IEE MATERIALS AND DEVICES SERIES 10

Series Editors: Professor D. V. Morgan
Professor K. Overshott

ELECTRICAL RESISTIVITY HANDBOOK

Other volumes in this series:

ELECTRICAL RESISTIVITY HANDBOOK

Edited by
G. T. Dyos and
T. Farrell

Peter Peregrinus Ltd. on behalf of the Institution of Electrical Engineers

Published by: Peter Peregrinus Ltd., London, United Kingdom

British Library Cataloguing in Publication Data

A CIP catalogue record for this book
is available from the British Library

ISBN 0 86341 266 1

Printed in England by Short Run Press Ltd., Exeter

Contents

Contents

Preface

It has long been a source of annoyance to one of the editors (G.T.D.) that there has not been a definitive compilation on the resistivity of materials as a function of temperature. So it was with this idea in mind and the close working relationship between the editors that the solidification of this handbook became reality. The handbook has been written for the needs of scientists and engineers who require a brief introduction to resistivity theory, the practical techniques of resistivity measurement and immediate access to resistivity data. No attempt has been made to include the effects of crystalline resistivity, pressure or magnetic effects, super-conductivity or irradiation damage.

If enough material can be gathered it is hoped that a further volume will be issued, particularly with respect to alloys, proprietary named alloys, steels, and materials. The editors would be glad to receive any comments, suggestions and information concerning data or suitable referenced material for inclusion in the next volume.

It became apparent towards the end of the compilation of this handbook that there is a lack of data available for certain polycrystalline elements, of known purity. The following table indicates the areas where there is a scarcity (*) of published data as a function of temperature bands.

Temperature ranges	0–300K	300–600K	>600K
Cadmium	*		
Cerium	*		
Europium	*		
Gadolinium		*	*
Gallium	*		
Indium	*	*	
Iridium	*		
Lead	*	*	
Manganese			*

Mercury	*	*	
Niobium			*
Osmium	*	*	
Rhodium	*	*	
Ruthenium	*	*	
Selenium	*	*	
Technetium	*	*	*
Tellurium	*	*	
Thallium	*	*	
Thorium	*	*	*
Tin	*	*	
Uranium		*	*
Ytterbium	*	*	*
Yttrium	*	*	*

Many of the large number of publications examined by the editors did not include the purity of the materials used; also many presented their data in an analysed form which rendered it almost impossible to extract the actual resistivity data relating to their samples. Regrettably the editors were not able to use this material.

The editors wish to thank Mr. J. Walker and Mr. J. Webster of EA Technology for the use of the reference library facilities.

Gordon T. Dyos
GT Innovations
Llandegla, Clwyd, UK

March 1992

Chapter 1
Introduction

1.1 Electronic conduction in solids

To aid the understanding of the electrical resistivity data on the materials presented in this handbook and to assist with its estimation for materials whose exact composition is not listed some fundamental concepts of electronic conduction in solids are required. These concepts are presented at the descriptive level; the reader who requires a more detailed quantitative approach should consult the standard text books [1–5] or the cited references.

Electronic conduction is concerned with the motion of electrons within the solid under the influence of an applied electric field. Of necessity, therefore, the electrons must be free to move; experience shows that metals are good conductors of electricity and thus some of the electrons in the metal can be regarded as 'free'. The earliest approach to quantifying the electrical conduction in metals, recognising that electrons were responsible for the transport of charge, was that of Drude [6], who treated the metal as an electron gas. When an electric field, E, is applied to the metal, the electrons of charge $-e$ are accelerated by the electrostatic force eE. After they have travelled for a certain average time, τ, the electrons suffer a collision with the atoms of the solid and are effectively arrested, transferring their kinetic energy to the solid. From simple mechanics, the mean drift velocity of the electrons, v_D, is given by

$$v_D = \frac{Ee\tau}{m}$$

where m is the mass of the electron. If the volume density of the electrons is n, then the electrical current per unit area, J, is given by

$$J = nev_D = \frac{n\,e^2\tau E}{m} \tag{1.1}$$

1

This is a statement of Ohm's law, where the electrical conductivity ρ is defined as $J = \sigma E$.

The electrical resistivity $\rho = 1/\sigma$ is therefore

$$\rho = \frac{m}{n\, e^2 \tau} \qquad (1.2)$$

To obtain an estimate of ρ from eqn. 1.2 the values of n and τ are required. Taking the 'relaxation time' τ to be given by λ/v_0 where λ is the electron mean free path and v_0 the thermal velocity of the electrons, then from the equipartition law $m v_0^2 / 2 = 3 k_B T / 2$ and

$$\rho = \frac{(3 k_B m T)^{1/2}}{n\lambda\, e^2} \qquad (1.3)$$

If it is assumed that each atom contributes one electron to the conduction process, n is the number density of atoms in the solid and if λ is taken to be the interatomic spacing, then eqn. 1.3 gives values of ρ in the 10^{-6} to 10^{-7} ohm metre range which covers the room temperature experimentally observed resistivities for most metals.

Although this simple classical Drude theory gives Ohm's law and taking the power dissipation to be approximately $m v_D^2 / \tau$ is consistent with Joule heating, (it can also be used to derive the relationship between the electrical and thermal conductivities of metals—the Wiedemann–Franz law), it cannot readily explain the differences in the electrical resistivity upon alloying; nor does it predict the correct variation of resistivity with temperature. The Drude theory is restricted to metals; it cannot explain the electrical resistivity of semiconductors. However, the important point to emerge from this approach is the concept of collisions or scattering of electrons and the significance of τ, the 'relaxation time'.

The quantum mechanics approach pioneered by Sommerfeld [7] assumes the free electrons of the metal to be moving in a constant internal potential and that the electrons in this potential well obey the laws of quantum mechanics rather than the laws of classical mechanics. Essentially this well consists of a quasi-continuum of energy levels and when the valence electrons are 'poured' into the well they progressively fill these levels up to a certain energy, E_F, known as the Fermi level. This value of E_F varies from metal to metal but is typically a few electron volts (1 electron volt $= 1 \cdot 6 \times 10^{-19}$ joules). In this quantum Fermi gas model, the number of electrons which take part in the conduction is restricted to those lying within the energy range $k_B T$ of the Fermi level where k_B is Boltzmann's constant. Thus, at room temperature, for a monovalent metal with $E_F \approx 5$ eV, the number of electrons is only 5×10^{-3} times the total number of valence electrons. The average drift velocity of the classical mechanics is replaced by the Fermi velocity, $v_F \approx (2 E_F / m)^{1/2}$, which for $E_F \approx 5$ eV is about $1 \cdot 3 \times 10^6$ m/s. If these mean values are substituted into eqn. 1.3, and the mean free path, λ, is again taken to be the interatomic spacing, then the Sommerfeld model predicts a resistivity which is several orders of magnitude larger than the experimental values. However, the Fermi gas model derives the Wiedemann–Franz law with remarkable agreement with experiment and successfully accounts for electronic specific heat. Thus the difficulty it encounters in describing conductivity must lie in the definition of the mean free path. In order to reconcile the prediction of the Fermi gas model with experiment, the electron mean free path must be larger than interatomic spacings.

The Fermi gas model, in its simplicity, treated the electrons as being in a constant potential well. Solids are comprised of crystals the size of which are large compared with atomic dimensions (with the exception of amorphous metals). Each ion within the crystal lattice has its own potential well and as the ions are arranged periodically, there is a periodic arrangement of potential wells.

If one atomic potential well is considered in isolation then there are discrete electron energy levels such as are encountered in the hydrogen atom. If two such wells are brought into close proximity, then from the Pauli exclusion principle which forbids any two electrons from occupying the same energy level (neglecting electron spin) the isolated discrete energy levels split into two levels. When a large number, N, of such wells are brought into close proximity, which is the case in a crystal lattice, then the original discrete energy level is broadened into a quasi-continuous energy band (actually comprising N energy levels). The gaps between the original isolated discrete energy levels may be preserved, but are smaller; in some cases the gaps no longer exist and the energy bands overlap. This 'Band Theory of Solids' provides the framework for understanding the classification of solids into metals, semi-conductors and insulators.

If the electron is considered to be a wave travelling perpendicular to a set of lattice planes then Bragg reflection will occur at certain wavelengths. For these wavelengths, λ_F, the wave is not propagated through the crystal and therefore they are forbidden to the conduction electrons of the crystal. The wave vectors, k, corresponding to the wavelengths ($k = 2\pi/\lambda_F$) occur at the boundaries of the Brillouin zone (the unit cell of the reciprocal lattice) and constitute energy gaps.

The Brillouin zone model relates the electron energy levels to the crystal structure. In regions close to the zone boundaries the variation of electron energy with its wave vector deviates from that where the periodic potential can be treated as constant, as a direct result of energy gaps. One consequence is that the mass of the electron is no longer that of the free electron; if the Fermi level occurs in such a region the mass of the electron should be replaced by an effective mass, m^*, and this can be quite different from m.

The quantum mechanical approach provides the basis for the understanding of solids in terms of their conduction behaviour. The band theory and Brillouin zone model are primarily concerned with electronic structure, whereas electron conduction is concerned with the motion of the electrons under the influence of an externally applied electric field within that structure. The solution to the vexing problem of the mean free path being considerably larger than atomic dimensions was provided by Bloch [8], who considered, via a Fourier analysis, the propagation of electron waves in the lattice structure. If the structure is perfectly periodic, no scattering occurs, whereas deviations from periodicity give rise to scattering and hence to electrical resistance.

If the potential wells at the lattice sites are identical and the lattice sites themselves are regularly spaced then the electron wave propagates almost undisturbed; in other words very little scattering occurs, which is equivalent to a mean free path which is much longer than interatomic distances. In fact, in a perfectly periodic structure, the electron wave is not disturbed at all and in such a structure the resistivity would be zero (this is not to be confused with superconductivity, where the resistivity is zero in non-periodic structures). A perfectly periodic structure is never realised in practice although it is approached in very pure single crystal solids at temperatures close to absolute zero.

Thus, in normal solid electronic conductors the electrical resistivity arises from departures from periodicity in the lattice potential wells and there are many ways in which this can occur; these include

(i) displacement of the atoms due to their thermal motion
(ii) imperfections in the crystal lattice, such as disolocations, vacancies and grain boundaries in polycrystalline solids
(iii) the substitution of foreign atoms at the lattice sites.

The departures from periodicity arising from (i) and (ii) are characteristic of pure metals whilst all three sources contribute to the resistivity of metallic alloys.

1.2 Pure metals

In a well annealed pure metal at low temperature, the density of the imperfections is low and their contribution to the electrical resistivity is small. Dislocations, which arise for example by cold working, give rise to a temperature independent resistivity which is usually negligible in comparison with the resistivity at high temperatures where phonon scattering dominates (see below) but may contribute significantly to the resistivity at cryogenic temperatures. For this reason, when studying pure metals it is preferable to use well annealed specimens. On the other hand vacancies are thermally activated; their number density increases with temperature. At room and cryogenic temperatures the vacancy contribution to the resistivity is negligibly small and only starts to become significant at temperatures approaching the melting point. However, this is somewhat academic in the sense that the vacancy contribution to the resistivity is inherent to the metal and temperature.

At finite temperatures the atoms vibrate about their mean positions. Effectively the interatomic spacing is no longer constant; as the temperature increases, so does the amplitude of the vibrations, giving rise to greater changes in the interatomic spacing and hence larger departures from periodicity. The result is a temperature variation of the mean free path and hence of the electrical resistivity.

The vibration of the atoms can be regarded as waves propagating through the crystal lattice. By treating the lattice vibrations as being quantised (the quantum of energy is known as a phonon) it can be shown that there is a cut-off phonon frequency; phonons of frequency larger than the cut-off frequency are not allowed. The temperature corresponding to this frequency is known as the Debye temperature (usually denoted by θ_D).

Now the scattering of the electrons is essentially the interaction of the electron waves (which are defined by the periodicity of the lattice) and the displacements of the lattice sites which can be described by lattice waves. The characteristic electron energy E_F (the Fermi energy) is equivalent to many thousands of degrees K, whilst the Debye temperature is typically a few hundred degrees K. Thus the effect of temperature on the electronic structure of a solid is small compared with its effect on the lattice and consequently, for scattering purposes, the effect of temperature on the lattice only need be considered.

At temperatures above θ_D, all lattice modes are excited and contribute to the scattering of the electrons. The phonon occupancy of the excited lattice modes increases linearly with temperature and consequently as the mean free path is inversely proportional to the number of scattering centres, the resistivity increases linearly with temperature. At temperatures much lower than θ_D not all lattice modes are excited; only those with frequency ν less than $k_B T/h$ (where h is Planck's constant) will be excited and at very low temperatures the number of excited modes falls off as $(T/\theta_D)^3$. Thus for temperatures lower than θ_D, the resistivity varies more rapidly than linear, and it was shown by Bloch [9] that at these very low temperatures the resistivity varies as the fifth power of temperature. Gruneisen [10] later developed the Bloch analysis to give an expression for the resistivity which covered the entire temperature range.

Some pure metals can exist with different crystal structures known as allotropes. Each allotrope has its own Brillouin zone and thus its own characteristic resistivity, which may exhibit a different temperature dependence. Such phase changes are more often encountered in metallic alloys and these are discussed later.

Whilst electron scattering by phonons is the dominant process in determining the temperature variation of the resistivity of most metals, other temperature-dependent scattering processes, such as electron–electron scattering, which gives a T^2 dependence for the resistivity at low temperatures, occur, and these are often seen in the magnetic and transition metals. Magnetic phase transformations occur in the ferromagnetic metals at the Curie temperature, T_c. These transitions are associated with a continuous decay in long range magnetic order which vanishes at T_c. As the temperature is increased and T_c is approached the resistivity rises at an increased rate reflecting the increase in disorder. At T_c, where magnetic disorder is complete, there is a change in slope of the resistivity/temperature curve.

The abrupt increase in the electrical resistivity of many metals at the melting point in passing from the solid to the liquid state and its steady increase with temperature in the liquid state, is associated with the absence of long range crystal structure and the increase in disorder in the liquid state.

1.3 Metallic alloys

1.3.1 Dilute alloys

When a 'foreign' atom is substituted for an atom of the host metal, the potential well at the substitution site is changed. The contributions to the change in the potential well due to the substitution arise from the change in valency and change in atomic volume. The valence electrons from the foreign atom merge with the conduction electron number of the host with the result that the potential well at the substitution site has a different charge from that which exists at the host sites; in the mathematical treatment the potential well has a different depth. The change in atomic volume causes the spatial dimensions of the impurity potential well to differ from that of the host.

When foreign atoms are randomly substituted into the host lattice and the crystal structure of the host is preserved, then provided that the foreign atoms do not interact with each other, the electrical resistivity increases in proportion to the number of foreign atoms added. Such a situation is realised in practice in dilute alloys where the alloy addition is typically less than 5%. Moreover, with such dilute alloys, the foreign atoms vibrate as though they were host atoms and consequently the additional contribution to the resistivity is temperature independent. This was first recognised experimentally in the 1860s by Matthiessen and Vogt and a simple rule, known as Matthiessen's rule, which applies to many dilute alloys, can be stated:

$$\rho_a(T) = \rho_p(T) + \rho(c) \qquad (1.5)$$

where $\rho_a(T)$ and $\rho_p(T)$ are the temperature variations of the resistivities of the dilute alloy and the host and $\rho(c)$ is the concentration dependent resistivity of the foreign atoms, which is almost linear in concentration. There are deviations from this rule even for very dilute alloys, but with the exception of transition metal alloys involving the magnetic metals the deviations are very small and for practical purposes can often be regarded as negligible.

1.3.2 Concentrated alloys

When the number of substitutional foreign atoms becomes large, for example when their concentration is greater than 5%, the assumption that they do not interact with each other is no longer valid and Matthiessen's rule breaks down. The distinction between host and foreign atom becomes obscure (with a 50 atomic percent binary alloy there is no distinction) and thus the $\rho_p(T)$ term of eqn. 1.4 loses its original significance. That is not to say the resistivity is temperature independent but rather the temperature dependence is not necessarily that of the major constituent of a concentrated alloy. As the concentration of foreign atoms increases, then so does $\rho(c)$ of eqn. 1.4 although it is no longer linear with concentration, and the result is that the contribution from the foreign atoms dominates.

To complicate matters further, according to the equilibrium phase diagram the alloy may exist as a single or multiphase, ordered or disordered solid solution depending on concentration and temperature. The data is presented in this handbook in terms of resistivity against temperature. Whilst this is the preferred means of presentation, it is sometimes easier to discuss the resistivity at a given temperature as a function of concentration.

The simplest situation in concentrated alloys is that of a binary system which exhibits complete solid solubility series at a given temperature. For such a disordered alloy Nordheim [11] showed that the resistivity was of the form $\rho \propto c(1-c)$, which can be written more generally as $\rho = A + c(1-Bc)$ where A and B are constants which depend on the alloy series and c is the concentration of one of the components of the alloy. The Nordheim rule specifically relates to the impurity resistivity of concentrated alloys. With such alloys impurity resistivity normally dominates and the rule can be applied to the total resistivity, especially for room temperature and below.

The above account relates to substitutional impurities which are randomly distributed in the host lattice. There are some alloy systems where for certain compositions the impurity atoms are located at distinct lattice positions, creating a long range 'ordered' state with a degree of periodicity which is absent in the random state. At these compositions, where such a 'superlattice' exists, the resistivity of the ordered state is somewhat less than that of the random state. Generally the random state is preferred at high temperatures and the long range ordered state only occurs below a critical temperature. Thus care should be exercised when using Nordheim's rule to estimate the resistivity of an alloy.

Phase-equilibrium diagrams of alloy systems indicate that the crystal structure of the alloy often changes with temperature. As mentioned above these allotropes have different electronic structures which give rise to differences in the electrical resistivity, and this is manifest as a discontinuity in the resistivity/temperature plot, the discontinuity occurring at the boundary of the phase field. In a particular phase field, Nordheim's rule can in principle be applied, but the constants A and B will differ from phase to phase.

Alloy systems do not necessarily exist in a single phase. Indeed the mechanical properties of many engineering alloys depend on the co-existence of two or more phases. Each phase has its own characteristic resistivity which may exhibit a different temperature variation. For the binary two-phase systems the resistivity may be regarded as lying between two limits corresponding to the two phases being elongated and aligned parallel to the current flux and where the two phases are platelets perpendicular to the current flow. If the volume fractions of the phases are known together with the characteristic resistivities of the individual single-phase alloys which bound the two phase fields then the limits can be evaluated by regarding them as parallel and series networks. There is no general rule which governs whether the series or parallel configuration will exhibit the greater resistivity and in any case it is highly unlikely that these two somewhat idealised situations will be encountered in practice; likewise, there is no general rule to indicate whether or not the resistivity of the two-phase alloy will be greater or less than that of one of the individual bounding single-phase alloys; however, it might be anticipated that the resistivity would be less than that of the single-phase disordered alloy of the same composition and this is often observed.

In alloy systems where magnetic transformations occur the change in long range order is reflected in the resistivity and is often manifest as a change in slope of the resistivity/temperature curve at the critical or transition temperature. In general, the magnitude of the effect of long range order on the resistivity is difficult to estimate. Ferromagnetic alloys usually contain one of the ferromagnetic pure metals whilst anti-ferromagnetic long range ordering can be found in many of the alloy systems involving transition and rare-earth metals.

1.4 Semiconductors

As mentioned above, the band theory of solids provides a framework for classifying solids into metals, semiconductors and insulators, the last of which are not relevant to this handbook. In metals there are continuous energy states above the Fermi energy due either to the bands not being completely filled or the bands overlapping in energy, whereas in semiconductors there exists an energy gap, E_g, between the valence and conduction bands with the Fermi energy lying in the gap. In the pure elemental semiconductors such as germanium and silicon, electrons must be thermally excited into the conduction band (and as a result an equal number of holes created in the valence band) for these materials to behave as intrinsic conductors. At temperatures of around room temperature and below, the thermal energy, $k_B T$ ($< 0 \cdot 03$ eV), is small compared with E_g ($\approx 0 \cdot 7$ eV for germanium, $\approx 1 \cdot 1$ eV for silicon) and as the number of carriers varies according to $\exp(-E_g/2k_B T)$ these materials are poor conductors; at absolute zero no thermal excitation can occur and they behave as insulators. Increasing the temperature causes the number of carriers to increase exponentially and hence the resistivity to fall. Small amounts of impurities can give rise to donor and acceptor levels within the energy gap which are very close, in energy terms, to the conduction band and valence band respectively; the energy separation, E_i, may be comparable to the room temperature thermal energy ($k_B T = 0.026$ eV). As the number of carriers now varies as $\exp(-E_i/2k_B T)$, the result is that many more electrons are excited from the donor level into the conduction band (or correspondingly many more holes created in the valence band) at lower temperatures, leading to n-type (electron) and p-type (hole) conductivity. This conductivity, which is proportional to the concentration of impurities, can be several orders of magnitude larger than the intrinsic conductivity. On the other hand the addition of some impurities to an already impure semiconductor can reduce the total number of carriers via compensation. Thus the influence of impurities on the resistivity is in marked contrast to the case of metals. Because impurity concentration is usually very low and these materials are normally used in single crystal form where the defect concentration is low, phonon scattering is dominant, but the variation of resistivity with temperature is most strongly influenced by the variation in the number of carriers. For these reasons only the high temperature resistivity of the elemental semiconductors is given in the handbook.

1.5 Summary

To summarise these fundamental concepts, the resistivity of the pure metals is well understood. Reliable experimental data exists over a wide range of temperature and this in general conforms to theoretical expectations. For dilute alloys (less than 5 atomic percent) provided the increase in residual resistivity per atomic percentage impurity addition is known the resistivity can be estimated with reasonable confidence from Matthiessen's rule. This rule holds remarkably well for many alloy systems although there are exceptions, notably when the host metal is one of the ferromagnets.

In concentrated alloys the situation is much more complicated. While Nordheim's rule can be used to obtain a fair estimate of the resistivity of some concentrated binary alloys it has limited applicability. This is due to the many factors which influence the resistivity; these include atomic and magnetic long range order, changes in crystal structure (or phase) as the alloy composition and temperature vary and co-existence of two or more phases. All these factors together make it almost impossible to make even a fair estimate of the resistivity of a given alloy, let alone its temperature dependence. If the resistivity of an alloy of near composition and similar heat treatment is known then by using the principles outlined above, a reasonable estimate of the resistivity of the alloy in question may be obtained.

The purpose of this handbook is to provide a comprehensive compilation of experimental data in graphical form of resistivity versus temperature which we hope will be of assistance in this respect. Failing this, one must resort to the experimental determination of the resistivity; the next few pages deal with some of the experimental methods of measuring resistivity.

Chapter 2
Measurement techniques

2.1 Introduction

There are many techniques available for the determination of electrical resistivity. The simplest and most widely used is the so-called four probe method in which the resistance of a portion of the sample is measured when a direct current passes through the sample. This technique involves two current contacts and two voltage contacts attached to the sample. The four probe technique is applicable to the vast majority of metals and alloys and a variation of the method, the Van der Pauw technique, is widely used for the determination of the resistivity of semiconductor wafers. The technique is applicable over a wide range of temperature. However, difficulties may be experienced with contacts if, for example, the material is brittle or chemically reactive (as it may be in the molten state). For such materials, methods which do not rely on making physical contact to the sample are preferable; the most widely used non-contacting techniques are those based on eddy currents induced in the sample when it is placed in a magnetic field which varies with time. In the following paragraphs the four probe technique and some of the eddy current techniques are briefly described.

The theoretical background to the four probe and eddy current methods can be demonstrated by the application of two of Maxwell's equations:

$$\mathrm{curl}\ \boldsymbol{E} = -\frac{\partial \boldsymbol{B}}{\partial t} = -\mu\mu_0\frac{\partial \boldsymbol{H}}{\partial t} \tag{2.1}$$

$$\mathrm{curl}\ \boldsymbol{H} = \boldsymbol{J} + \frac{\partial \boldsymbol{D}}{\partial t} = \frac{\boldsymbol{E}}{\rho} + \epsilon\epsilon_0\frac{\partial \boldsymbol{E}}{\partial t} \tag{2.2}$$

In these equations, $\boldsymbol{E}$ is the electric field, $\boldsymbol{D}$ is the electric displacement, $\boldsymbol{H}$ the magnetic field, $\boldsymbol{B}$ the magnetic induction, ϵ_0 and μ_0 are the permittivity and permeability of free space, ϵ and μ are the relative permittivity and permeability of the sample and J is the current in unit cross sectional area of the sample of resistivity ρ.

For a conducting sample, the displacement current $\partial D/\partial t$ is negligible in comparison with J and thus eqn. 2.2 becomes

$$\text{curl } \boldsymbol{H} = \boldsymbol{J} = \frac{\boldsymbol{E}}{\rho} \tag{2.2a}$$

Eqn. 2.2a is the basis for the four probe method. The expression $\boldsymbol{J} = \boldsymbol{E}/\rho$ is simply a statement of Ohm's law. Thus writing $J = I/A_S$ where I is the current in the sample of uniform cross-section A_S and V the voltage measured by two impedance probes separated by a distance L, the resistance R is

$$R = \rho \frac{L}{A_S} \tag{2.3}$$

If the sample is cylindrical of radius a, the resistance is

$$R = \left(\frac{\rho}{a^2}\right) \frac{L}{\pi} \tag{2.3a}$$

The basis for the eddy current methods is seen by substituting eqn. 2.1 into eqn. 2.2a:

$$\text{curl curl } \boldsymbol{H} = \nabla^2 \boldsymbol{H} = \frac{1}{\rho} \text{ curl } \boldsymbol{E} = -\frac{\mu\mu_0}{\rho} \frac{\partial H}{\partial t} \tag{2.4}$$

For a sample having isotropic magnetic and electrical properties, which is true of most polycrystalline materials, and where the magnetic field is aligned along the principal sample dimension (for example along the axis of a cylinder, say the z direction) the vector eqn. 2.4 becomes scalar and since $B = \mu\mu_0 H$ and the magnetic flux, ϕ, in the sample is $\phi = A_S B$, we have

$$\frac{\partial \phi}{\partial t} = -\frac{\rho}{\mu\mu_0} \frac{\partial^2 \phi}{\partial z^2} \tag{2.4a}$$

Eqn. 2.4a is the standard diffusion equation where the flux diffusivity (which has the dimensions of $(\text{length})^2/\text{time}$) is $\rho/\mu\mu_0$. It governs how the magnetic flux in the sample changes with a time dependent magnetic field. Solutions to the diffusion equation for various geometrical arrangements and different boundary conditions (for example, the time dependence of the magnetic field) are given in standard texts [12]. By way of illustration, consider a cylindrical sample of radius a which experiences a uniform magnetic flux, ϕ_0, which is suddenly removed at time $t=0$; the average flux at time t is given by

$$\phi = \frac{4\phi_0}{a^2} \sum_{n=1}^{\infty} \frac{1}{\beta_n^2} \exp\left(-\beta_n^2 \left(\frac{\rho}{\mu\mu_0 a^2}\right) t\right) \tag{2.5}$$

where the coefficients β_ν are the positive roots of $J_0(\beta) = 0$, J_0 being the zero order Bessel function. Thus, by observing the time dependence of the flux decay, the resistivity can be deduced. Details of this method along with other diffusion eddy current based methods involving periodic magnetic fields are discussed below. It is seen that in both eqns. 2.3 and 2.5 a term involving ρ/a^2 appears. The same term appears in other eddy current methods, which being diffusion based, must have a term involving $(\text{length})^2$—or cross sectional area—and this is important when considering the accuracy of the determination of ρ.

2.2 Four probe methods

These methods rely on the measurement of the resistance of the portion of the sample between the two voltage probes. For eqn. 2.3 to be applicable the cross-sectional area of the sample should be uniform and the voltage probes should be connected to the sample in such a way that the length, L, can be accurately determined.

The preferred method of measuring the resistance is to use a Wheatstone bridge, which compares the unknown resistance of the sample with that of three standard resistors, one of which is a variable standard. As the bridge is a null method, calibration of the detector is not necessary. The accuracy with which the null point can be determined depends not only on the sensitivity of the detector, but also on the relative magnitudes of the resistances of the arms of the bridge. A good general rule for optimum bridge operation is that the resistances in the four arms should be of the same order. With a well designed bridge and sensitive detector the bridge sensitivity can be very high and errors in the determination of resistivity arising from the resistance measurement can be regarded as being negligible.

Alternatively, the sample resistance can be measured using a constant current and measuring the probe voltage difference using a potentiometer or digital voltmeter; the latter has the advantage that it can be calibrated to give a direct reading of the resistance. To avoid zero errors it is usual practice to incorporate current reversal and take the average value of the two readings. Zero errors can arise not only from an incorrectly adjusted voltmeter but also from thermoelectric effects if a temperature gradient exists along the sample and the voltage leads, usually copper, have a different thermoelectric power from that of the sample being measured.

The accuracy with which the resistivity is determined from the resistance measurements can be dominated by the errors involved in the measurements of the sample cross-section and the separation of the voltage probes. This depends on the magnitude of A_S and L, which in turn is influenced by the magnitude of the resistivity through the need to have an accurate measurement of the resistance. For a pure metal with a resistivity of 5×10^{-8} ohm m, through which a current of 1 A passes and where the voltage is determined to an accuracy of 10 nV, then for a resistance measurement to an accuracy of $0 \cdot 01\%$, the voltage between the probes should be 100 μV, giving a resistance of 10^{-4} ohm. From eqn. 2.3, L/A_S should be 2×10^3/m and thus can be satisfied realistically with a cylindrical sample of diameter 8 mm and with $L = 100$ mm. The power dissipated in the sample is $0 \cdot 1$ mW and this will not cause any significant heating.

Thus, for the majority of metals and alloys, the resistivity at temperatures above room temperature, cylindrical samples measuring about 10 mm in diameter and with L of about 50–100 mm are ideal shapes. Clearly if this demands too large a quantity of material, the sample dimensions should be scaled down accordingly, but this will have the effect of increasing the errors in the measurement of L and A_S. At lower temperatures, particularly with annealed pure metals the resistivity can be as low as 2×10^{-11} ohm m and this usually requires the sample to be in the form of a thin wire.

The measurement of L usually presents no problems. If mechanical clamps embodying 'knife-edge' contacts are used as the voltage probes then the separation of appropriate datum marks on the clamps can be determined to plus or minus $0 \cdot 02$ mm with a travelling microscope; thus the error introduced in the measurement of L is less than $0 \cdot 1\%$.

The determination of A_S is more demanding. Firstly A_S should be uniform, and this can be checked by measuring the sample diameter (for the cylindrical sample) at various positions along L. It is preferable to use a non-contacting method for this measurement (if a micrometer is used, the contact pressure can lead to an erroneously low value for the diameter). Optical projection techniques and photographic enlargements can be used to get an accuracy of better than $0 \cdot 1\%$ in A_S, provided, of course, the sample is prepared to this accuracy.

Thus, provided care is taken in the measurement of L and A_S, the resistivity can readily be determined with an absolute accuracy of around $0 \cdot 1\%$ at and above room temperature. At cryogenic temperatures, in particular at liquid helium temperatures and with pure metals, the error in the determination of the resistance is maintained at negligible proportions by employing voltage detection equipment which involves superconducting elements and has much higher sensitivity.

2.3 Temperature variation of resistivity

With the four probe technique, the temperature variation of the resistivity can be determined quite simply by thermally attaching the sample to a heat sink, the temperature of which is controlled. The sample temperature can be measured using a suitable thermocouple attached to one of the voltage probes and any temperature gradients can be measured using a differential thermocouple with its two junctions thermally anchored to the two voltage probes; in this way the mean sample temperature can be determined. The design of the furnace is not considered here but for high temperature work, a vacuum furnace is usually used to reduce the thermal load and prevent oxidation.

To overcome problems with thermoelectrically generated voltages, the voltage leads should be continuous to the wall of the vacuum vessel; likewise for accurate temperature measurements it is preferable to use continuous thermocouple leads.

When varying the temperature, the change in sample dimensions due to thermal expansion should be taken into account, especially since L and A_S are usually carefully determined at room temperature. To illustrate the magnitude of the error which may be introduced by the neglect of thermal expansion consider a cylindrical sample to which eqn. 2.3a applies. At any temperature T, the resistance is

$$R_T = \rho_T \frac{L_0}{\pi a_0{}^2} \frac{(1 + \alpha \Delta T)}{(1 + \alpha \Delta T)^2} = \rho_T \frac{L_0}{\pi a_0{}^2} (1 + \alpha \Delta T)^{-1}$$

where a is the thermal expansion coefficient, R_T is the measured resistance and ρ_T is the resistivity. Thus the percentage error introduced is $\alpha \Delta T \times 100$. Taking a representative value of $\alpha = 15 \times 10^{-6}$ for most metals and alloys then it is readily appreciated that the error can be comparable to the accuracy in the determination of the resistivity for a ΔT of 100 K.

2.4 Eddy current methods

Several techniques which rely on induced eddy currents for the determination of resistivity have evolved, the simplest of which corrrespond to the situations where the applied magnetic field is periodic, for example varying sinusoidally with time, and where the field is a step function being suddenly switched from zero to a preset constant value or conversely switched from a steady value to zero. Whilst these elegant techniques overcome the need for electrical contact to the sample they do have characteristic features which can give misleading results.

Firstly, it is important to recognise that the eddy currents are not uniformly distributed throughout the sample when a periodic magnetic field is applied. They decay exponentially from the sample surface with a measure of the decay being the 'skin depth'. The skin depth, δ, is given by

$$\delta = \left(\frac{\pi f \mu \mu_0}{\rho} \right)^{-1/2}$$

and is the depth at which the eddy currents have fallen to $1/e$ of their surface value. If the whole of the sample is to be examined, then, in the case of the periodically varying magnetic field, the frequency, f, should be chosen so that δ is not significantly less than the sample dimension, which, in the case of a cylinder with the magnetic field parallel to the cylinder axis, is the cylinder radius. If δ is significantly less than the sample dimensions then surface features such as roughness and contaminant films can affect the resistivity measurements. It must be emphasised that the only situation where uniform current density occurs is in the direct current measurements on uniform cross-section samples as discussed in the four probe method.

Secondly, the magnetic field can directly influence the motion of the electrons within the sample (in addition to 'indirect' influence via induction which gives rise to the eddy currents), leading to additional resistance over that caused by scattering. This 'magnetoresistance' is dependent on the magnitude of the applied magnetic field and its occurrence can be checked by varying the field strength.

These techniques are well suited to the determination of the resistivity of weakly magnetic materials where the relative permeability, μ, can safely be regarded as unity. However some caution should be exercised when dealing with ferromagnetic and strongly paramagnetic materials or materials which exhibit magnetic transformations where μ is neither constant nor unity.

The most convenient means of applying a magnetic field is to place the conducting sample in a coil. With the step function methods the resistivity is determined by observing the change of flux in the sample as a function of time after the field has suddenly been applied or removed; this is described for the case of field removal by eqn. 2.5. For periodically varying fields the difference between the impedance of the loaded and empty coil is measured.

2.5 Step function methods

The basis of this method has already been mentioned. A convenient arrangement is to insert the sample into a coil, apply a direct current to the coil for sufficient time to ensure complete flux penetration into the sample and then de-energise the coil by opening a switch. A simple circuit is shown in Fig. 2.1. The flux decay is detected by a secondary coil wound over the central region of the energising coil. The time dependence of the average flux, ϕ, in a cylindrical sample after removing the magnetic field is given by eqn. 2.5. If there are N turns on the second coil then

16

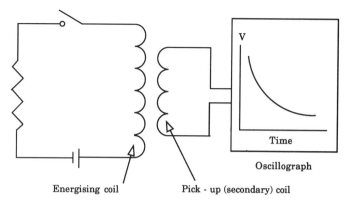

Figure 2.1 *Simple circuit for the step function method*

the voltage induced in that coil is

$$-N\frac{\partial\phi}{\partial t}$$

Using the relationship between ϕ and magnetic field H

$$\phi = \pi a^2 \mu\mu_0 H$$

and differentiating eqn. 2.5 with respect to time and multiplying by N, the secondary coil voltage is obtained:

$$V = 4\pi N\rho H_0 \sum_{n=1}^{\infty} \exp\left(-\beta_n^{2}\left(\frac{\rho}{\mu\mu_0 a^2}\right)t\right)$$

The voltage decays as a sum of exponential factors. The first three coefficients β_n^{2} have the values $5\cdot784$, $30\cdot470$ and $74\cdot892$; thus for longer times, the decay is a simple exponential governed by β_1^{2}. To illustrate this, consider the exponential term involving β_2 to contribute a maximum of $0\cdot1\%$ of the contribution from β_1. For this condition to apply:

$$\exp-\left(24\cdot676\left(\frac{\rho}{\mu\mu_0 a^2}\right)t\right) \leq 10^{-3}$$

Thus, for a material with $\rho = 5\times10^{-8}$ ohm m, $a = 5\times10^{-3}$ m and $\mu = 1$, the time should be greater than $1\cdot75\times10^{-4}$ seconds in which case the sum of the exponential factors is approximately $0\cdot2$. The measured voltage at this time is $1\cdot25\times10^{-5}$ H_0 volts and after $t = 5\times10^{-4}$ seconds the voltage has decayed to $6\cdot25\times10^{-7}$ H_0 volts. The voltage/time variation can be measured and stored using a suitable oscilloscope.

2.6 Periodically varying fields

The impedance, Z_L, of a magnetic field coil into which a conducting sample is placed (a loaded coil) comprises resistive and reactive components:

$$Z_L = R_L + iX_L$$

The resistance R_L and reactance X_L of the loaded coil are given by

$$R_L = R_0 + R_1$$

$$X_L = X_0 + X_1$$

where R_0 and X_0 refer to the empty solenoid and R_1 and X_1 are the resistive and reactive contributions from the eddy currents induced in the sample. In fact R_1 and X_1 measure the out of phase and in phase flux in the sample.

In general

$$R_1 = A'X_0F_R(y)$$

$$X_1 = A'X_0F_X(y)$$

$$\text{or} \quad \frac{X_1}{R_1} = \frac{F_X(y)}{F_R(y)} = F_{XR}(y)$$

where the constant A' is a measure of the ratio of the cross-sectional area of the sample to that of the coil and y is a parameter involving the skin depth and the dimension of the sample. Analytical expressions of the coil functions $F_X(y)$, $F_R(y)$ and hence $F_{XR}(y)$ can be derived for the simpler geometrical shapes such as 'infinite' cylinders, in which case y is given by

$$y = \frac{\sqrt{2}a}{\delta} = (2\pi\mu\mu_0)^{1/2}\left(\frac{fa^2}{\rho}\right)^{1/2}$$

where a is the radius, as before, and the coil 'filling factor', A', is simply A_S/A_C, where A_C is the cross-sectional area of the coil. The coil functions for cylindrical geometry are shown in Figs. 2.2 and 2.3 and for such a geometry these Figures can be used directly to obtain ρ once X_0, X_1 and R_1 have been measured.

Samples of irregular geometry, for which analytic expressions cannot be derived, can be used but the coil/sample system should be calibrated over a range of frequencies using a 'standard' specimen of known resistivity, ρ_S, having identical shape to that to be determined. The experimentally determined R_S/X_0, X_S/X_0 and X_S/R_S give $A'F_R(y)$, $A'F_X(y)$ and $F_{XR}(y)$ respectively (X_S and R_S refer to the calibration with the standard resistivity specimen). There are advantages in using $F_{XR}(y)$ as this eliminates the need to know the coil filling factor which, for non-simple geometries, is difficult to calculate.

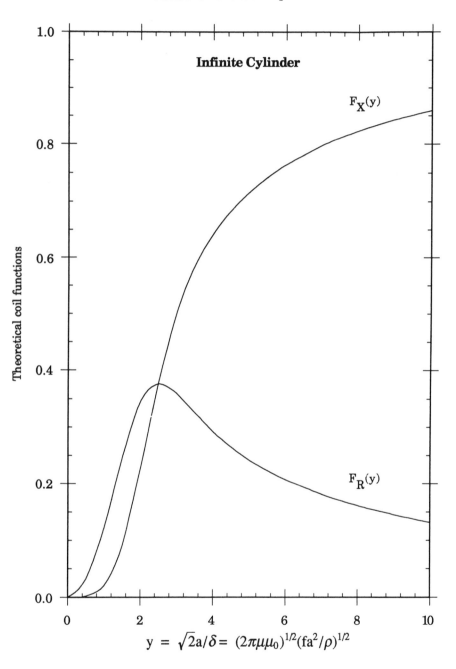

Figure 2.2 *Coil functions $F_R(y)$ and $F_X(y)$ for an infinite cylinder*

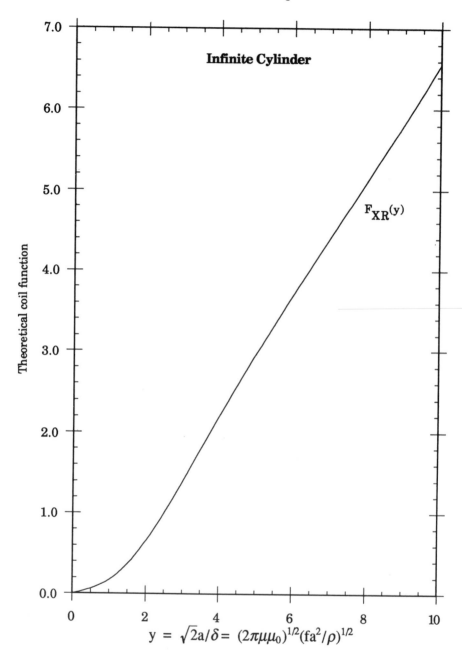

Figure 2.3 *Coil function $F_{XR}(y)$ for an infinite cylinder*

Having determined the empty coil parameter and obtained the coil functions, a single frequency measurement is sufficient to determine the resistivity of the unknown sample. The choice of frequency is important for two reasons. Firstly, as mentioned beforehand, it is desirable that $\delta > a$ and secondly, since the resistivity determination involves the subtraction of the unloaded coil measurements from the loaded coil measurements, it is desirable to have a large difference between the two sets of measurements. This can be illustrated by reference to Fig. 2.2 and again taking a cylindrical sample of radius 5 mm and resistivity of 5×10^{-8} ohm m. $F_X(y)$ and $F_R(y)$ are both greater than $0 \cdot 2$ in the range from $y = 2$ to $5 \cdot 5$ and this corresponds to $f = 1$ to $7 \cdot 5$ kHz. At $f = 1 \cdot 0$ kHz the skin depth in the above material is $3 \cdot 6$ mm whilst at $f = 7 \cdot 5$ kHz the skin depth is $1 \cdot 3$ mm. If the criterion of $\delta = a = 5$ mm is used, the frequency is 500 Hz giving $y = 1 \cdot 4$; $F_X(y)$ and $F_R(y)$ are reduced to $0 \cdot 07$ and $0 \cdot 22$ respectively. This implies that the maximum value of X_L/X_0 is $1 \cdot 07$ (taking $A' = 1$) which imposes a more stringent demand on the experimental determination of X_L and X_0. At higher frequencies ($7 \cdot 5$ kHz) the experimental demands are relaxed but skin depth considerations may affect the result.

2.7 Inductive techniques

2.7.1 Self inductance

The coil should be longer than the sample to avoid end effects, thus ensuring uniformity of the magnetic flux in the sample. Coil design details are beyond the scope of this account, but taking into consideration such factors as low thermal dissipation, maximum coil filling factor, small magnetic fields, etc. then for a cylindrical sample of diameter about 10 mm, R_0 and $2\pi f X_0$ ($f = 10^3$ Hz) can each be a few ohms, giving a coil impedance of around 10 ohms.

The loaded and unloaded coil impedances are measured using an appropriate variable frequency bridge. Two self-inductance bridge networks are shown in Fig. 2.4. The Anderson bridge, Fig. 2.4b, has the advantage that it avoids the use of variable standard inductors or capacitors. Taking the coil current to be in the milliampere range (for low thermal dissipation and low magnetic fields) and following the general rule for optimum bridge operation, the driving voltage of the audio-frequency oscillator is in the $0 \cdot 1$ to $1 \cdot 0$ V range. Using a tuned detector, the sensitivity of the bridge can be higher than $0 \cdot 005\%$ and with an appropriately chosen frequency and good coil filling factor, R_1 and $2\pi f X_1$ can be around $0 \cdot 15 \times 2\pi f X_0$ implying that R_1 and X_1 can be measured to an accuracy of $0 \cdot 05\%$. The error in y is that associated with the accuracy to which a can be measured and this has been discussed in connection with the four probe method.

The uncertainty in ρ arises from the uncertainties in y^2 (and hence a^2)

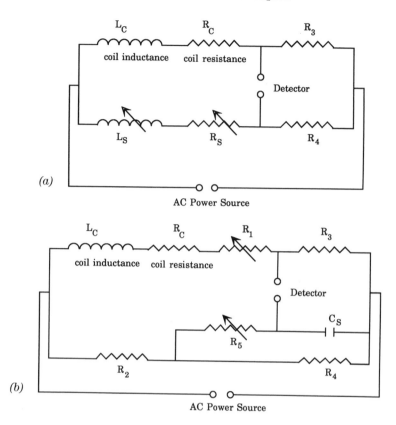

Figure 2.4 *Typical bridges for the determination of self inductance*

    (a) Typical bridge using standard inductors
    (b) Anderson bridge

and if $F_X(y)$ and $F_R(y)$ are used the uncertainties in A', which involve both a^2 and d^2 (d is the coil diameter); if $F_{XR}(y)$ is used the uncertainties are reduced as A' is not required. If a non-simple shape is used which necessitates experimental determination of $F_X(y)$ and $F_R(y)$ the errors are compounded by the dimensions of both the standard and unknown samples. Taking all these factors into consideration the accuracy in ρ when determined using the self-inductance technique is likely to be in the region of $0 \cdot 1\%$ with the lowest uncertainties being achieved with a uniform cross-section cylindrical sample.

2.7.2 Mutual inductance

The use of a secondary coil wound over the central region of the primary coil has the advantage of eliminating end effects and offers the prospect of examining regions of the sample along its length.

The analysis of the coils for mutual inductance is quite complex. As an approximation the reactance and resistance of a suitable unloaded coil system is scaled down from the self-inductance and resistance of the primary coil by the turns ratio. Thus, a typical system will have a mutual inductance in the region of 100 µH (as opposed to a few millihenries for the single coil self-inductance).

The determination of mutual inductance is most readily achieved by making a direct comparison with a variable standard mutual inductance, connecting the primary windings of both standard and unknown in series to an audio-frequency oscillator and connecting the secondary windings in series opposition so that the induced voltages oppose each other. In principle, the detector which is in series with the secondaries shows a null reading when the mutual inductances and unknown inductors are equal. In practice, because a resistance element exists, it is almost impossible to obtain a balance. The so-called Hartshorn bridge, shown in Fig. 2.5 overcomes this difficulty. The technique requires a balance to be obtained in the unloaded system by adjusting M_S and R. The sample is then inserted and the balance is restored by varying M_S and R by ΔM_S and ΔR from the unloaded condition. Recognising that R_1^+ and X_1^+ are

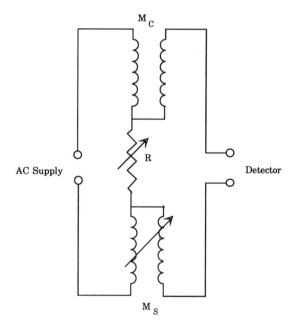

Figure 2.5 *The Hartshorn mutual inductance bridge*

$M_C =$ *mutual inductance of the coil*
$M_S =$ *variable standard mutual inductance*

measures of the out-of-phase and in-phase flux associated with the eddy currents, R_1^+ is given by ΔM_S and X_1^+ by $\Delta R/2\pi f$. As with the self inductance technique the value of y is then obtained from the calibrated coil function or, if cylindrical samples are used, from the theoretical coil functions shown in Figs. 2.2 and 2.3.

The mutual inductance technique relies on exactly the same parameters as the self inductance method; the uncertainties involved are again those associated with the measurement of sample and coil dimensions and hence the achievable accuracy in the determination of the resistivity is similar.

2.7.3 Inductive techniques and temperature variations

The primary purpose of using eddy current techniques is to avoid physical contact with the sample. When determining the resistivity as a function of temperature, the sample should be in an isothermal enclosure with the temperature of the enclosure monitored with an appropriate thermometric device such as a thermocouple, resistance thermometer or pyrometer. The use of an isothermal enclosure virtually eliminates thermoelectrically generated voltages.

When the highest accuracy is sought, the effects of thermal expansion must be taken into account and these effects enter through expansion of the sample and the coils (in the four probe method only the expansion of the sample needs to be considered).

The reactance of the unloaded coil is related to its dimension via $X_0 = \text{constant} \times d_c^2/l_c$ where d_c and l_c are the coil diameter and length respectively. The effect of an increase in temperature, ΔT, is to increase X_0 by $\alpha \Delta T \times 100\%$ where α may be taken to be the coefficient of expansion of the coil windings. The magnitude of the change is similar to that discussed in connection with the variation of resistance with temperature in the four probe method.

Neglecting the lead resistances, R_0 is proportional to l_w/d_w^2, where l_w and d_w are the wire length and diameter. Assuming the coil is tightly wound then $d_w \approx B_c l_c$ and $l_c \approx C_c l_c$ where B_c and C_c are constants involving the number of turns on the coil and the number of layers of windings. R_0 is therefore proportional to d_c/l_c^2 and the effect of a temperature increase of ΔT is to decrease R_0 by $\alpha \Delta T \times 100\%$. Again, the magnitude of the effect can easily exceed the accuracy with which the resistivity can be determined and corrections for thermal expansion should be made.

The sample dimensions enter through the parameter y which is proportional to $(r/a^2)^{1/2}$; thus if the thermal expansion behaviour of the sample is known, a simple correction can be applied for y.

However, it is not just a simple matter of correcting the impedance measurements for thermal expansion as is the case with the resistance measurements in the four probe method. Since R_0 and L_0 change with temperature, the bridge balance conditions also change. Thus, for the most reliable results, the implication is that the equipment should have provision for withdrawing the sample from the coil to enable R_0 and L_0 as well as R_L and X_L to be measured at each temperature.

The coil filling factor is also affected by thermal expansion. If F_X, F_R is used to determine the resistivity then a simple correction for thermal expansion (sample/coil) can be applied; this is not necessary if F_{XR} is used.

This account of experimental techniques concludes with a few remarks about an interesting method of resistivity measurement which involves induced eddy currents and as such avoids the use of electrical contact. Unlike the eddy current techniques discussed above it does not rely on electrical measurements. The technique involves a rotating uniform magnetic field such as that produced by the stators of a polyphase induction motor. When a cylindrical sample is placed in the field, the induced eddy currents interact with the flux causing the sample to experience a torque, Γ, given by

$$\Gamma = \frac{\pi \omega_r \mu^2 {\mu_0}^2 l a^4 H^2}{4\rho}$$

where ω_r is the relative angular velocity of H and the sample. If the sample's motion is restrained by a torsion wire, it will rotate to a fixed position where Γ is balanced by the reaction in the torsion wire, enabling the above expression to be used with ω_f, the angular velocity of the field replacing ω_r. Measurement of the deflection by the incorporation of a mirror into the torsion wire suspension and the use of optical levers gives an accurate measurement of Γ. The resistivity, ρ, is then determined using the above expression. This technique is more sensitive to the sample dimensions than the electrical techniques and is thus inherently less accurate. Furthermore, when using this method to determine the temperature variation of the resistivity it is important that corrections for thermal expansion be applied.

Chapter 3
Explanation of graphs

3.1 Data analysis

The majority of the data presented in this handbook has been gleaned from material published over the last two decades. The data has been extracted from the original published paper or book. The interpretation of the data is that of the original author except where an obvious error has occurred in the publishing of tabular material. Otherwise, every effort has been made to ensure the accuracy of the transcripted original data.

3.2 Catalogue system for materials

The catalogue system used in this book is based on two levels. Firstly the materials are listed in alphabetical order; where a material has several elements, they are listed in order of decreasing atomic percentage. Should a material have elements all having equal atomic percentage then they are listed in alphabetical order.

Data which has been presented in the original paper in graphical form and extracted is indicated by a '*' against the reference number in question. Data presented, in the original form, in tabular format is indicated by the absence of an asterisk against the reference number. Data from various sources which has been surveyed by an author to determine a best fit of that data is indicated by an 'R' after the reference. Data included in that survey is not represented in this handbook.

If the material is a disordered crystalline compound the letter 'D' is placed after the reference number and similarly the letter 'A' is used if the material is amorphous. It should be noted that the vast majority of authors do not indicate whether an allowance for thermal expansion has been applied to the resistivity of the material. Where the purity of an element has not been stated in the original paper, but the information greatly extends the range of resistivity, this has been denoted by '%na'.

3.3 Material composition

The purity of the elements in this book are expressed in terms of the weight percentage of the elements present.

All the compounds listed in this handbook are expressed in terms of atomic percentages, that is, the number of atoms of an element expressed as a percentage of the total number of atoms present in one molecule of the material.

Where compounds have been listed in the original reference as weight percentages, these have been converted into atomic percentages.

Text references

1. POLLOCK, D.D. (1985): 'Electrical conduction in solids: An introduction' (American Society for Metals, Ohio, USA)
2. ROSSITER, P.L. (1987): 'The electrical resistivity of metals and alloys' (Cambridge University Press, UK)
3. SCHRODER, K. (1983): 'Handbook of electrical resistivities of binary metallic alloys' (CRC Press, Florida, USA)
4. MEADEN, G.T. (1963): 'The electrical resistance of metals' (Heywood Books, UK)
5. KITTEL, C. (1976): 'Introduction to solid state physics' (John Wiley & Sons, New York, USA)
6. DRUDE, P. (1900):'Zur elektronentheorie der metalle', *Ann. Phys.*, **1**, p. 566
7. SOMMERFELD, A. (1928): 'Zur elektronentheorie der metalle auf grund der Fermischen statistik', *Z. Phys.*, **47**, p. 1
8. BLOCH, F. (1928): 'Uber die quantenmechanik der elektronen in kristallgittern', *Z. Phys.*, **52**, p. 555
9. BLOCH, F. (1930): 'Zum elektrischen widerstandsgesetz bie tiefen temperaturen', *Z. Phys.*, **59**, p. 208
10. GRUNEISEN, E. (1933): 'Die abhangigkeit das elektrischen widerstandes reiner metalle van der temperatur', *Ann. Phys.*, **16**, p.530
11. NORDHEIM, L. (1931): 'Zur elektronentheorie der metalle', *Ann. Phys.*, **9** (5), pp. 607–678
12. CARSLAW, H. S., and JAEGER, J.S. (1959): 'Conduction of heat in solids' (Oxford University Press, UK)

Index of materials

Material resistivity graphs

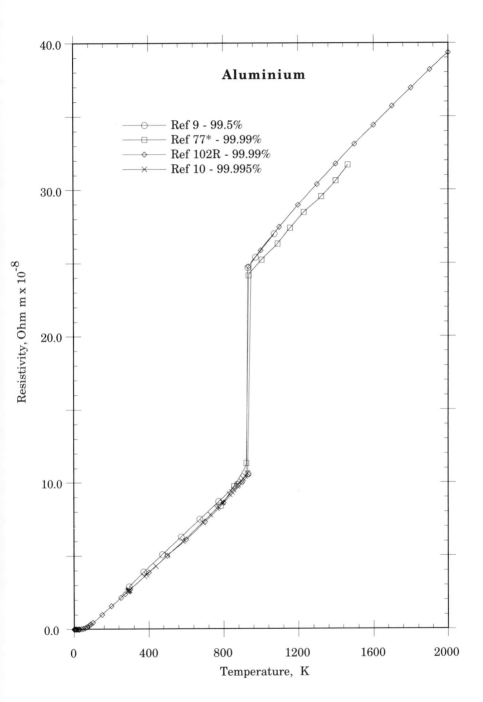

Aluminium

Ref 9 - 99.5%
Ref 77* - 99.99%
Ref 102R - 99.99%
Ref 10 - 99.995%

Resistivity, Ohm m x 10^{-8}

Temperature, K

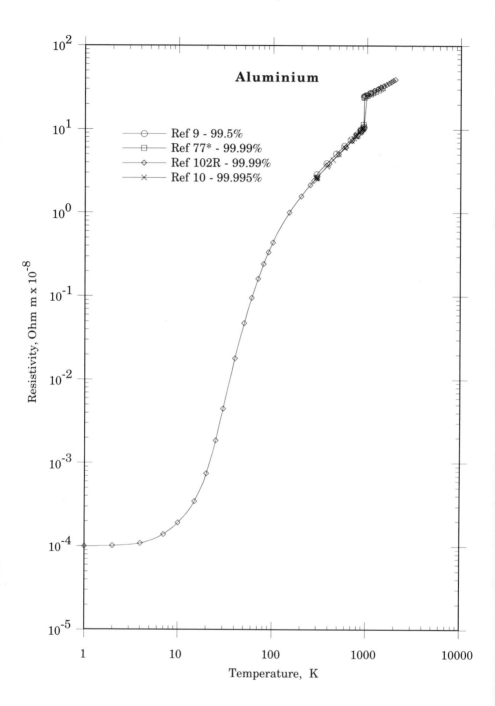

Aluminium

Ref 9 - 99.5%
Ref 77* - 99.99%
Ref 102R - 99.99%
Ref 10 - 99.995%

Resistivity, Ohm m x 10^{-8}

Temperature, K

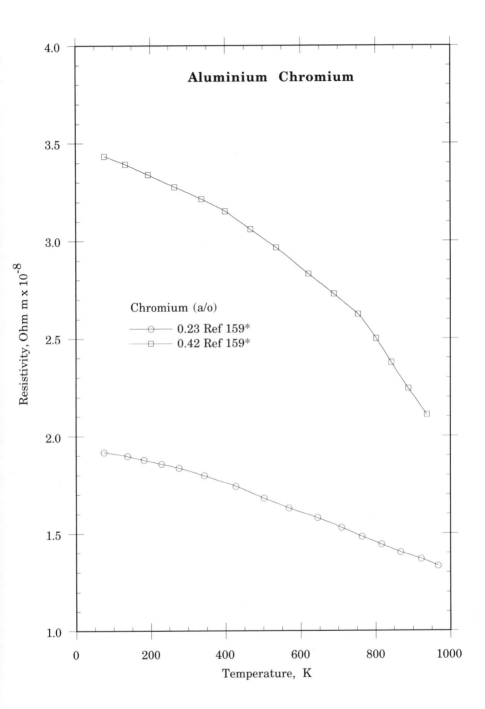

Aluminium   Chromium

Chromium (a/o)
— ⊙ — 0.23 Ref 159*
— ☐ — 0.42 Ref 159*

Resistivity, Ohm m x 10^{-8}

Temperature, K

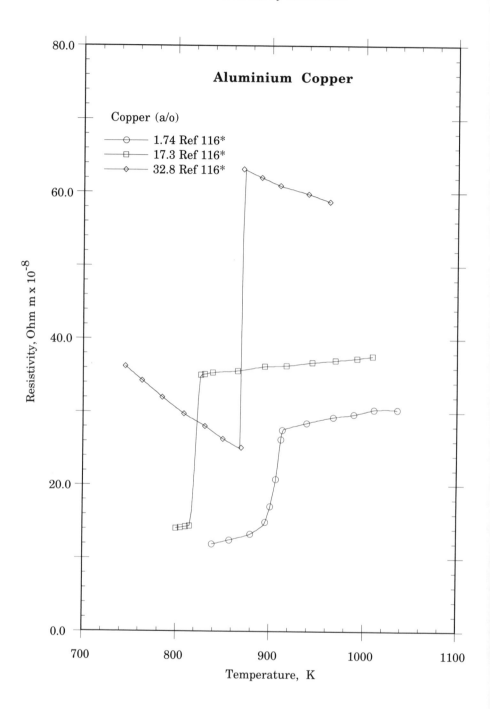

Aluminium Copper

Copper (a/o)

○ 1.74 Ref 116*
□ 17.3 Ref 116*
◇ 32.8 Ref 116*

Resistivity, Ohm m x 10^{-8}

Temperature, K

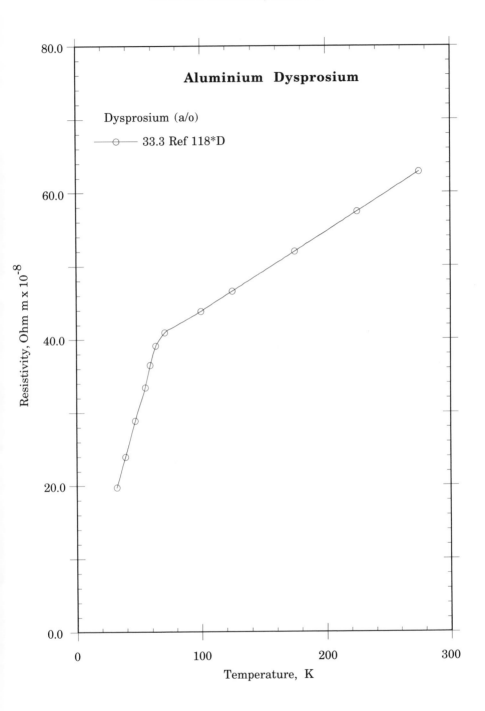

Aluminium Dysprosium

Dysprosium (a/o)

—⊙— 33.3 Ref 118*D

Resistivity, Ohm m x 10⁻⁸

Temperature, K

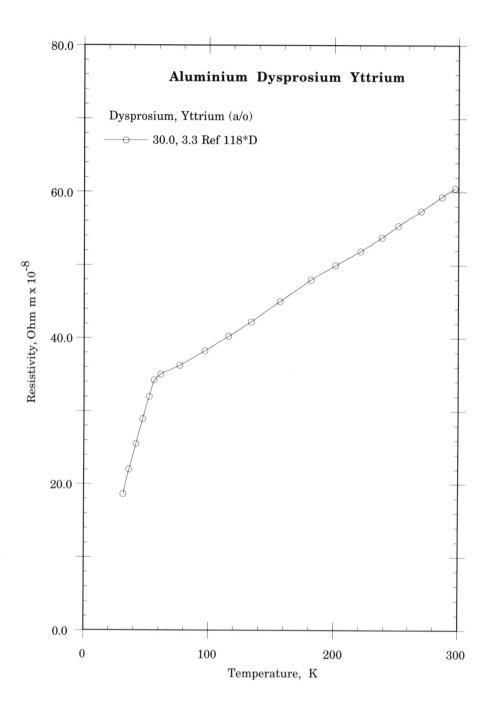

Aluminium Dysprosium Yttrium

Dysprosium, Yttrium (a/o)

30.0, 3.3 Ref 118*D

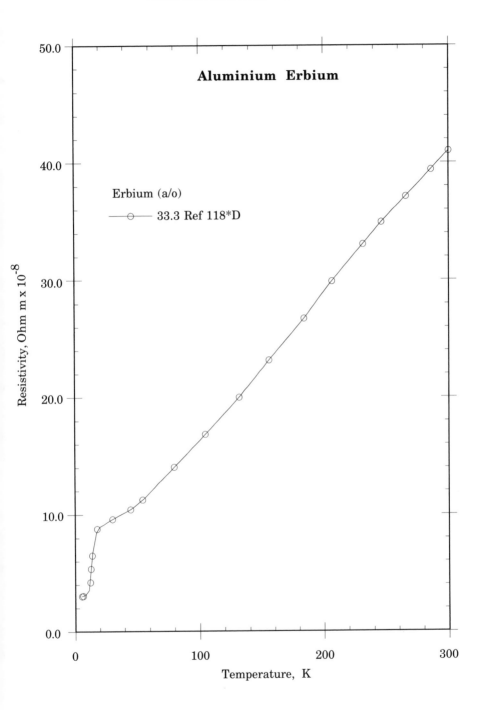

Aluminium Erbium

Erbium (a/o)

—⊙— 33.3 Ref 118*D

Resistivity, Ohm m x 10^{-8}

Temperature, K

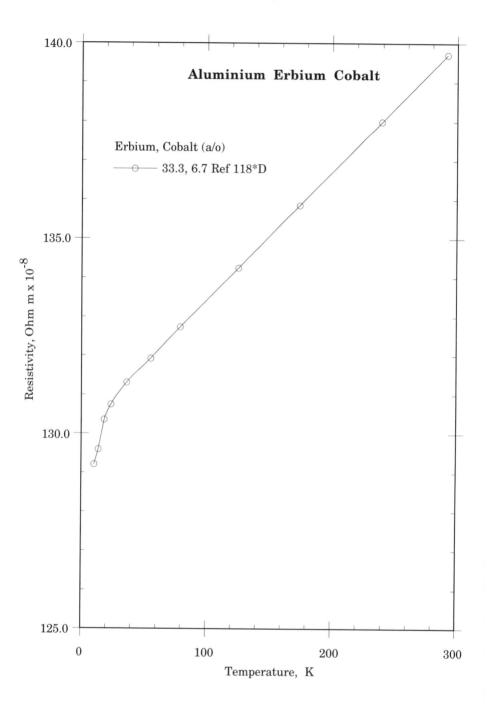

Aluminium Erbium Cobalt

Erbium, Cobalt (a/o)

——⊙—— 33.3, 6.7 Ref 118*D

Resistivity, Ohm m x 10^{-8}

Temperature, K

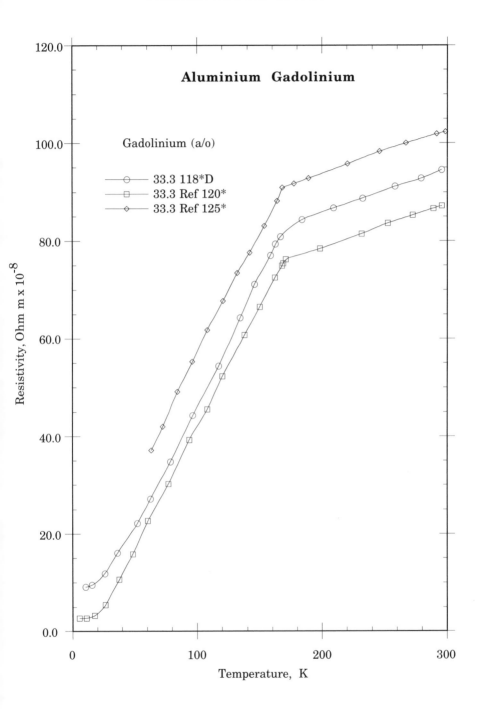

Aluminium Gadolinium

Gadolinium (a/o)

○ 33.3 118*D
□ 33.3 Ref 120*
◇ 33.3 Ref 125*

Resistivity, Ohm m x 10^{-8}

Temperature, K

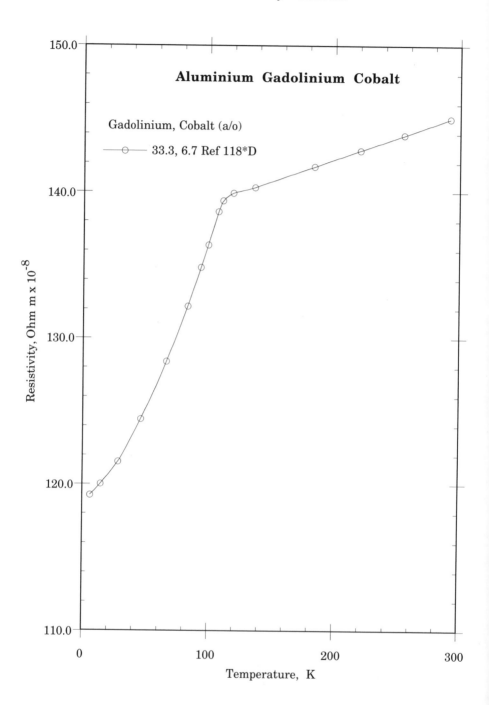

Aluminium Gadolinium Cobalt

Gadolinium, Cobalt (a/o)

——⊙—— 33.3, 6.7 Ref 118*D

Resistivity, Ohm m x 10^{-8}

Temperature, K

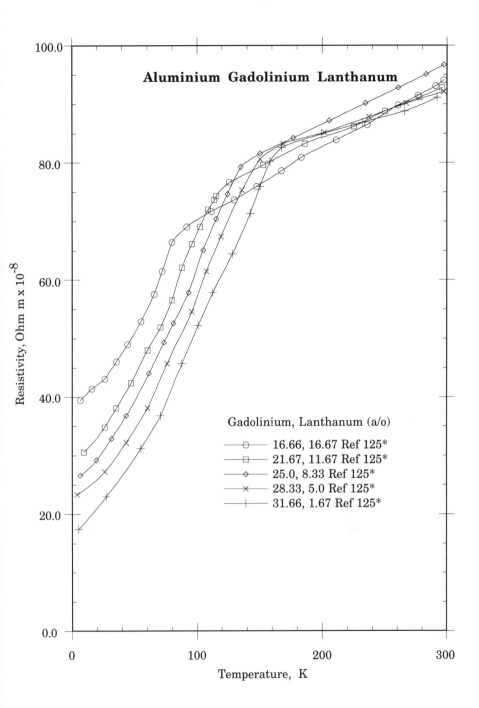

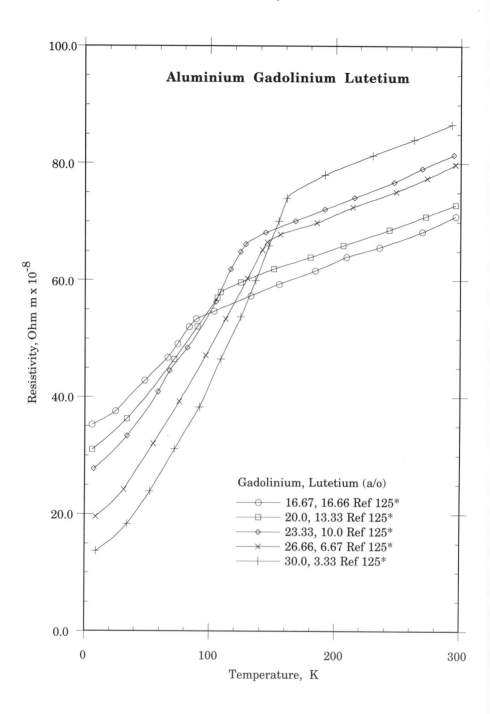

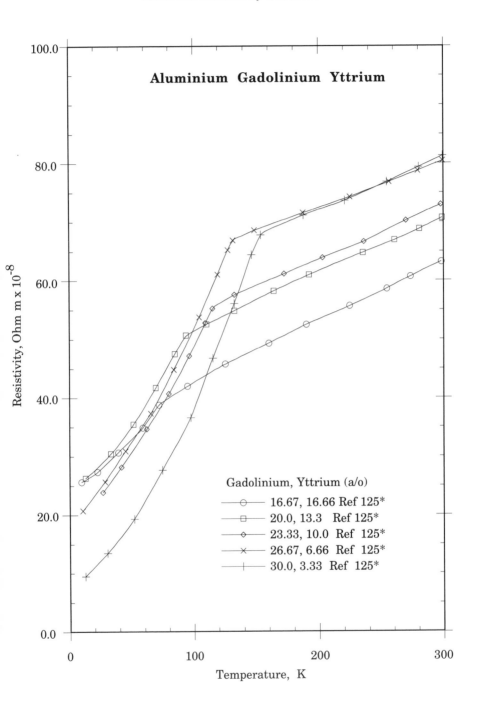

Aluminium   Gadolinium   Yttrium

Resistivity, Ohm m x 10^{-8}

Temperature, K

Gadolinium, Yttrium (a/o)
- —o— 16.67, 16.66 Ref 125*
- —□— 20.0, 13.3   Ref 125*
- —◇— 23.33, 10.0 Ref 125*
- —×— 26.67, 6.66 Ref 125*
- —+— 30.0, 3.33 Ref 125*

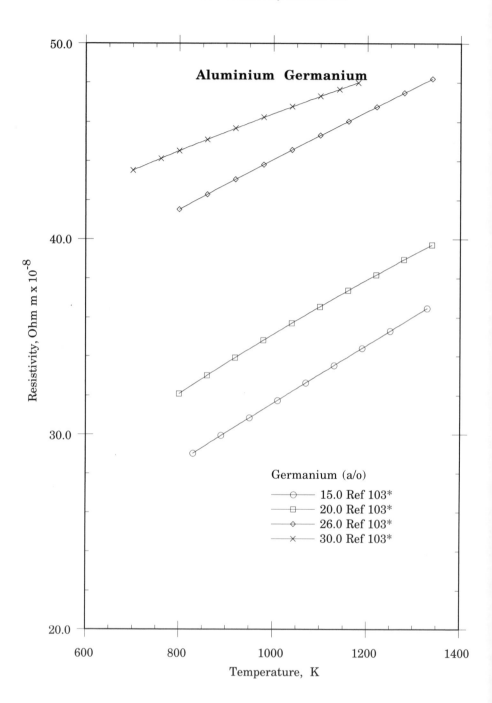

Aluminium Germanium

Resistivity, Ohm m x 10⁻⁸

Temperature, K

Germanium (a/o)
— ○ — 15.0 Ref 103*
— □ — 20.0 Ref 103*
— ◇ — 26.0 Ref 103*
— × — 30.0 Ref 103*

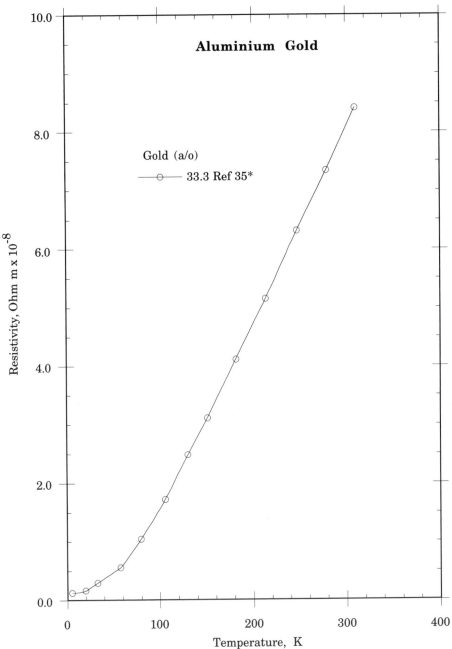

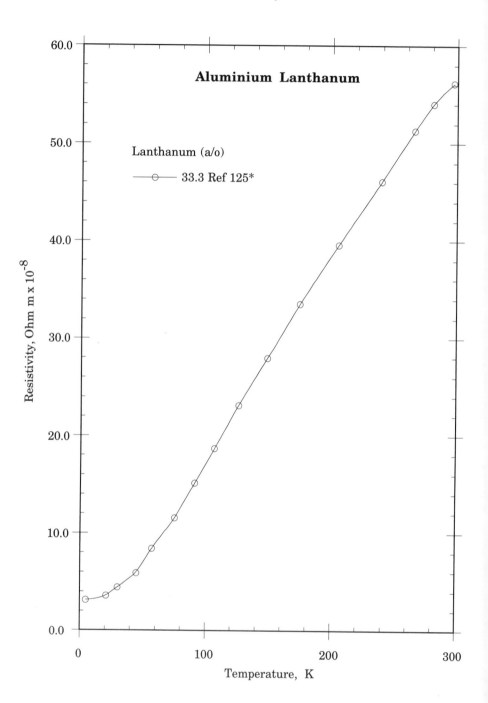

Aluminium Lanthanum

Lanthanum (a/o)

—○— 33.3 Ref 125*

Resistivity, Ohm m x 10^{-8}

Temperature, K

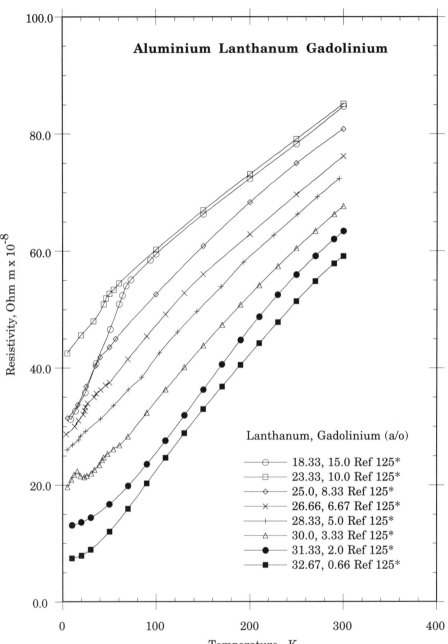

Aluminium Lanthanum Gadolinium

Resistivity, Ohm m x 10^{-8}

Temperature, K

Lanthanum, Gadolinium (a/o)

- ○ 18.33, 15.0 Ref 125*
- □ 23.33, 10.0 Ref 125*
- ◇ 25.0, 8.33 Ref 125*
- × 26.66, 6.67 Ref 125*
- + 28.33, 5.0 Ref 125*
- △ 30.0, 3.33 Ref 125*
- ● 31.33, 2.0 Ref 125*
- ■ 32.67, 0.66 Ref 125*

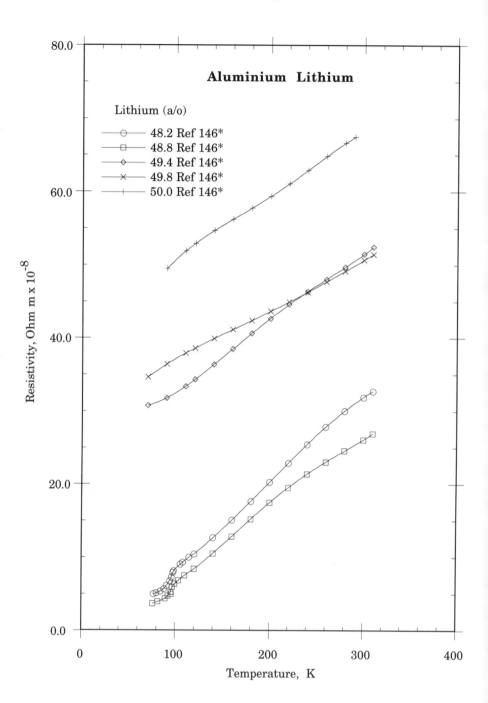

Aluminium Lithium

Lithium (a/o)

- ○— 48.2 Ref 146*
- □— 48.8 Ref 146*
- ◇— 49.4 Ref 146*
- ×— 49.8 Ref 146*
- +— 50.0 Ref 146*

Resistivity, Ohm m x 10^{-8}

Temperature, K

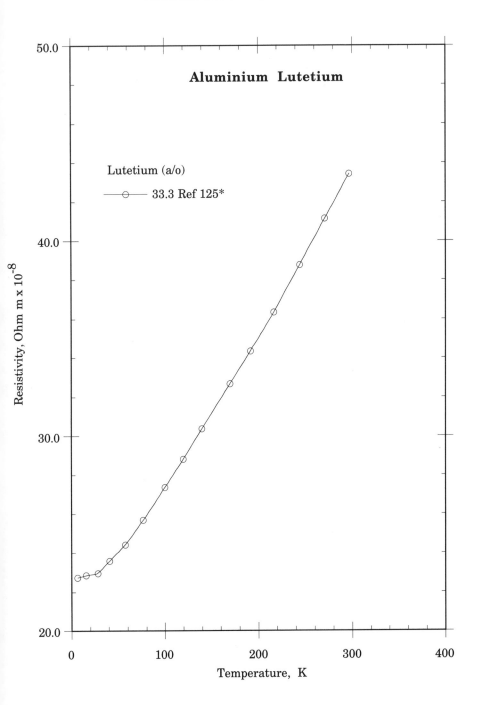

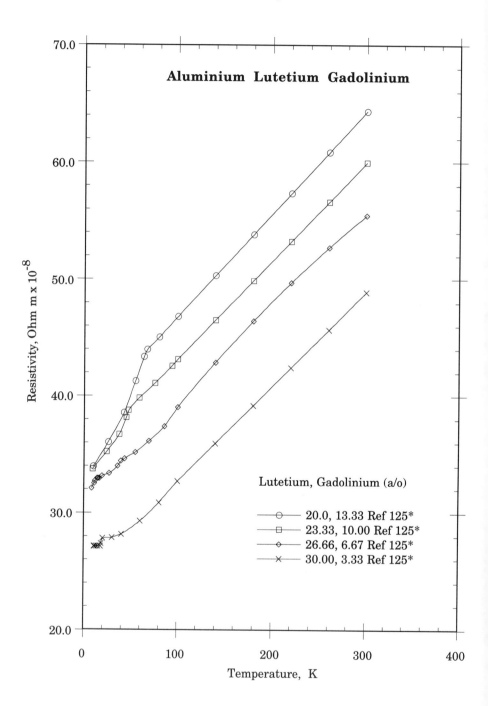

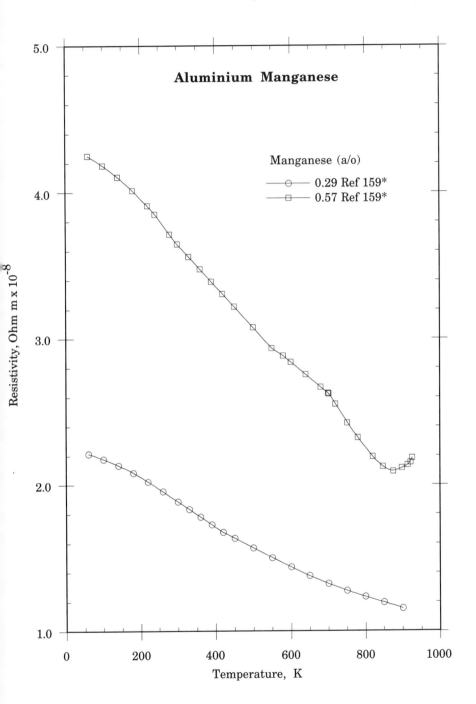

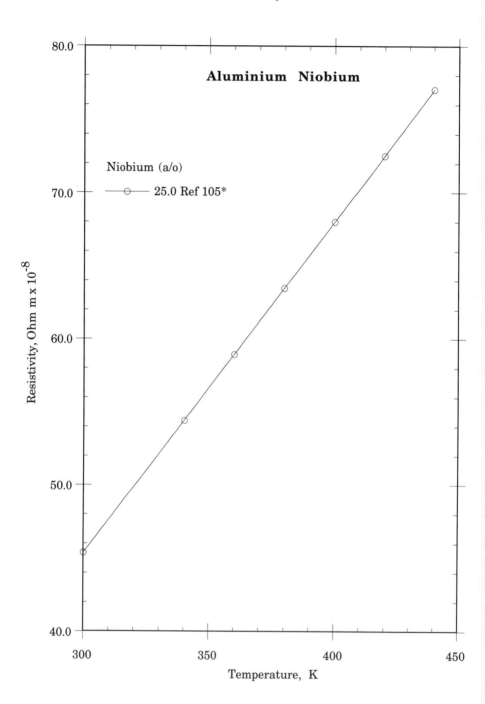

Aluminium   Niobium

Niobium (a/o)

25.0 Ref 105*

Resistivity, Ohm m x 10^{-8}

Temperature, K

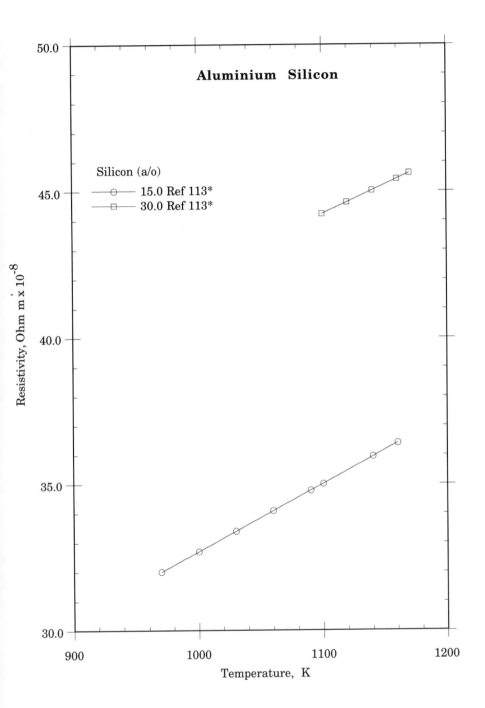

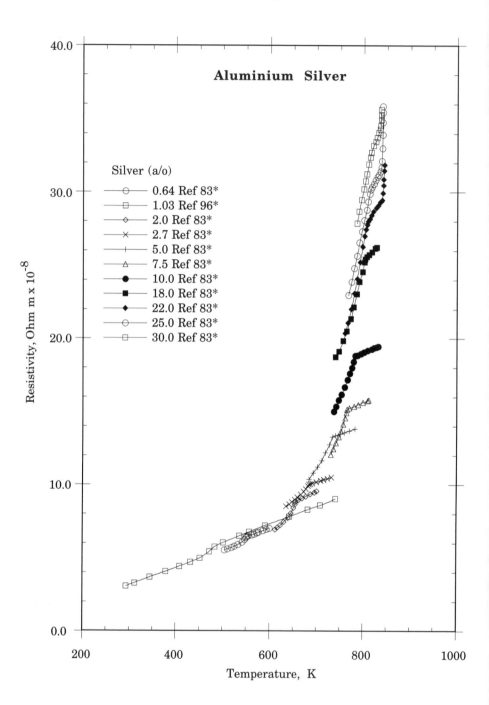

Aluminium Silver

Silver (a/o)
- ○— 0.64 Ref 83*
- □— 1.03 Ref 96*
- ◇— 2.0 Ref 83*
- ×— 2.7 Ref 83*
- +— 5.0 Ref 83*
- △— 7.5 Ref 83*
- ●— 10.0 Ref 83*
- ■— 18.0 Ref 83*
- ◆— 22.0 Ref 83*
- ○— 25.0 Ref 83*
- □— 30.0 Ref 83*

Resistivity, Ohm m x 10^{-8}

Temperature, K

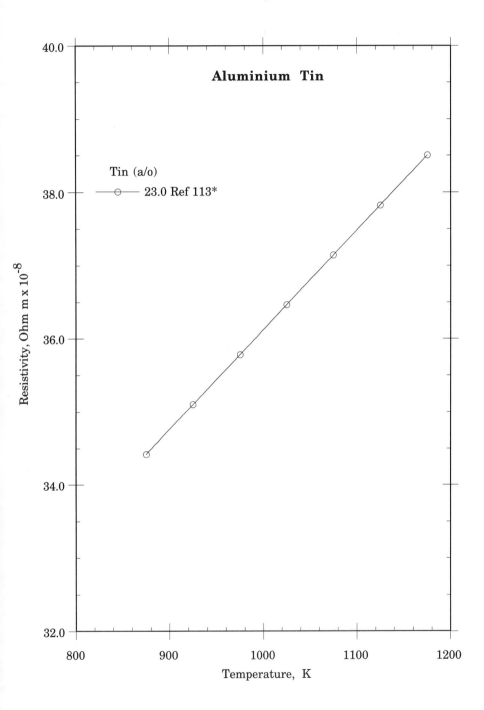

Aluminium Tin

Tin (a/o)
—◦— 23.0 Ref 113*

Resistivity, Ohm m x 10^{-8}

Temperature, K

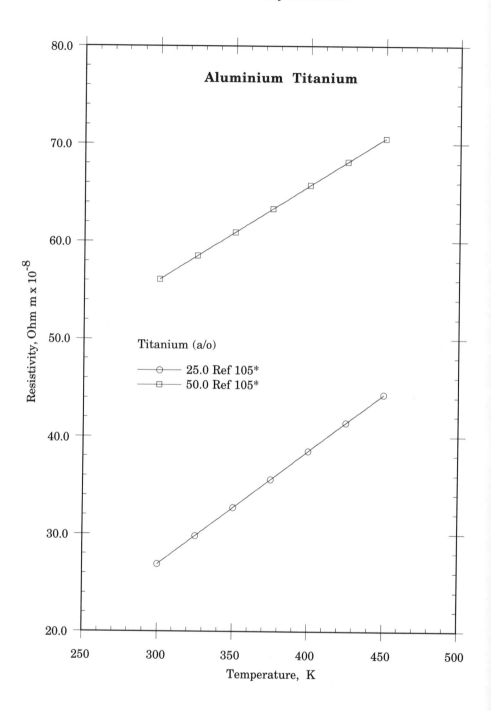

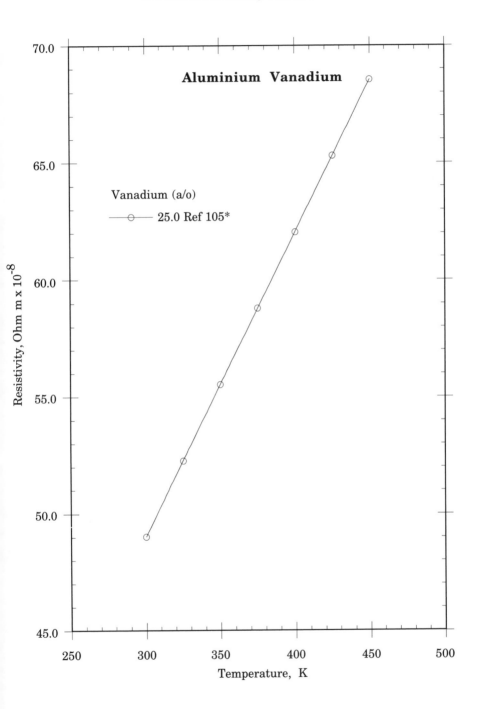

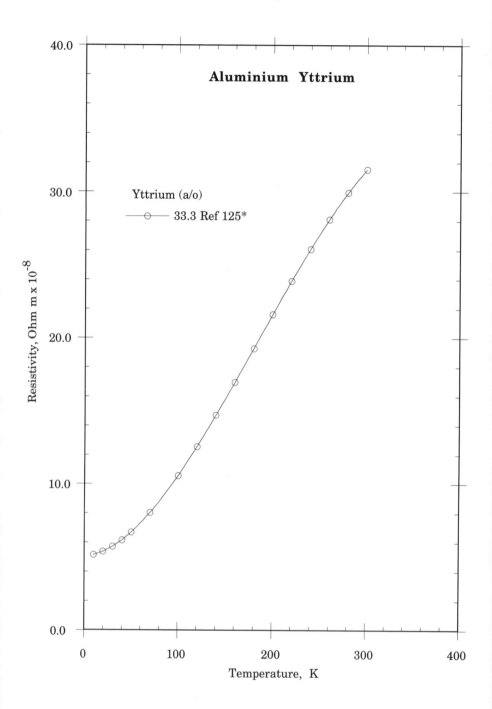

Aluminium Yttrium

Yttrium (a/o)
—○— 33.3 Ref 125*

Resistivity, Ohm m x 10⁻⁸

Temperature, K

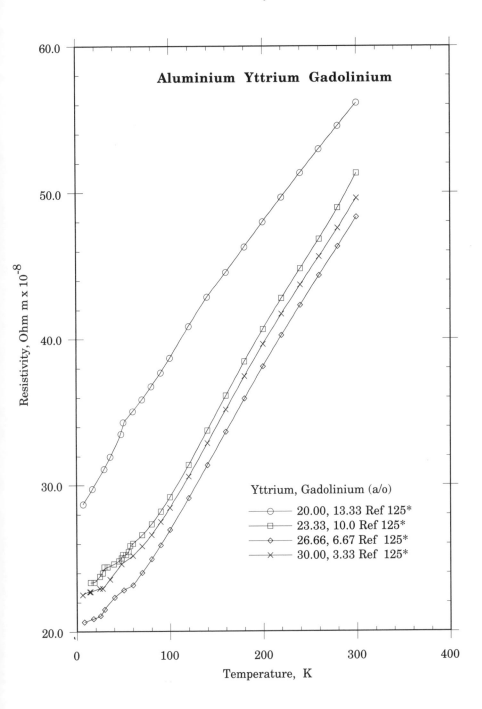

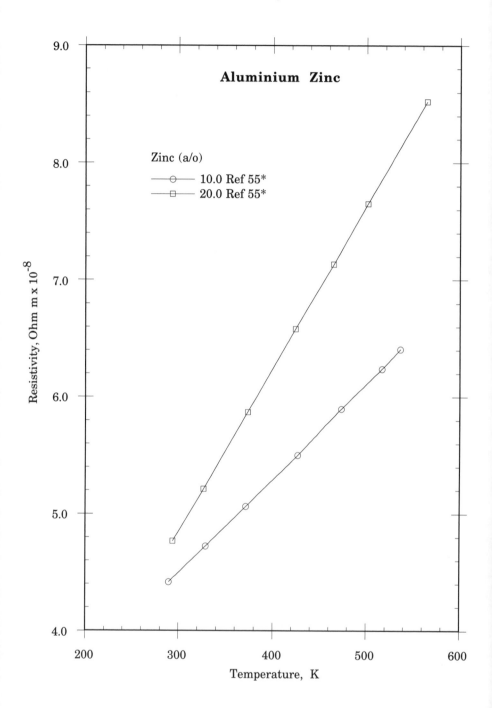

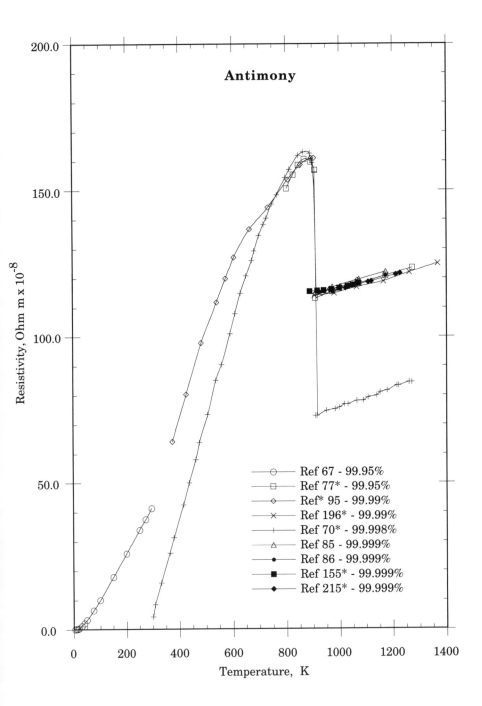

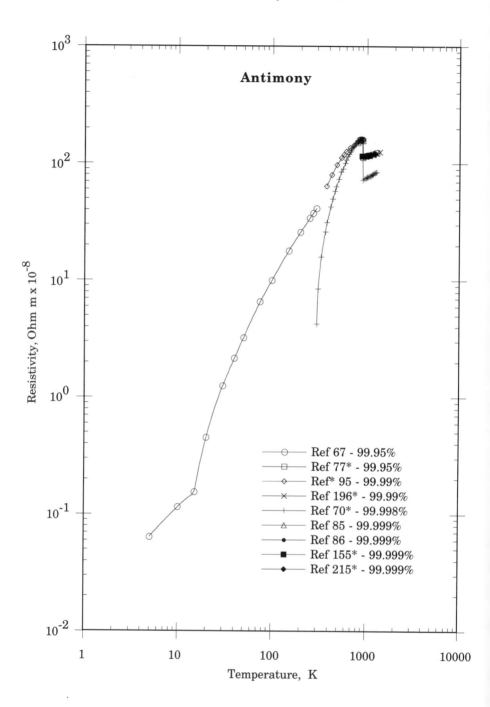

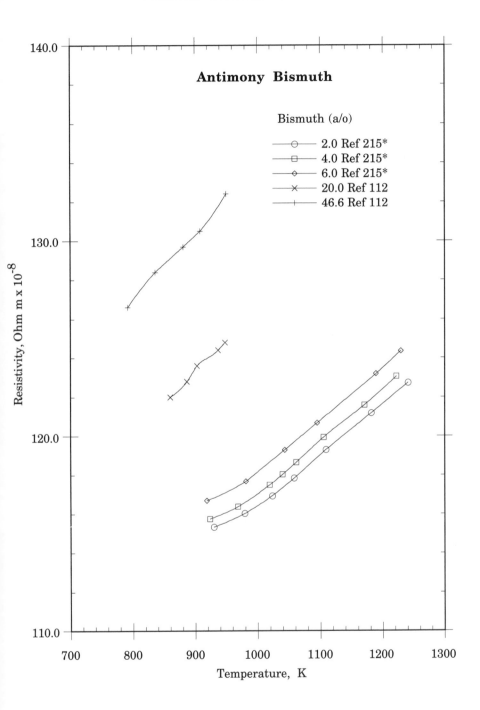

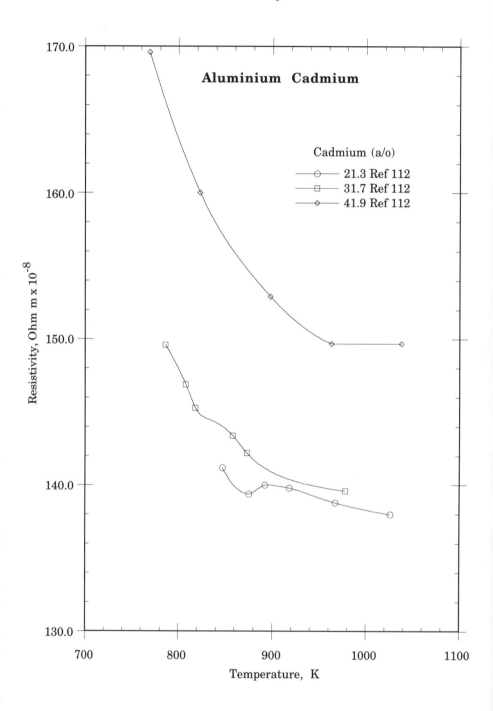

Aluminium Cadmium

Cadmium (a/o)

21.3 Ref 112
31.7 Ref 112
41.9 Ref 112

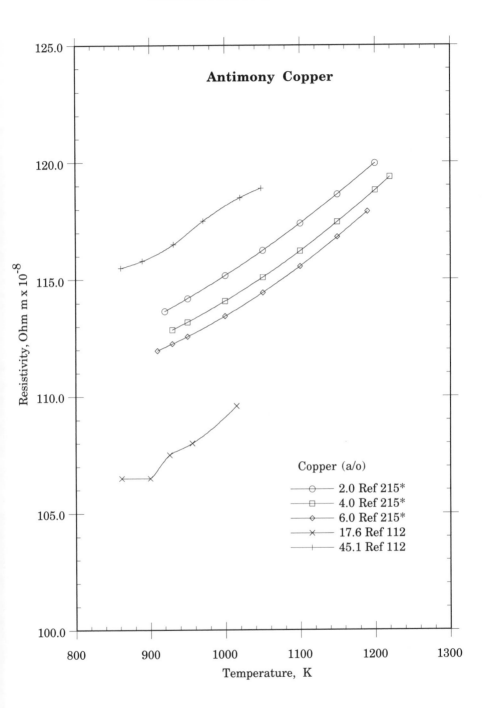

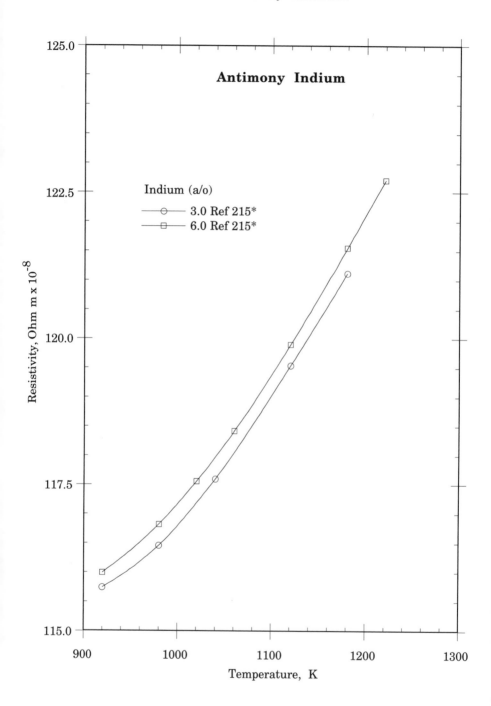

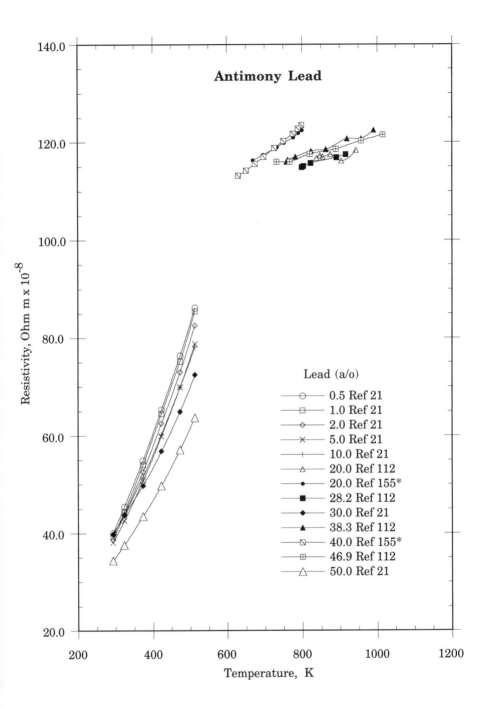

Antimony Lead

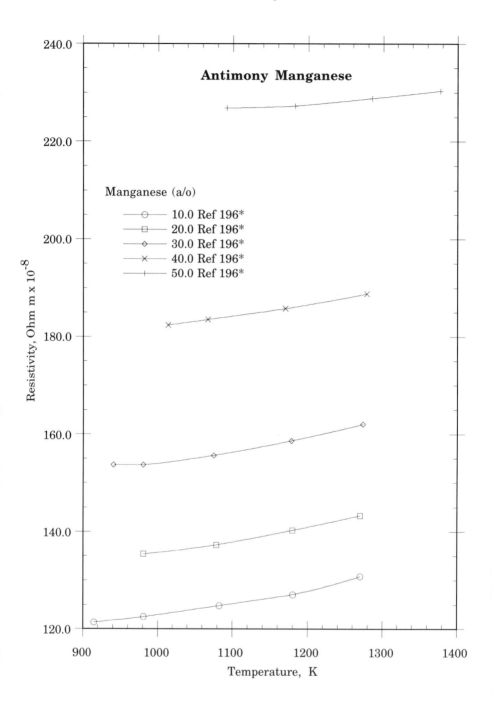

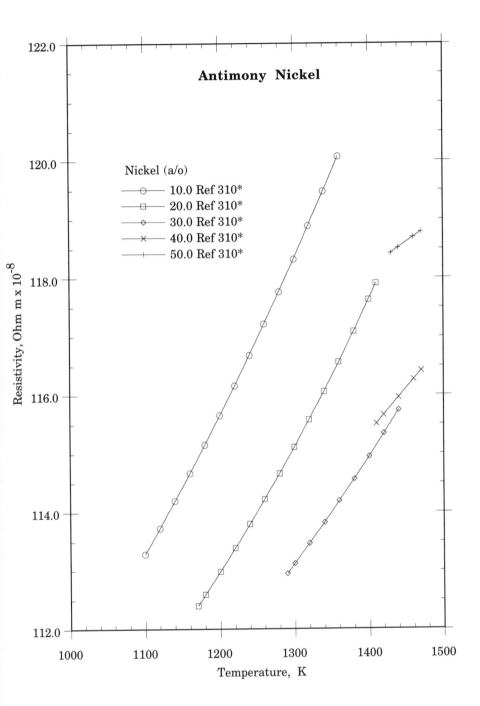

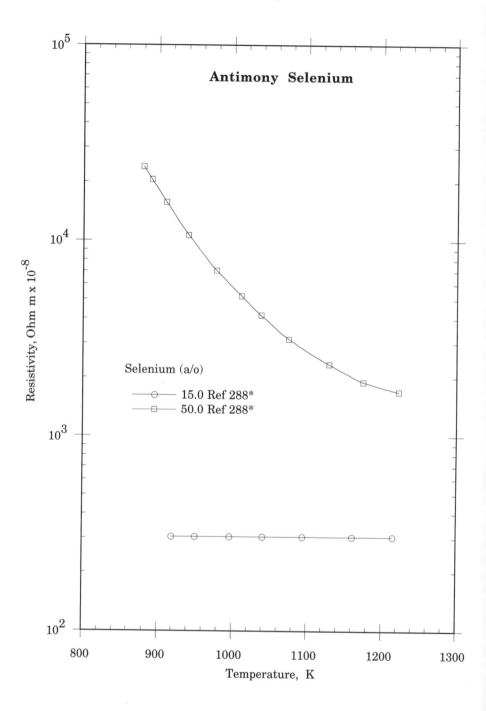

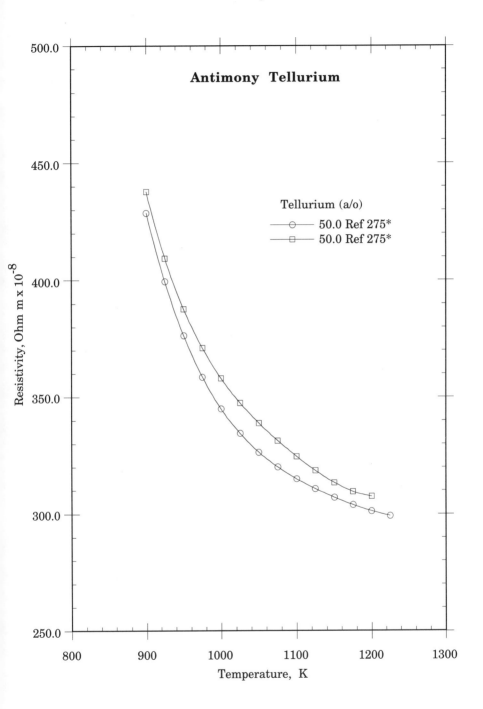

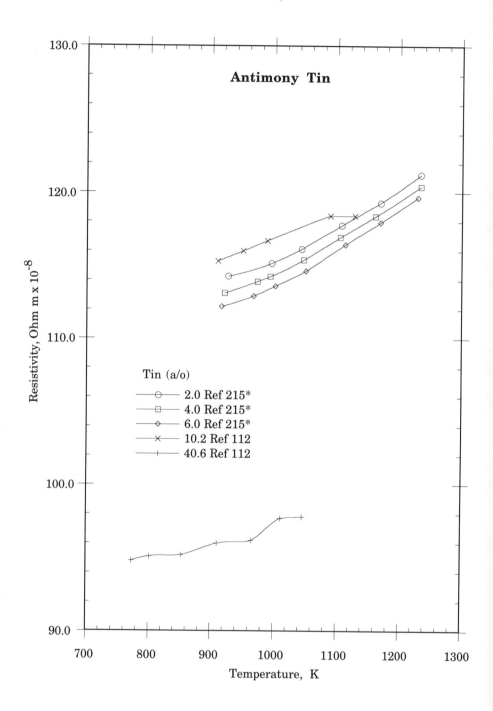

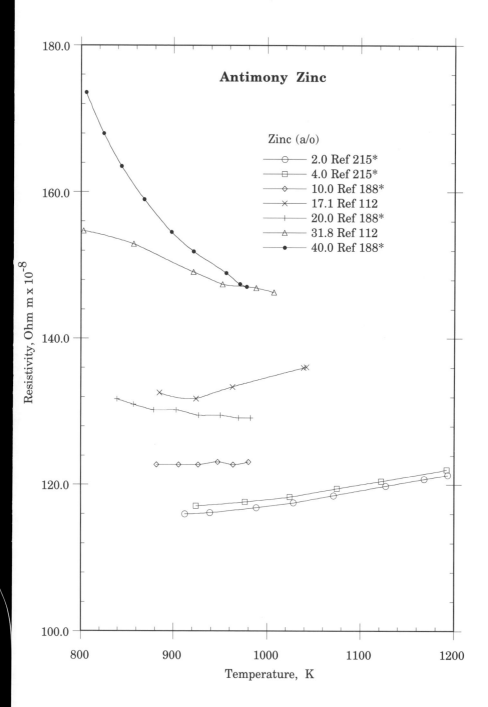

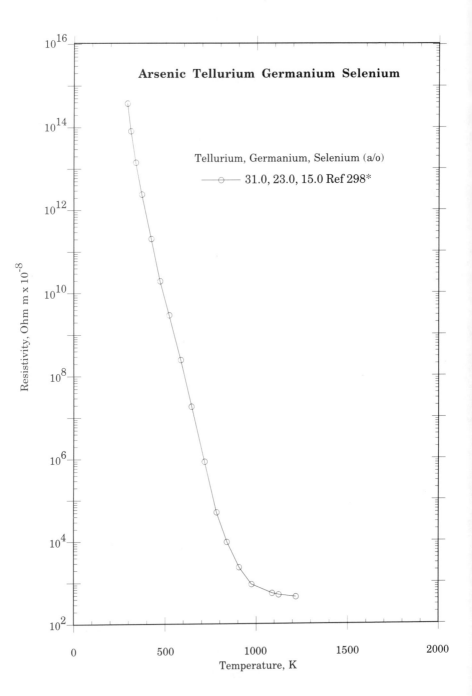

Arsenic Tellurium Germanium Selenium

Tellurium, Germanium, Selenium (a/o)
—o— 31.0, 23.0, 15.0 Ref 298*

Resistivity, Ohm m x 10⁻⁸

Temperature, K

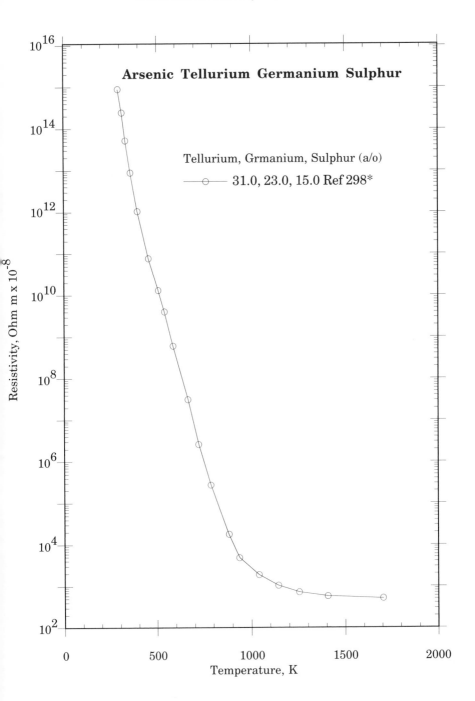

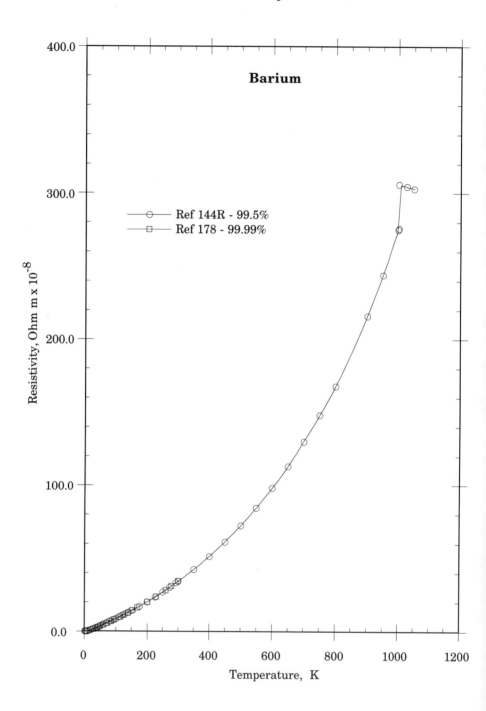

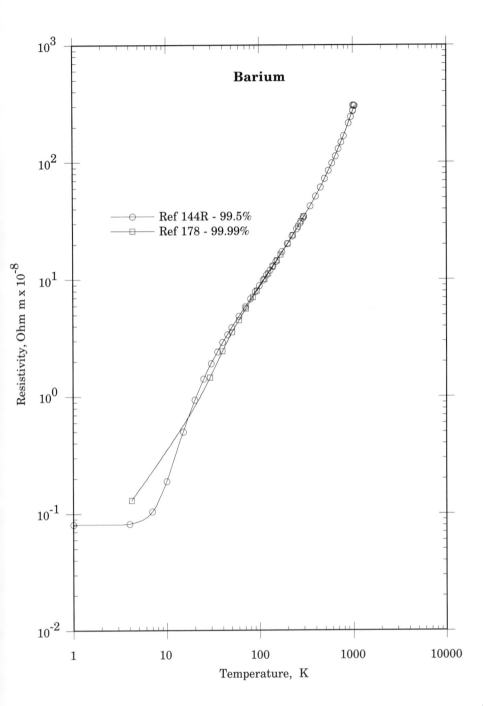

Barium

Ref 144R - 99.5%
Ref 178 - 99.99%

Resistivity, Ohm m x 10⁻⁸

Temperature, K

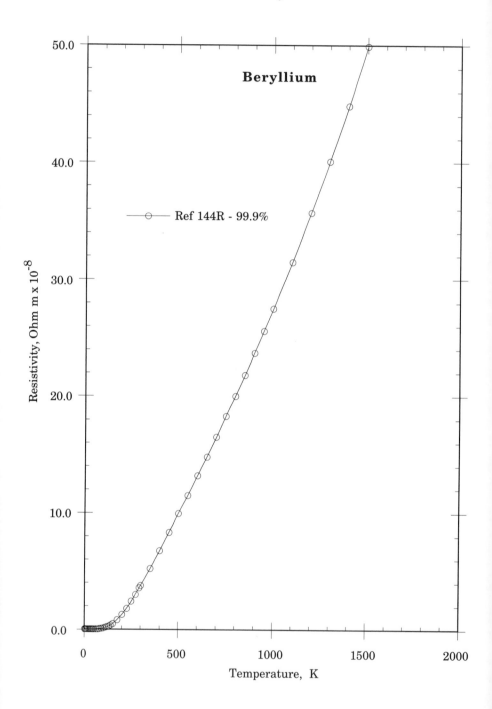

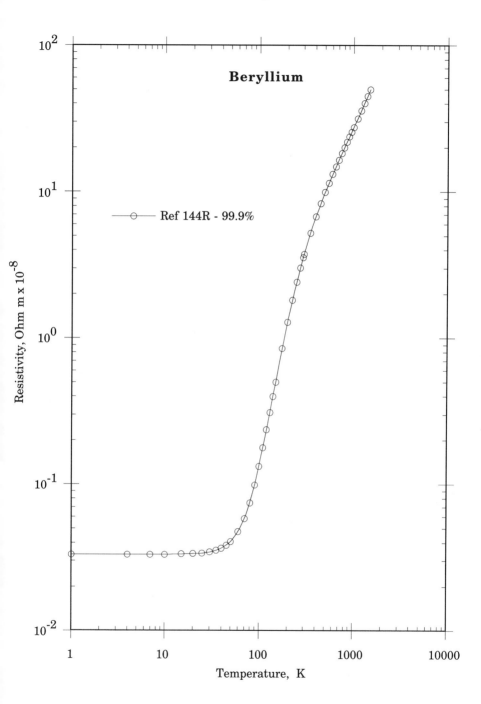

Beryllium

Ref 144R - 99.9%

Resistivity, Ohm m x 10^{-8}

Temperature, K

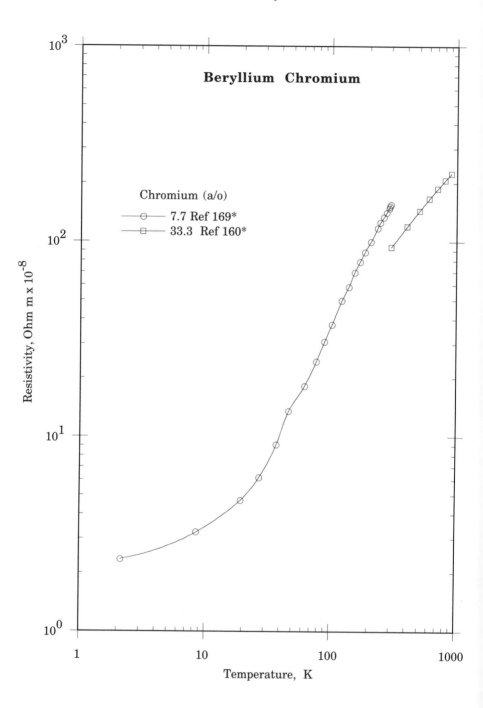

Beryllium   Chromium

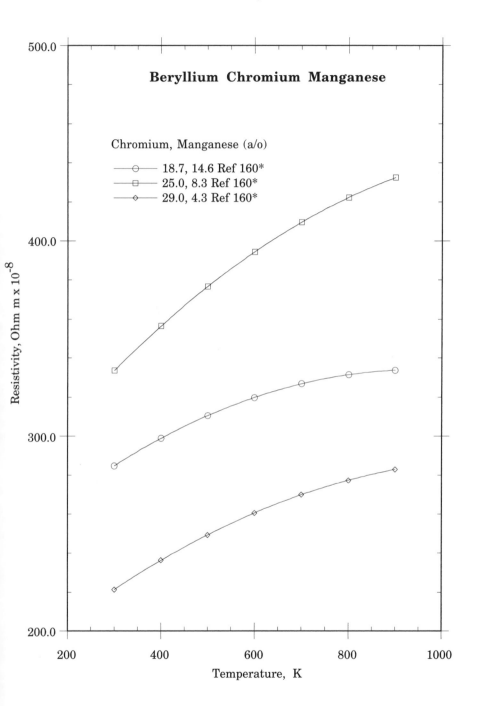

Beryllium Chromium Manganese

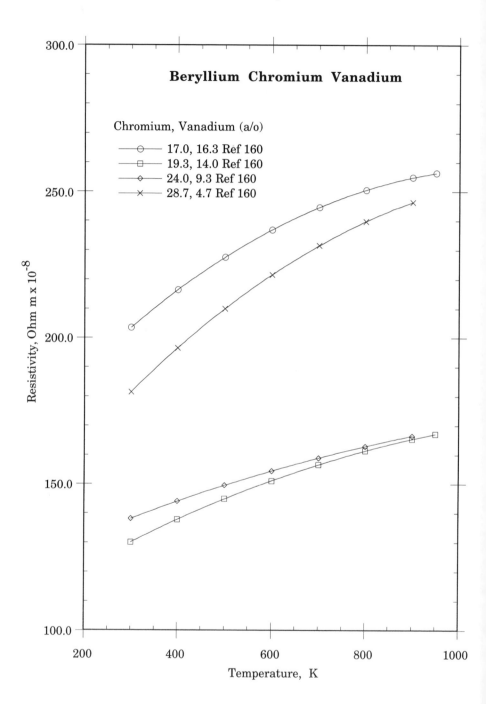

Beryllium   Chromium   Vanadium

Chromium, Vanadium (a/o)
- ⊖ 17.0, 16.3 Ref 160
- □ 19.3, 14.0 Ref 160
- ◇ 24.0, 9.3 Ref 160
- × 28.7, 4.7 Ref 160

Resistivity, Ohm m x 10^{-8}

Temperature, K

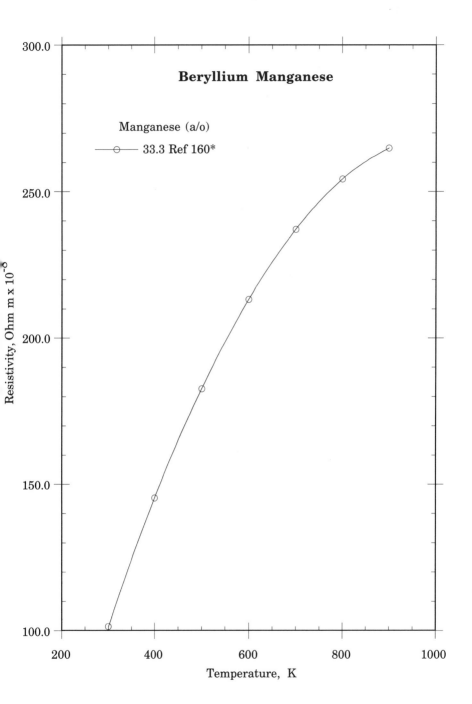

Beryllium Manganese

Manganese (a/o)

—⊖— 33.3 Ref 160*

Resistivity, Ohm m x 10^{-6}

Temperature, K

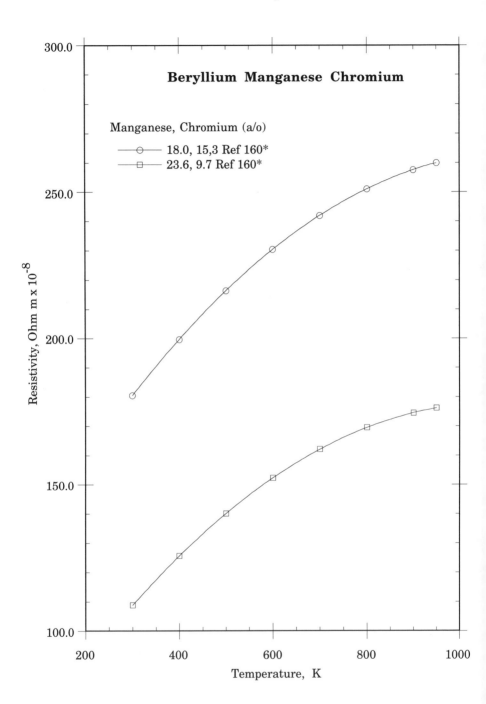

Beryllium Manganese Chromium

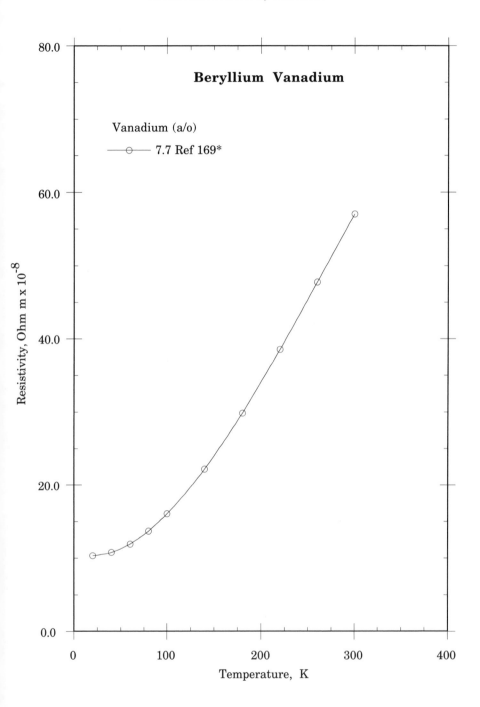

Beryllium Vanadium

Vanadium (a/o)
—○— 7.7 Ref 169*

Resistivity, Ohm m x 10^{-8}

Temperature, K

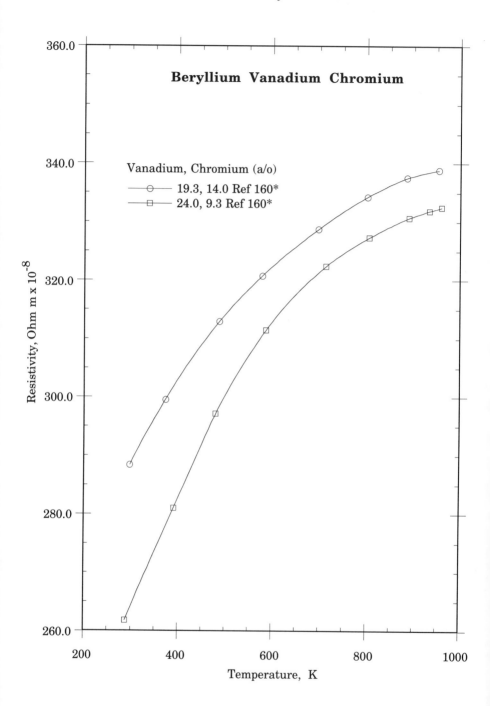

Beryllium Vanadium Chromium

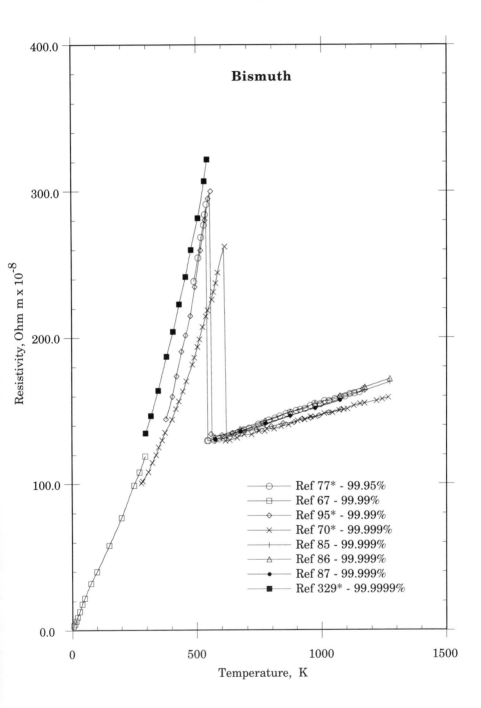

Bismuth

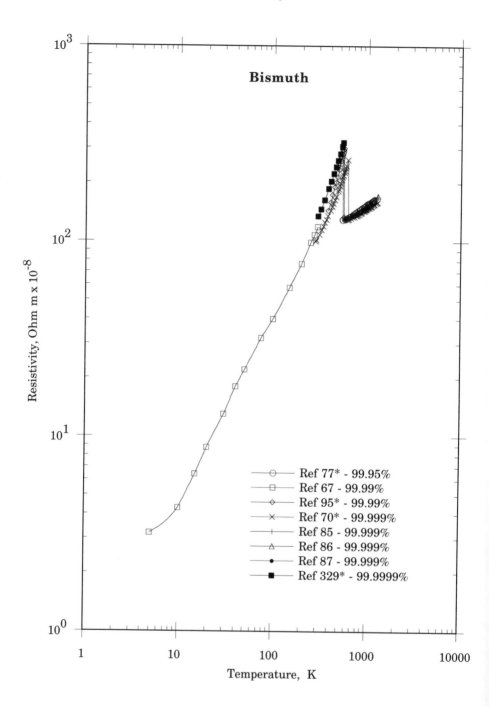

Bismuth

Resistivity, Ohm m x 10^{-8}

Temperature, K

Ref 77* - 99.95%
Ref 67 - 99.99%
Ref 95* - 99.99%
Ref 70* - 99.999%
Ref 85 - 99.999%
Ref 86 - 99.999%
Ref 87 - 99.999%
Ref 329* - 99.9999%

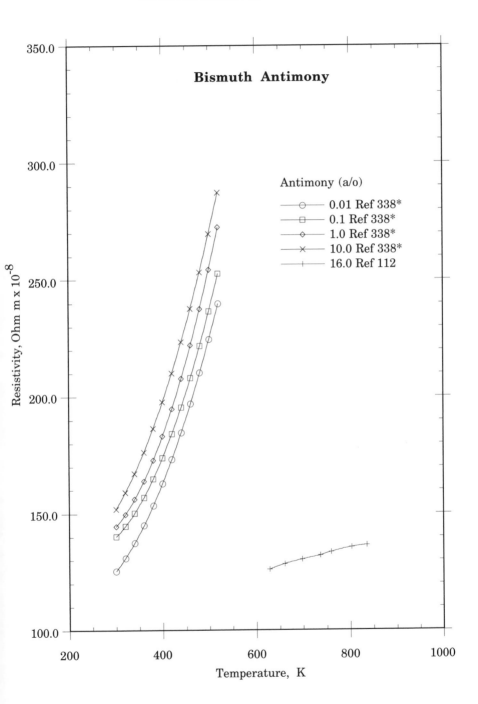

Bismuth Antimony

Antimony (a/o)

- —○— 0.01 Ref 338*
- —□— 0.1 Ref 338*
- —◇— 1.0 Ref 338*
- —×— 10.0 Ref 338*
- —+— 16.0 Ref 112

Resistivity, Ohm m x 10⁻⁸

Temperature, K

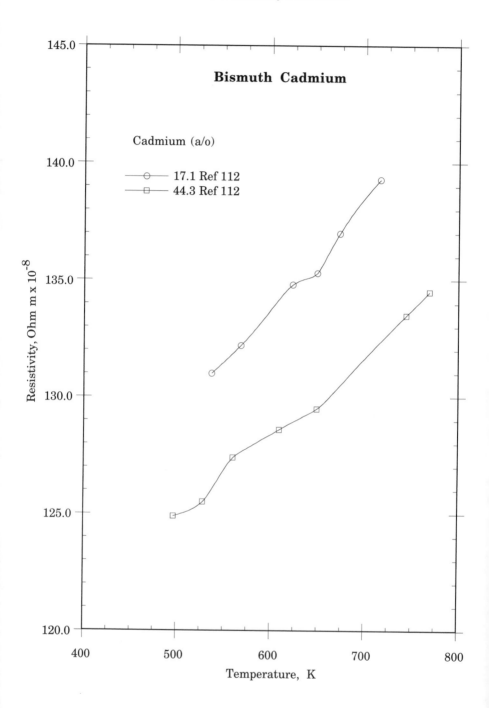

Bismuth Cadmium

Cadmium (a/o)

17.1 Ref 112
44.3 Ref 112

Resistivity, Ohm m x 10^{-8}

Temperature, K

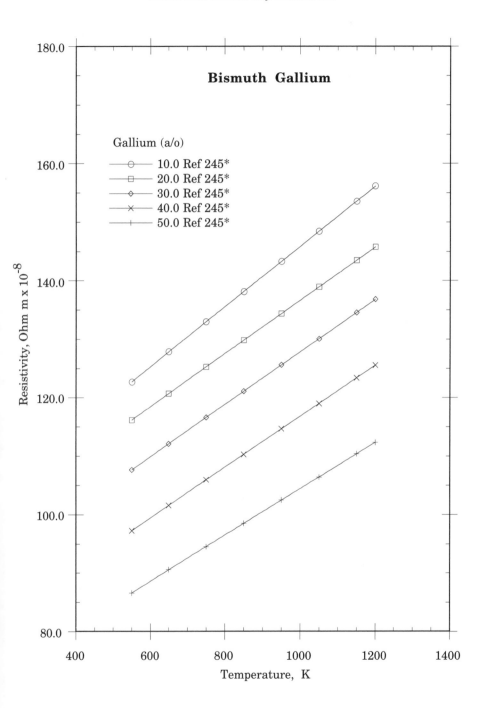

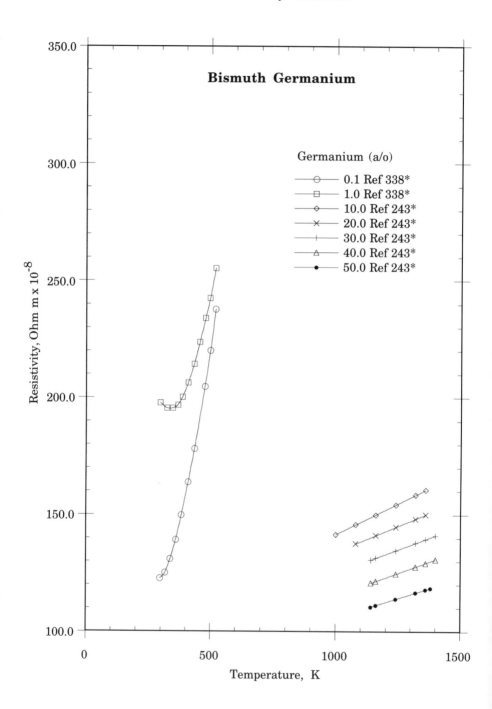

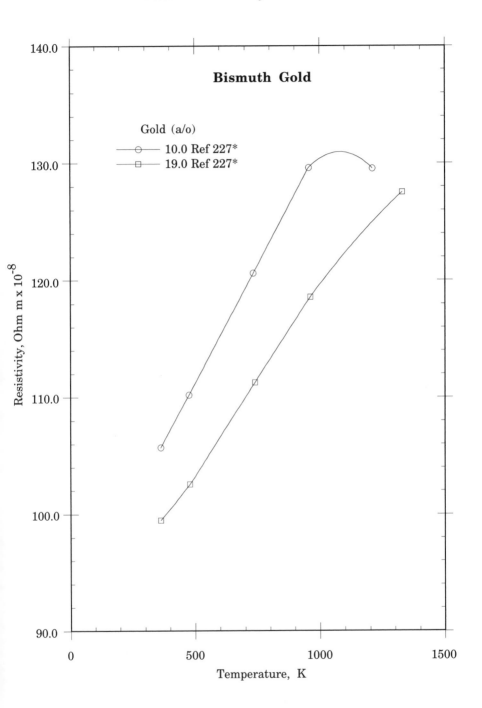

Bismuth Gold

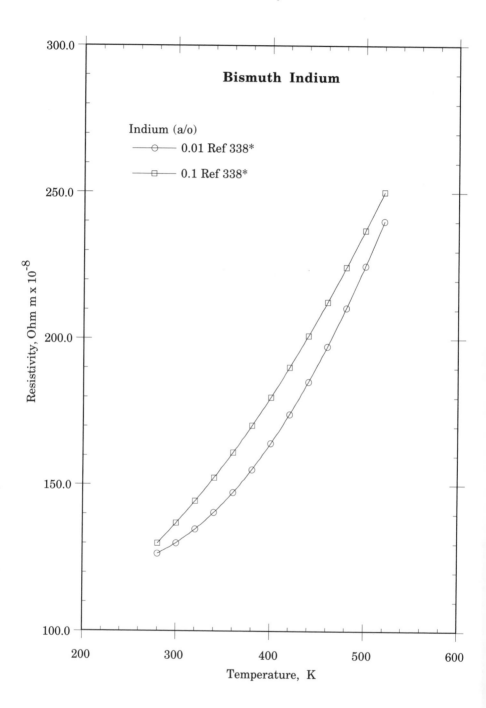

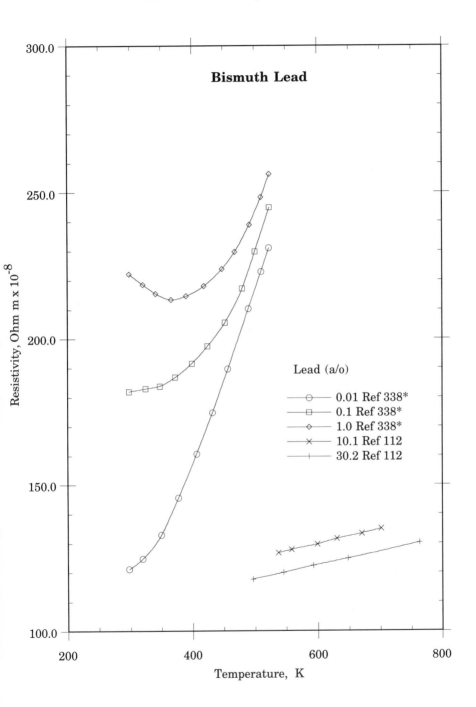

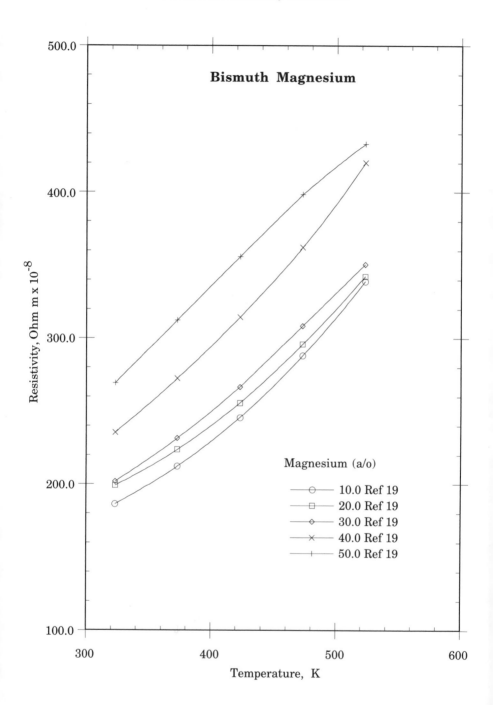

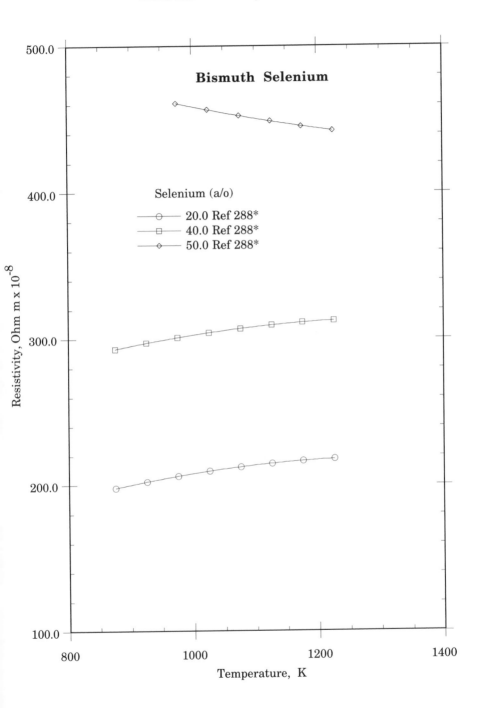

Bismuth Selenium

Selenium (a/o)

— ◦ — 20.0 Ref 288*
— □ — 40.0 Ref 288*
— ◇ — 50.0 Ref 288*

Resistivity, Ohm m x 10^{-8}

Temperature, K

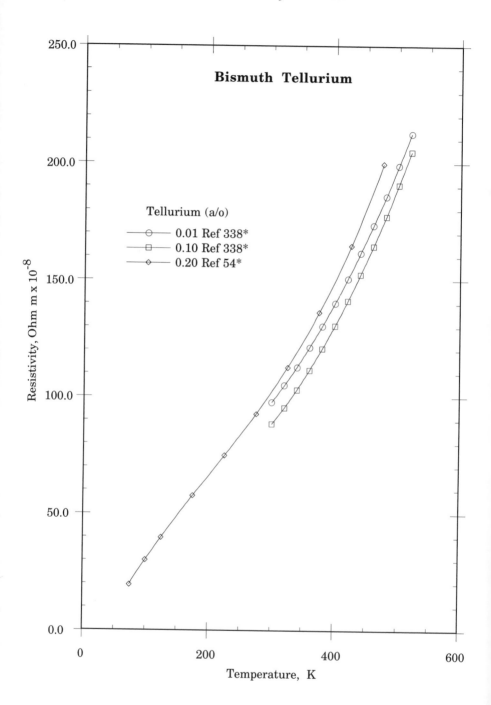

Bismuth Tellurium

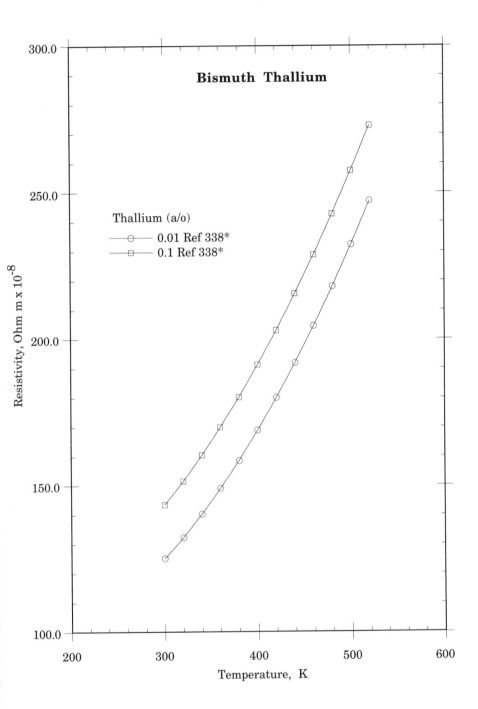

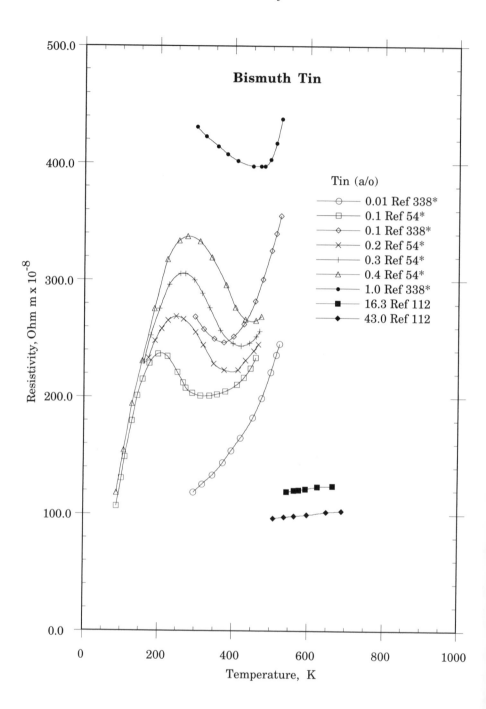

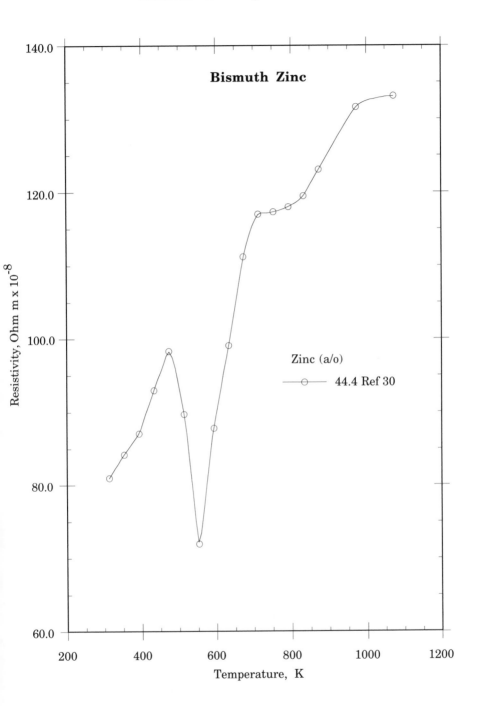

Bismuth Zinc

Zinc (a/o)

⊖ 44.4 Ref 30

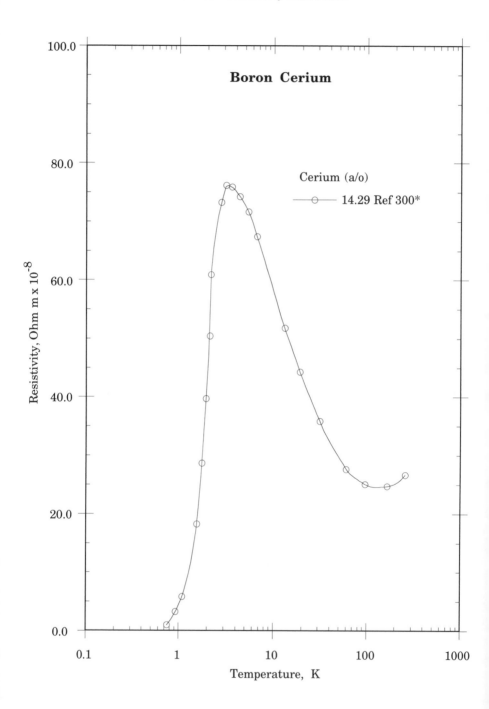

Boron Cerium

Cerium (a/o)

14.29 Ref 300*

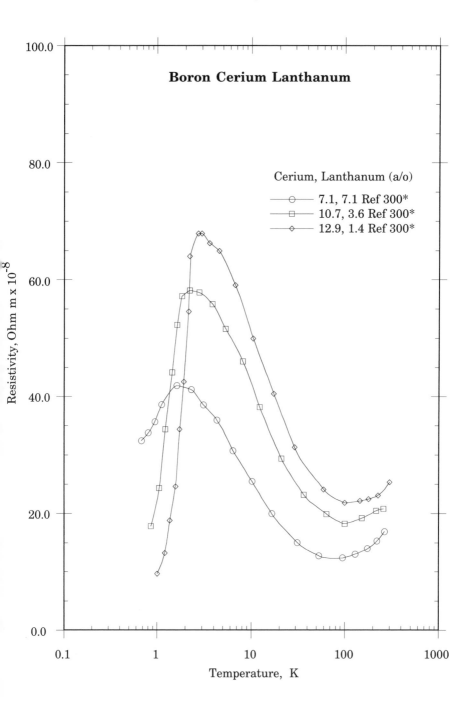

Boron Cerium Lanthanum

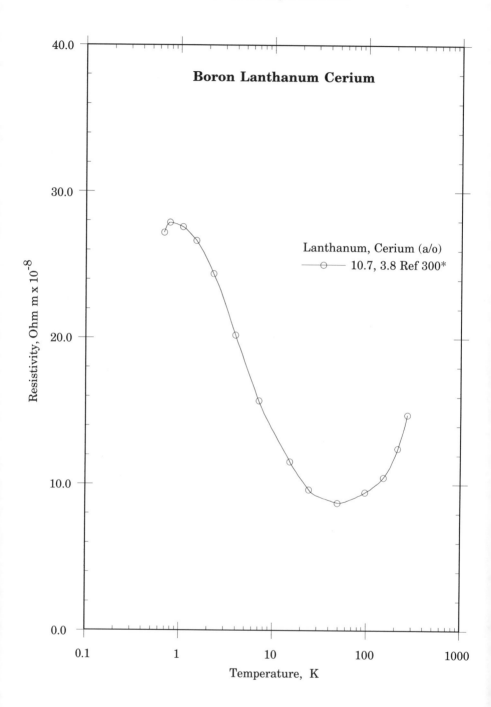

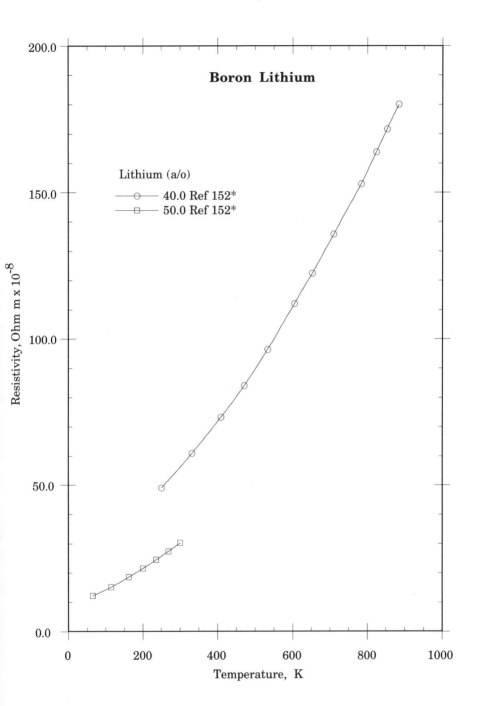

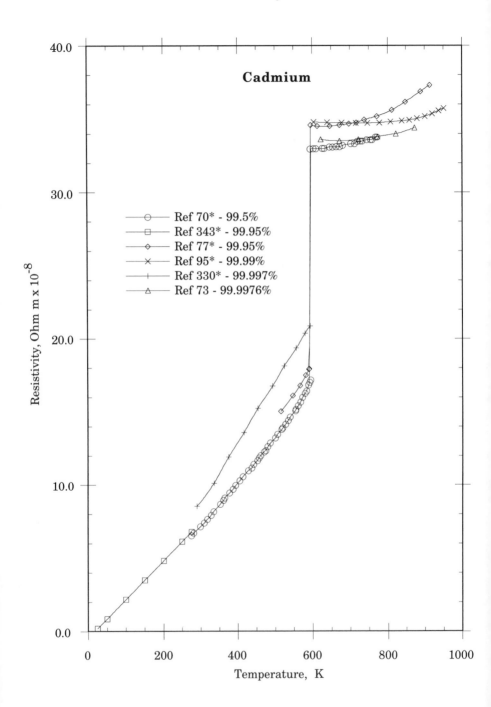

Cadmium

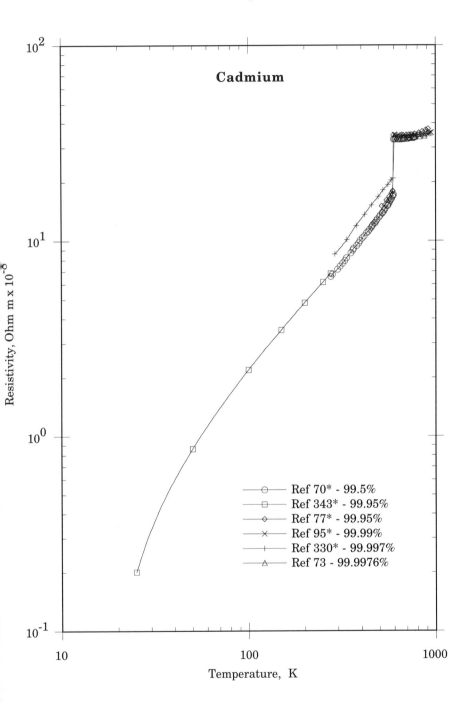

Cadmium

Resistivity, Ohm m x 10^{-8}

Temperature, K

- ○ — Ref 70* - 99.5%
- □ — Ref 343* - 99.95%
- ◇ — Ref 77* - 99.95%
- ✕ — Ref 95* - 99.99%
- + — Ref 330* - 99.997%
- △ — Ref 73 - 99.9976%

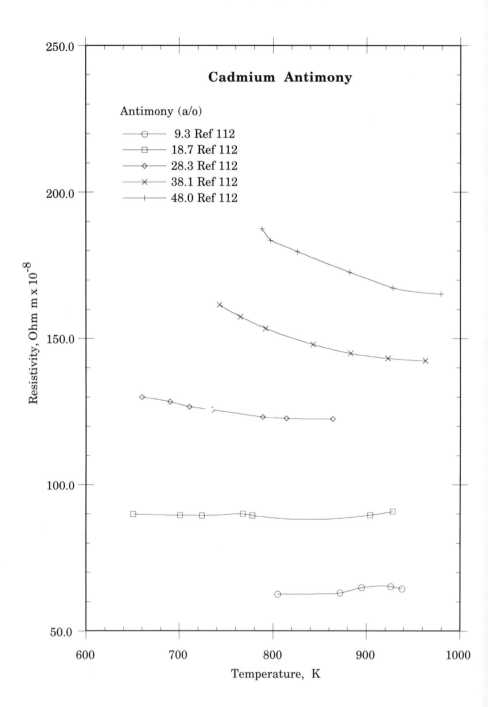

Cadmium Antimony

Antimony (a/o)
- ◦ 9.3 Ref 112
- □ 18.7 Ref 112
- ◇ 28.3 Ref 112
- ✕ 38.1 Ref 112
- ＋ 48.0 Ref 112

Resistivity, Ohm m x 10^{-8}

Temperature, K

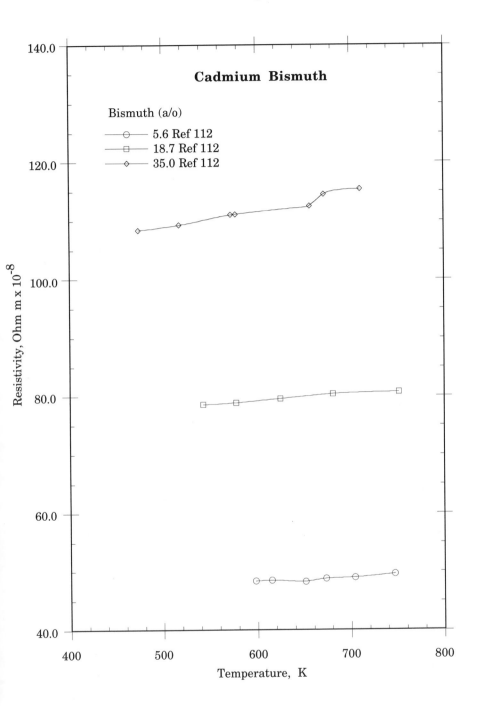

Cadmium Bismuth

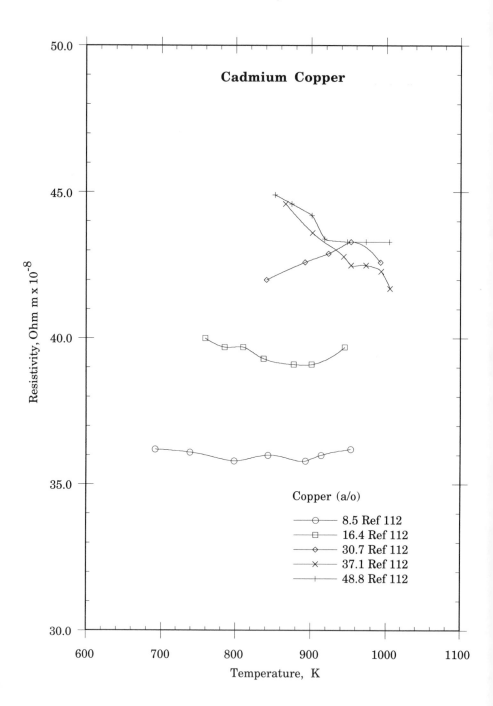

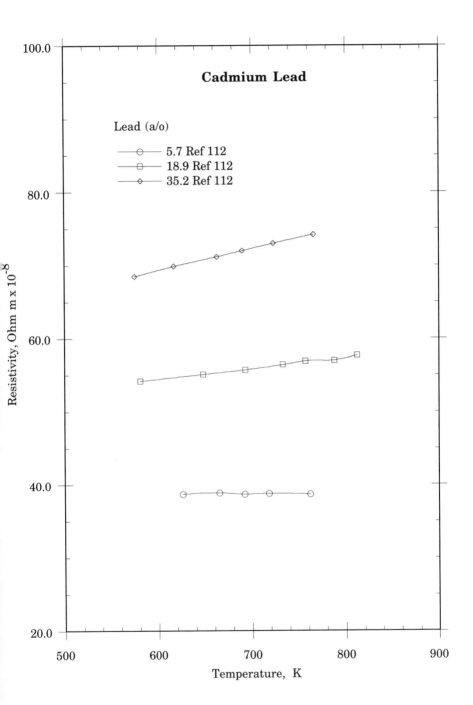

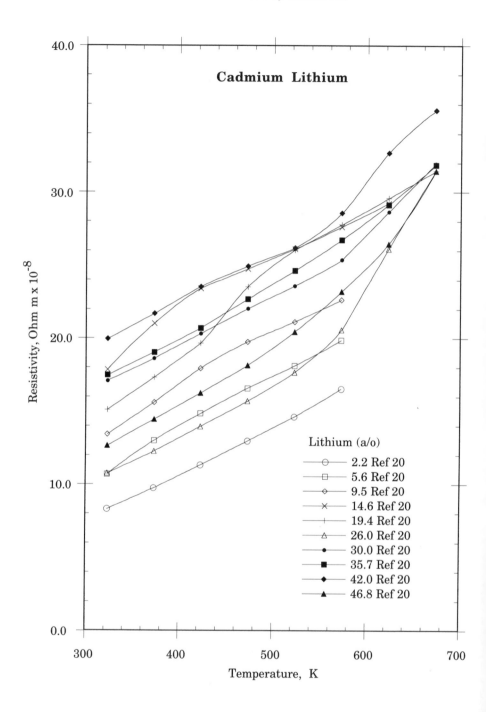

Cadmium Lithium

Resistivity, Ohm m x 10^{-8}

Temperature, K

Lithium (a/o)
2.2 Ref 20
5.6 Ref 20
9.5 Ref 20
14.6 Ref 20
19.4 Ref 20
26.0 Ref 20
30.0 Ref 20
35.7 Ref 20
42.0 Ref 20
46.8 Ref 20

120

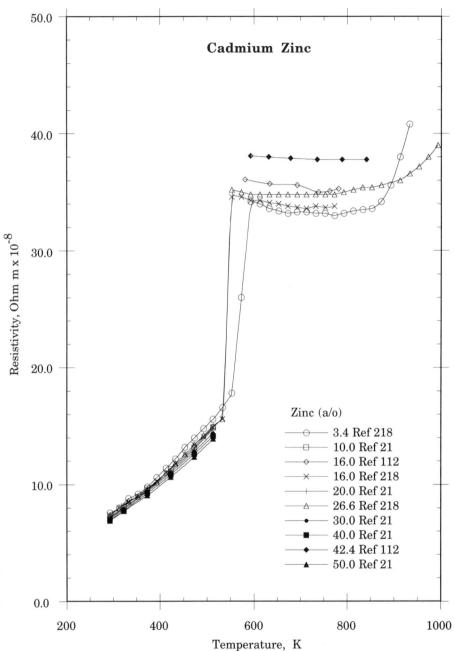

Cadmium Zinc

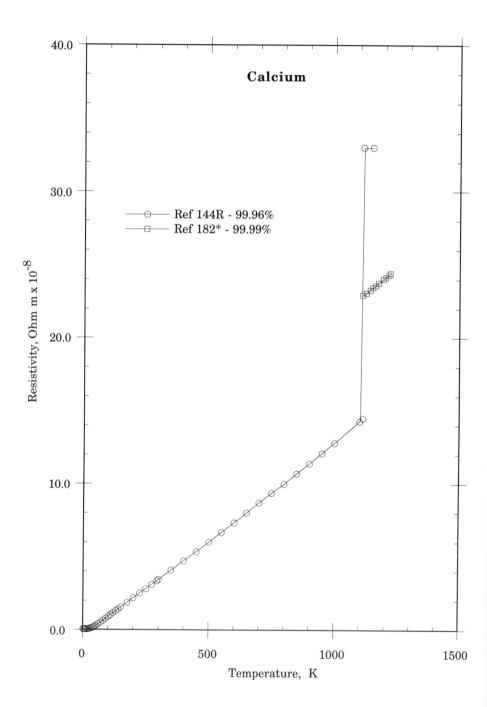

Calcium

Ref 144R - 99.96%
Ref 182* - 99.99%

Resistivity, Ohm m x 10^{-8}

Temperature, K

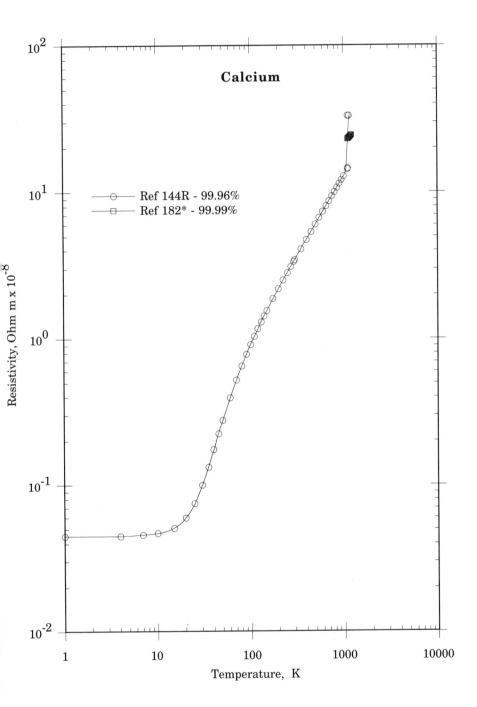

Calcium

Resistivity, Ohm m x 10^{-8}

Temperature, K

Ref 144R - 99.96%
Ref 182* - 99.99%

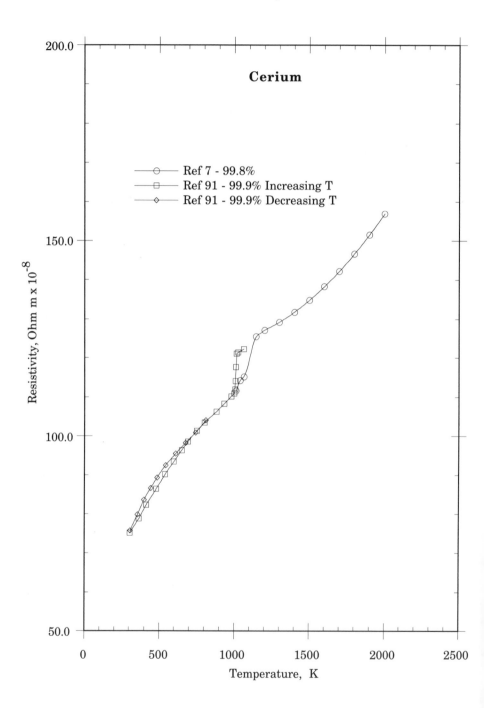

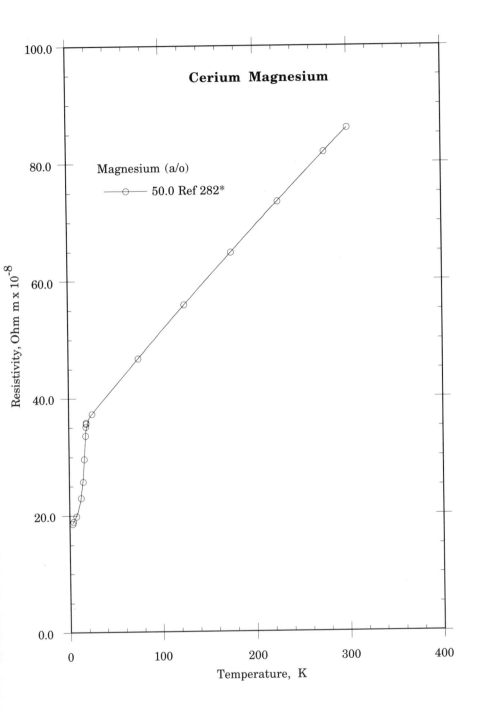

Cerium Magnesium

Magnesium (a/o)

———o——— 50.0 Ref 282*

Resistivity, Ohm m x 10^{-8}

Temperature, K

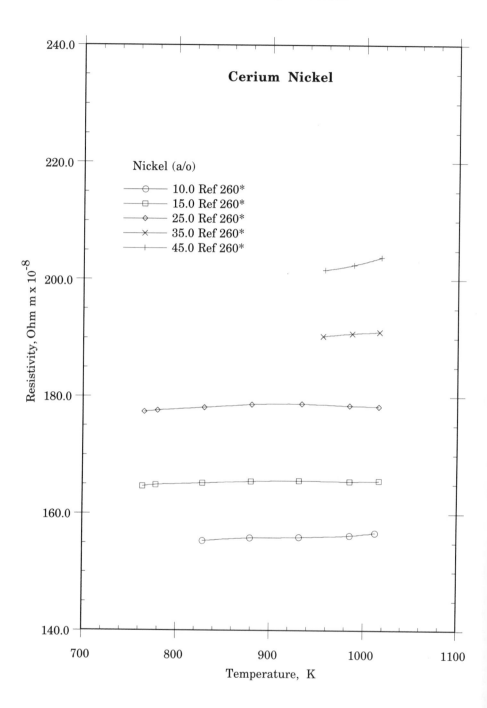

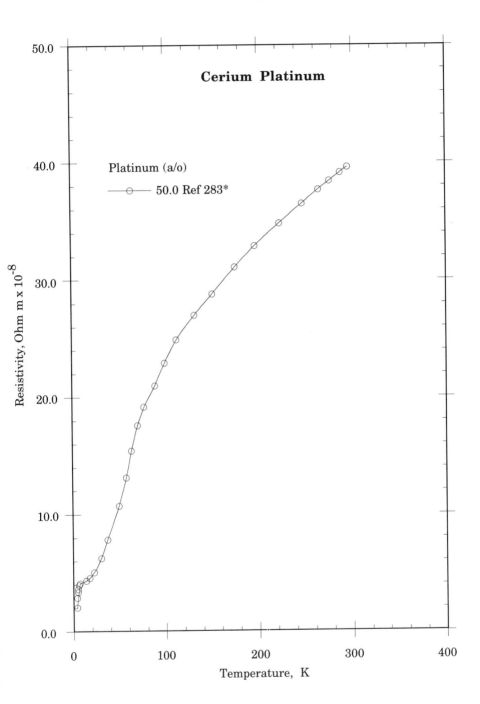

Cerium Platinum

Platinum (a/o)

50.0 Ref 283*

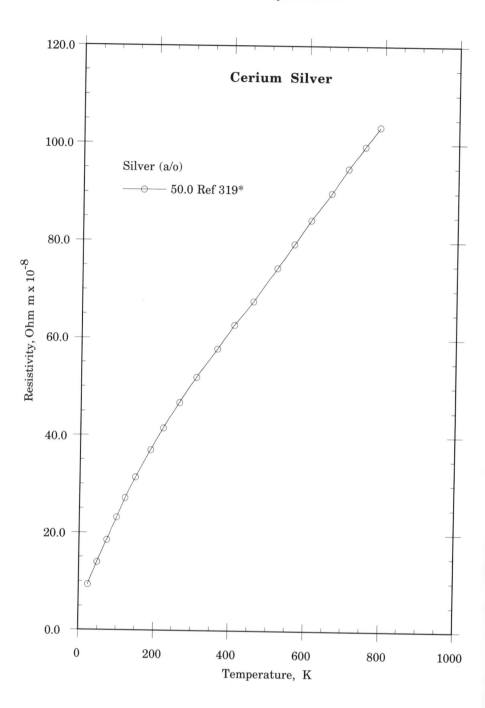

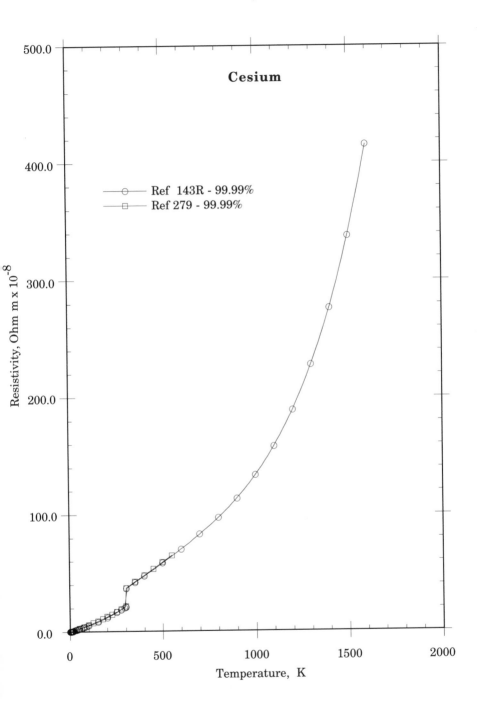

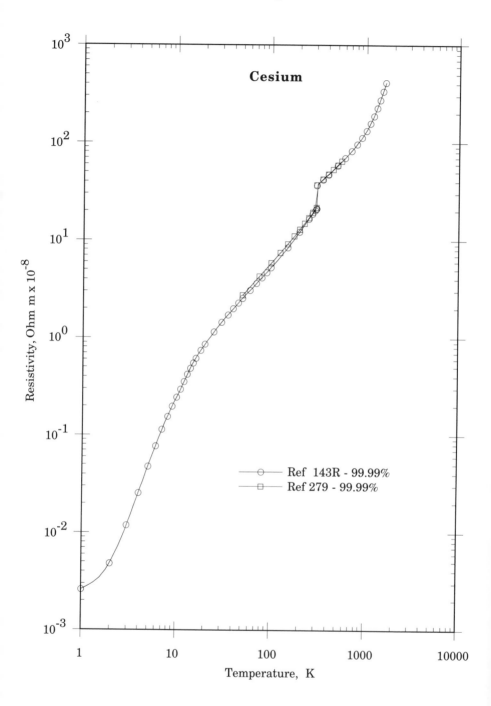

Cesium

Resistivity, Ohm m x 10^{-8}

Temperature, K

Ref 143R - 99.99%
Ref 279 - 99.99%

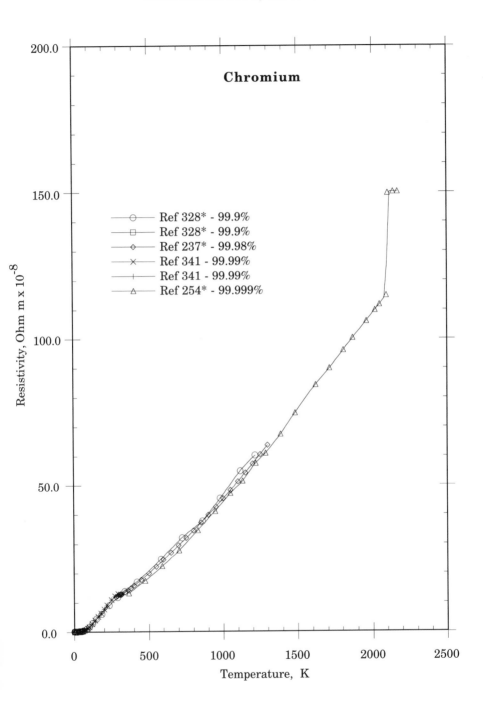

Chromium

Resistivity, Ohm m x 10^{-8}

Temperature, K

Ref 328* - 99.9%
Ref 328* - 99.9%
Ref 237* - 99.98%
Ref 341 - 99.99%
Ref 341 - 99.99%
Ref 254* - 99.999%

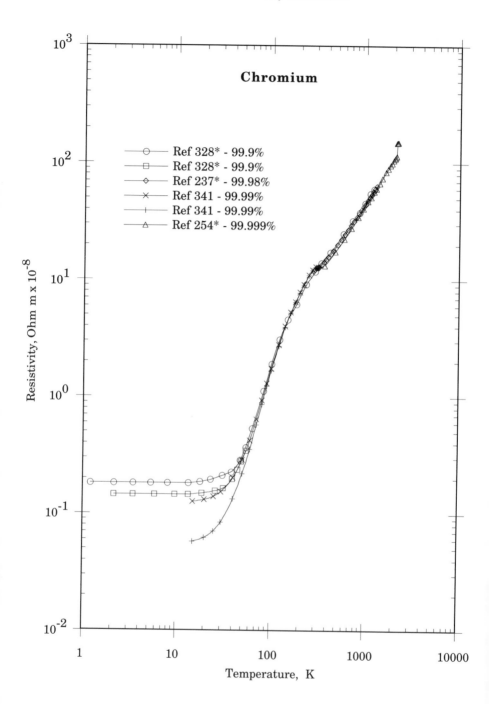

Chromium

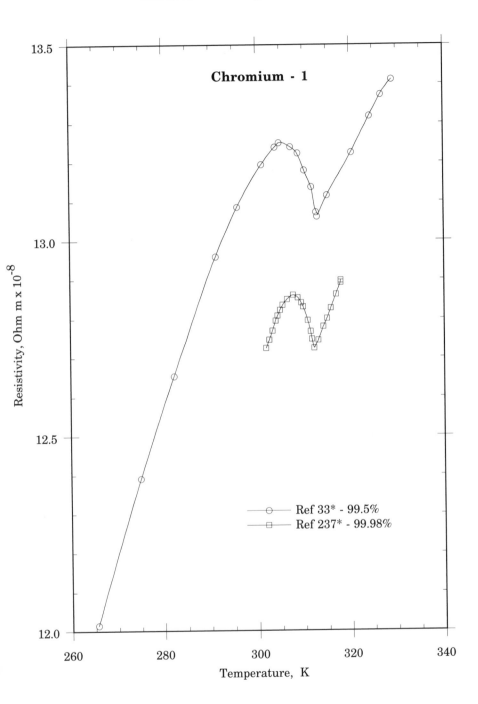

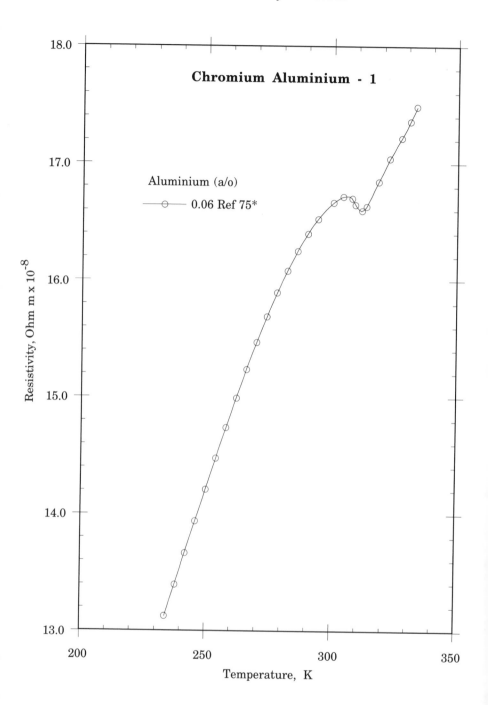

Chromium Aluminium - 1

Aluminium (a/o)

0.06 Ref 75*

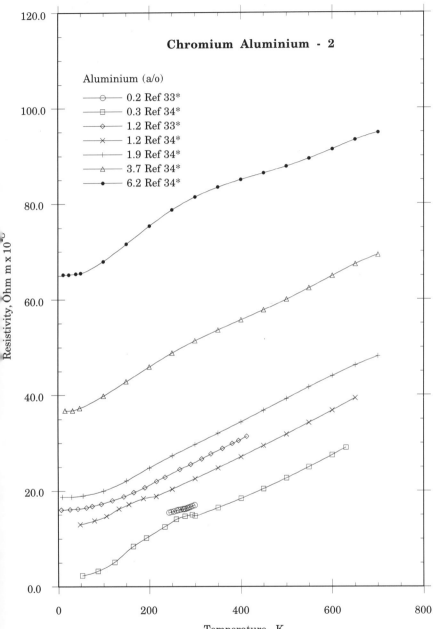

Chromium Aluminium - 2

Aluminium (a/o)

— ○ — 0.2 Ref 33*
— □ — 0.3 Ref 34*
— ◇ — 1.2 Ref 33*
— × — 1.2 Ref 34*
— + — 1.9 Ref 34*
— △ — 3.7 Ref 34*
— ● — 6.2 Ref 34*

Resistivity, Ohm m x 10⁻⁸

Temperature, K

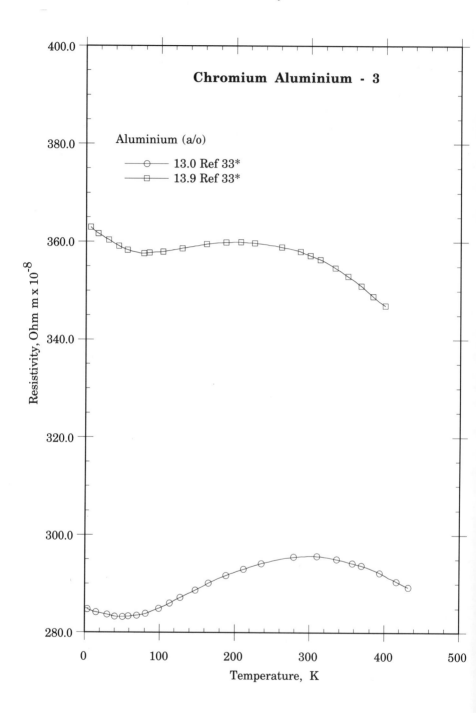

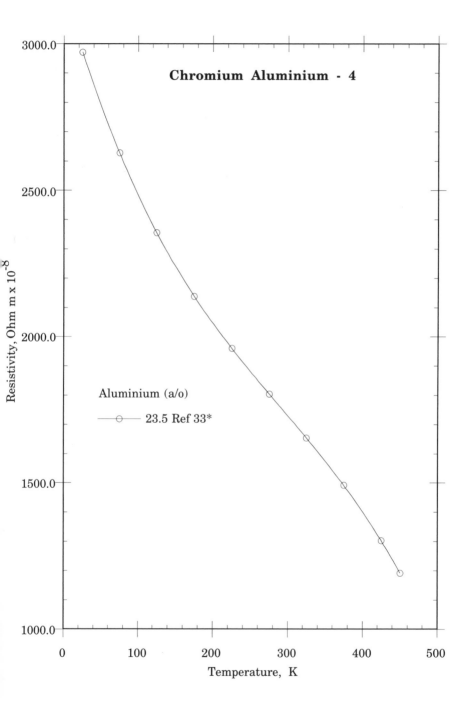

Chromium Aluminium - 4

Aluminium (a/o)
23.5 Ref 33*

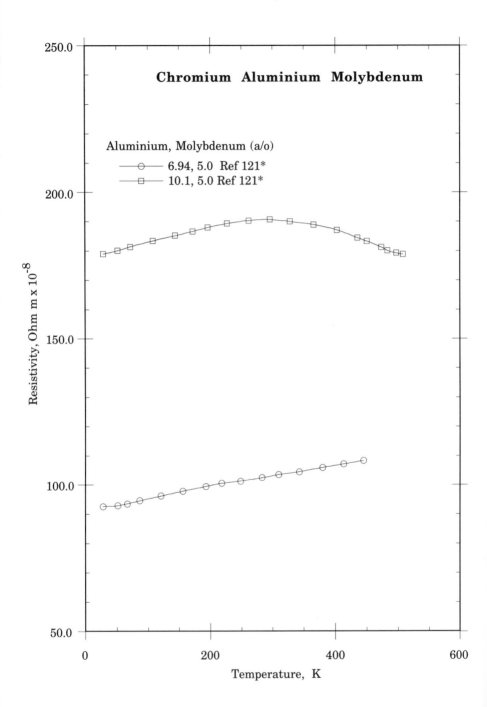

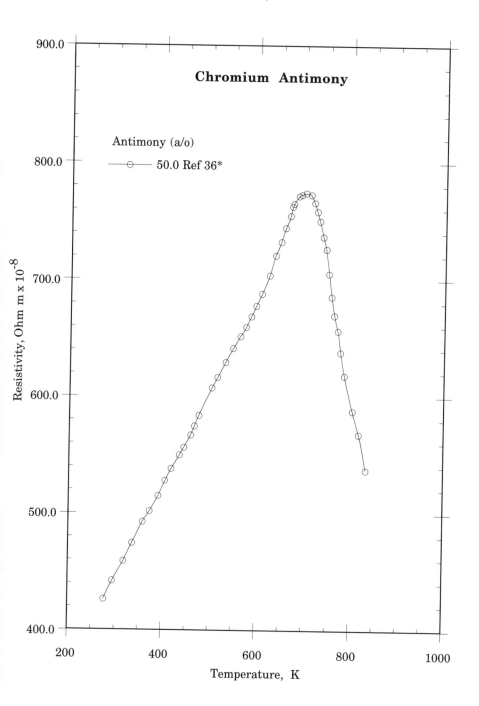

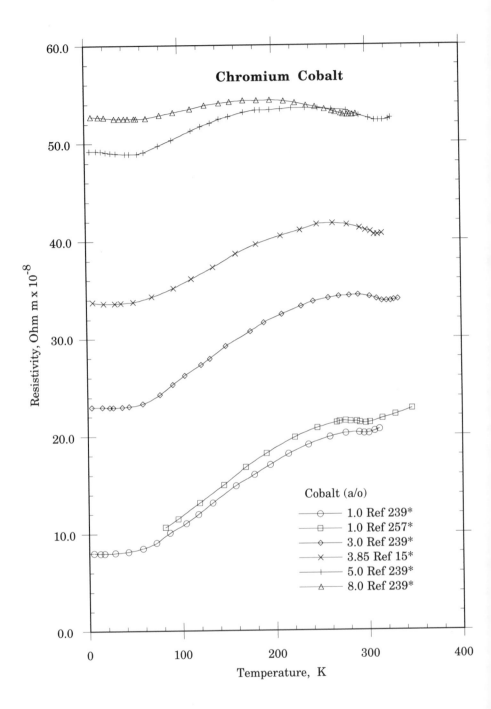

Chromium Cobalt

Resistivity, Ohm m x 10^{-8}

Temperature, K

Cobalt (a/o)
- ⊖ 1.0 Ref 239*
- ☐ 1.0 Ref 257*
- ◇ 3.0 Ref 239*
- ✕ 3.85 Ref 15*
- + 5.0 Ref 239*
- △ 8.0 Ref 239*

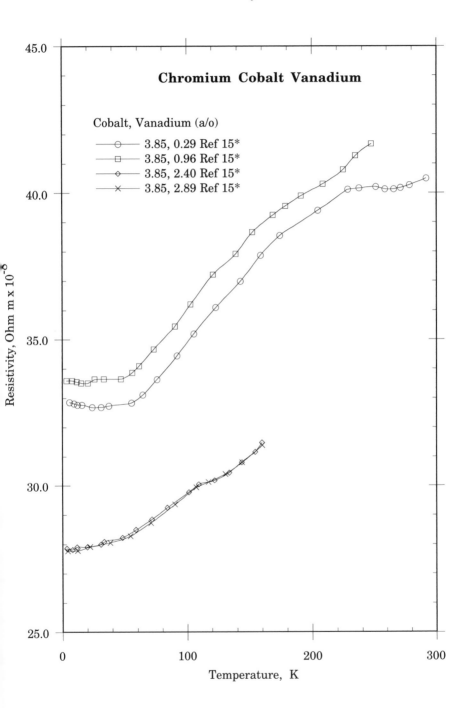

Chromium Cobalt Vanadium

Cobalt, Vanadium (a/o)
- ⊶ 3.85, 0.29 Ref 15*
- ⊟ 3.85, 0.96 Ref 15*
- ⬦ 3.85, 2.40 Ref 15*
- ⤫ 3.85, 2.89 Ref 15*

Resistivity, Ohm m x 10^{-8}

Temperature, K

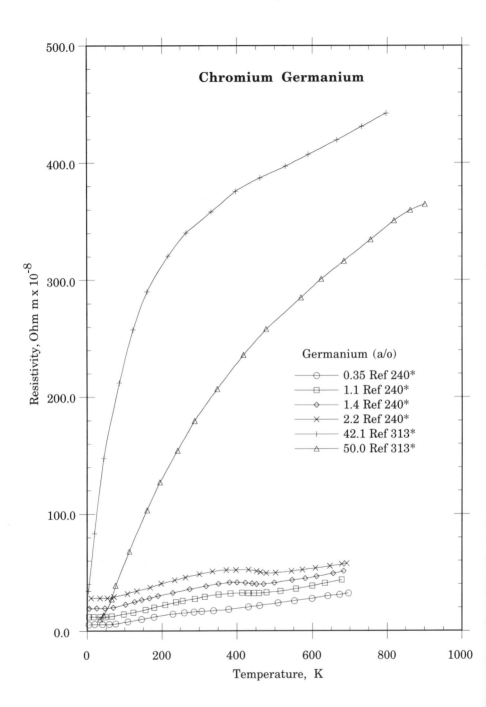

Chromium Germanium

Resistivity, Ohm m x 10^{-8}

Temperature, K

Germanium (a/o)

- ⊙ 0.35 Ref 240*
- □ 1.1 Ref 240*
- ◇ 1.4 Ref 240*
- ✕ 2.2 Ref 240*
- + 42.1 Ref 313*
- △ 50.0 Ref 313*

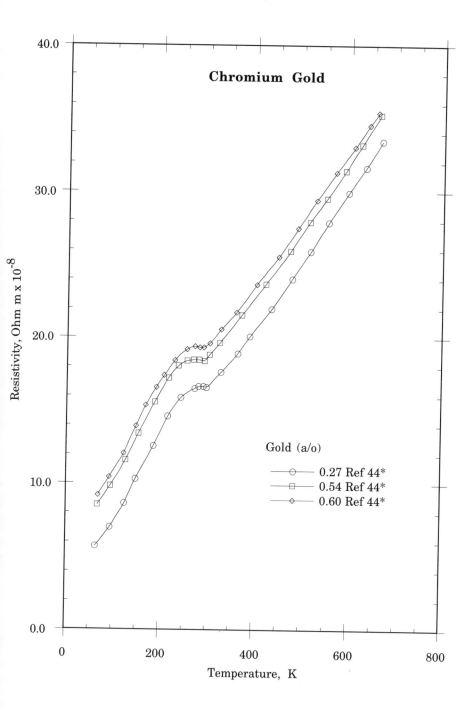

Chromium Gold

Resistivity, Ohm m x 10^{-8}

Temperature, K

Gold (a/o)

—⊙— 0.27 Ref 44*
—□— 0.54 Ref 44*
—◇— 0.60 Ref 44*

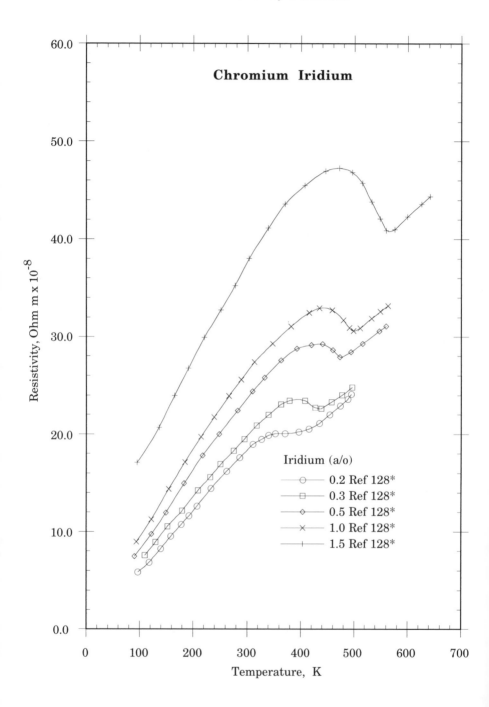

Chromium Iridium

Iridium (a/o)
- ⊖ 0.2 Ref 128*
- ⊟ 0.3 Ref 128*
- ◇ 0.5 Ref 128*
- ✕ 1.0 Ref 128*
- ＋ 1.5 Ref 128*

Resistivity, Ohm m x 10^{-8}

Temperature, K

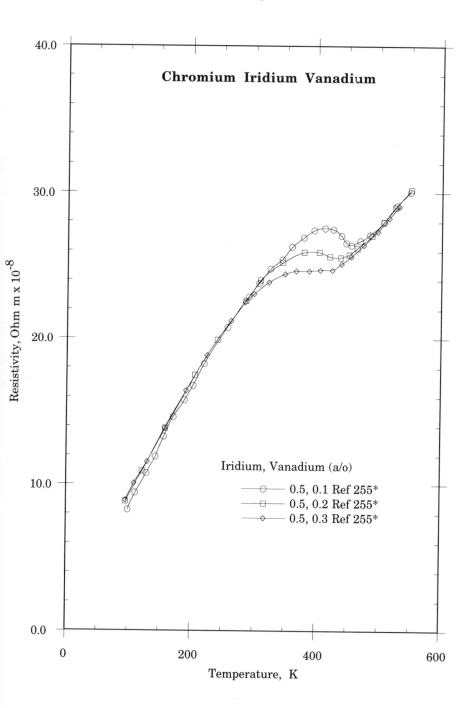

Chromium Iridium Vanadium

Iridium, Vanadium (a/o)

—⊖— 0.5, 0.1 Ref 255*
—□— 0.5, 0.2 Ref 255*
—◇— 0.5, 0.3 Ref 255*

Resistivity, Ohm m x 10^{-8}

Temperature, K

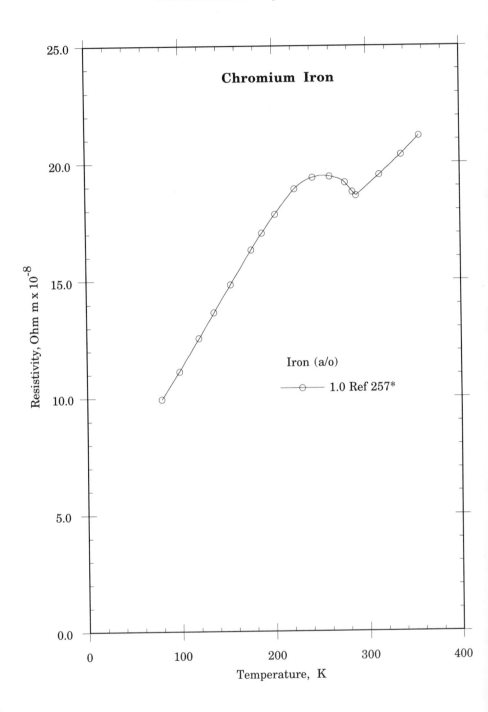

Chromium Iron

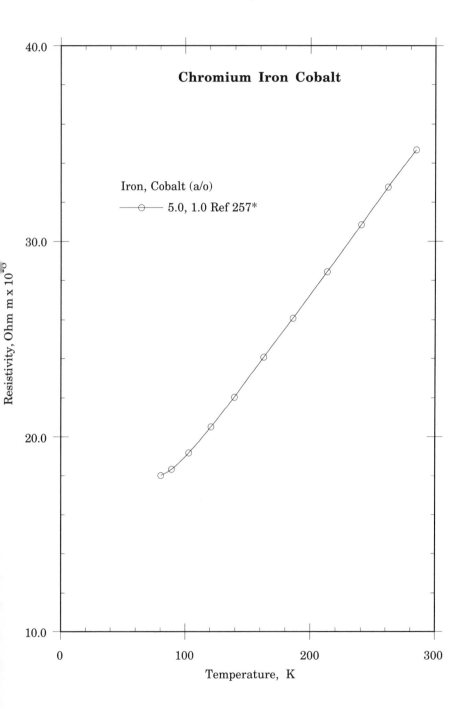

Chromium Iron Cobalt

Iron, Cobalt (a/o)
—⊙— 5.0, 1.0 Ref 257*

Resistivity, Ohm m x 10⁻⁸

Temperature, K

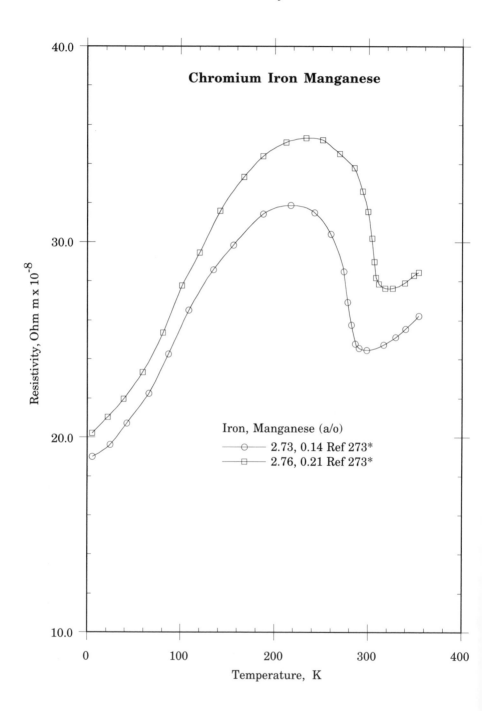

Chromium Iron Manganese

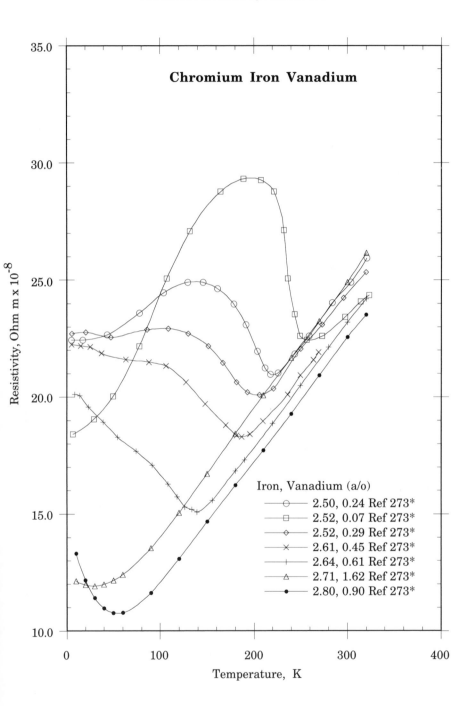

Chromium Iron Vanadium

Iron, Vanadium (a/o)
—⊙— 2.50, 0.24 Ref 273*
—□— 2.52, 0.07 Ref 273*
—◇— 2.52, 0.29 Ref 273*
—×— 2.61, 0.45 Ref 273*
—+— 2.64, 0.61 Ref 273*
—△— 2.71, 1.62 Ref 273*
—●— 2.80, 0.90 Ref 273*

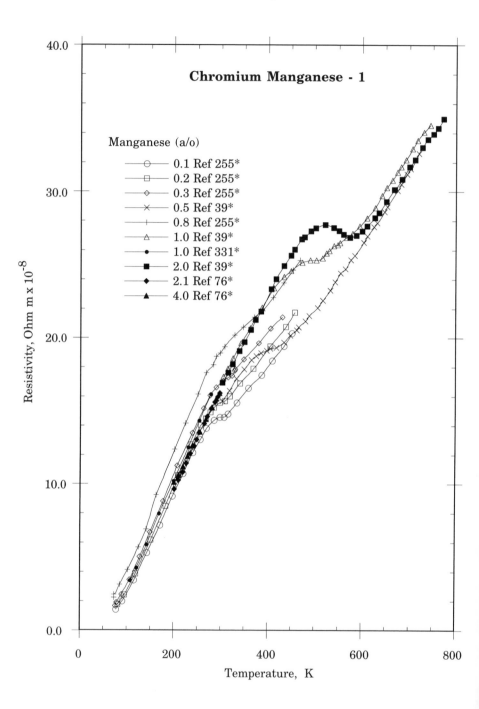

Chromium Manganese - 1

Manganese (a/o)
- ⊙ 0.1 Ref 255*
- ⬚ 0.2 Ref 255*
- ◇ 0.3 Ref 255*
- ✕ 0.5 Ref 39*
- + 0.8 Ref 255*
- △ 1.0 Ref 39*
- ● 1.0 Ref 331*
- ■ 2.0 Ref 39*
- ◆ 2.1 Ref 76*
- ▲ 4.0 Ref 76*

Resistivity, Ohm m x 10⁻⁸

Temperature, K

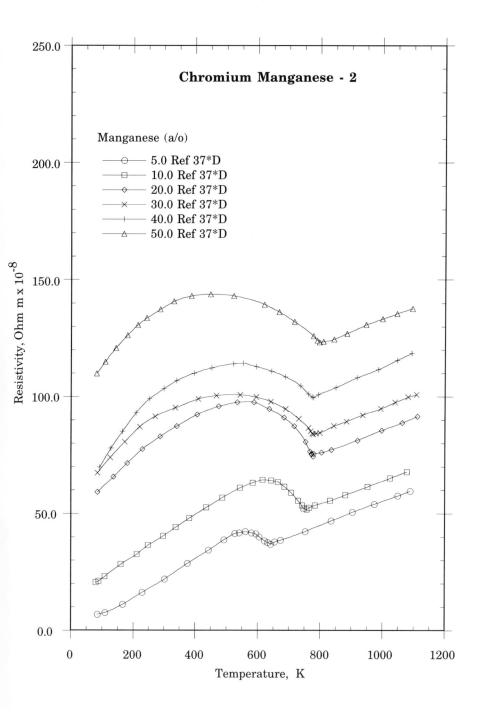

Chromium Manganese - 2

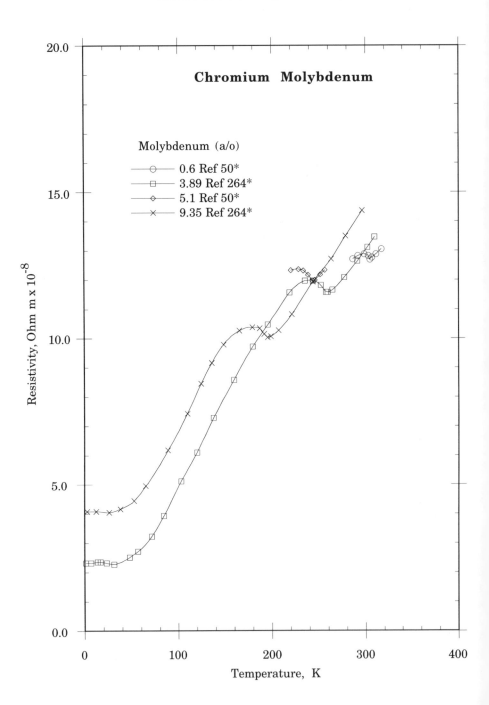

Chromium Molybdenum

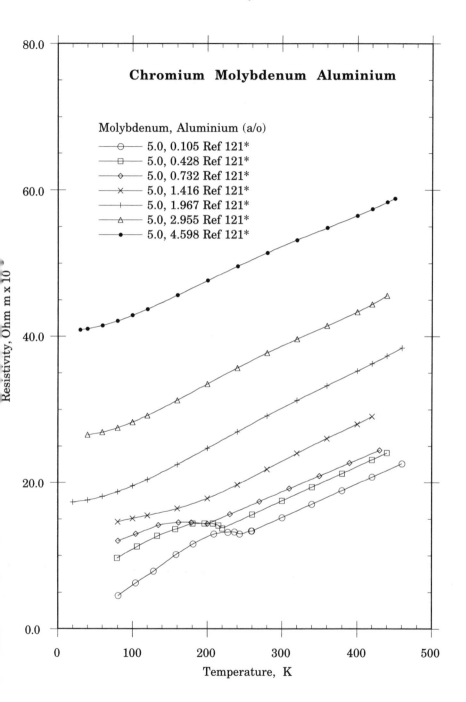

Chromium Molybdenum Aluminium

Molybdenum, Aluminium (a/o)
- ⊖ 5.0, 0.105 Ref 121*
- ▭ 5.0, 0.428 Ref 121*
- ◇ 5.0, 0.732 Ref 121*
- ✕ 5.0, 1.416 Ref 121*
- + 5.0, 1.967 Ref 121*
- △ 5.0, 2.955 Ref 121*
- ● 5.0, 4.598 Ref 121*

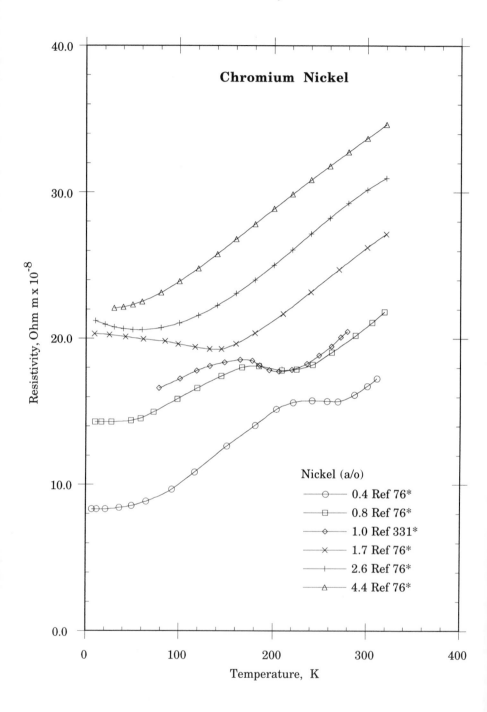

Chromium Nickel

Nickel (a/o)
- ⊙ 0.4 Ref 76*
- ▫ 0.8 Ref 76*
- ◇ 1.0 Ref 331*
- ✕ 1.7 Ref 76*
- + 2.6 Ref 76*
- △ 4.4 Ref 76*

Resistivity, Ohm m x 10⁻⁸

Temperature, K

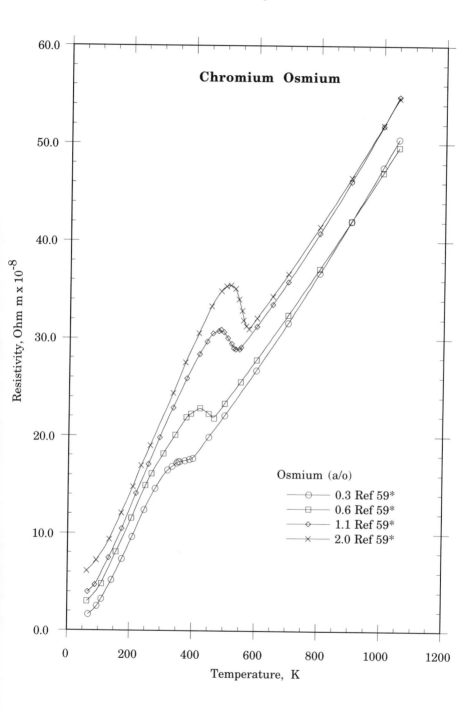

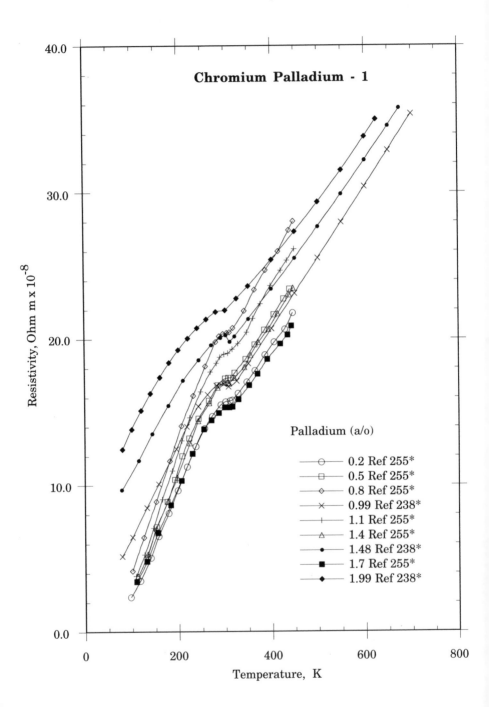

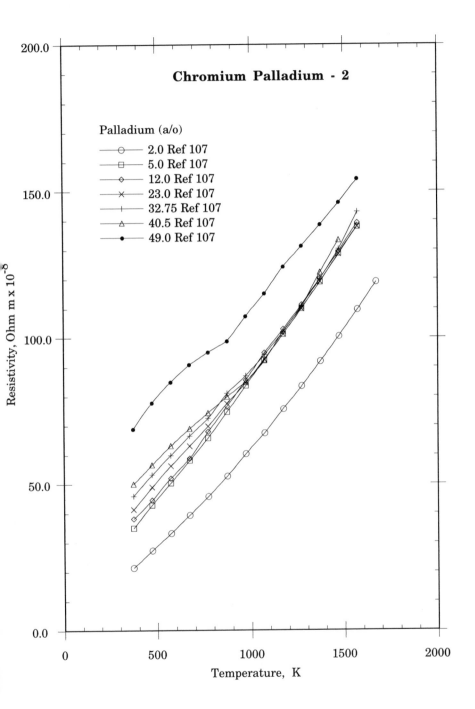

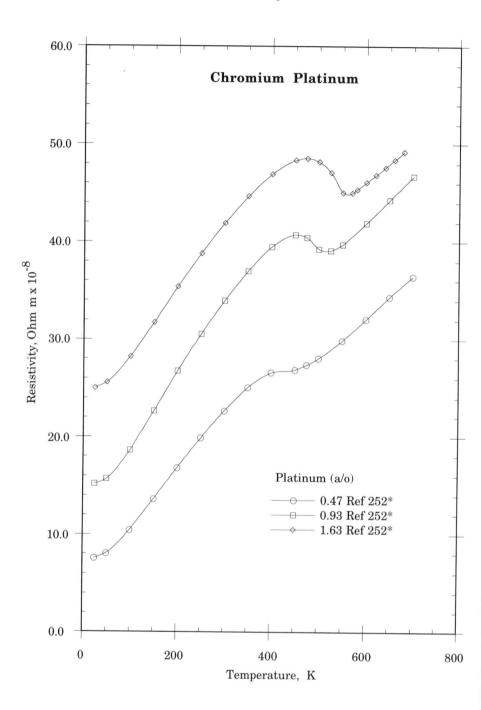

Chromium Platinum

Resistivity, Ohm m x 10^{-8}

Temperature, K

Platinum (a/o)

— ○ — 0.47 Ref 252*
— □ — 0.93 Ref 252*
— ◇ — 1.63 Ref 252*

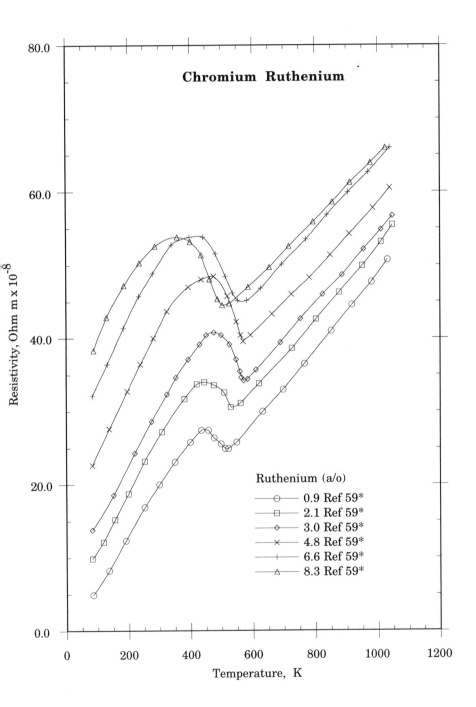

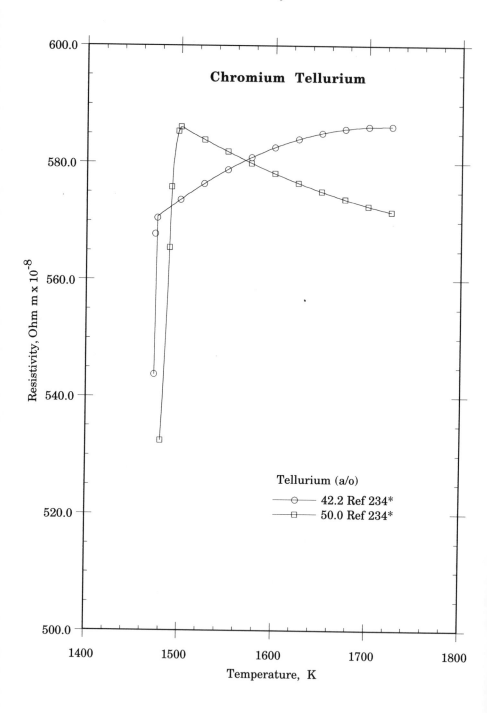

Chromium Tellurium

Resistivity, Ohm m x 10^{-8}

Temperature, K

Tellurium (a/o)
42.2 Ref 234*
50.0 Ref 234*

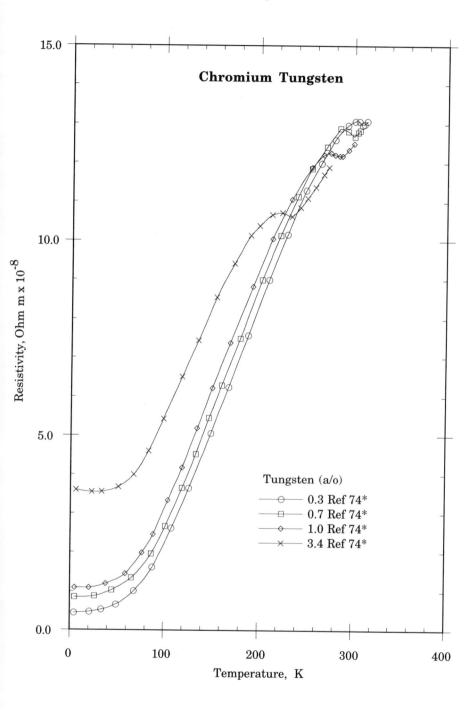

Chromium Tungsten

Resistivity, Ohm m x 10^{-8}

Temperature, K

Tungsten (a/o)

—⊙— 0.3 Ref 74*
—□— 0.7 Ref 74*
—◇— 1.0 Ref 74*
—✕— 3.4 Ref 74*

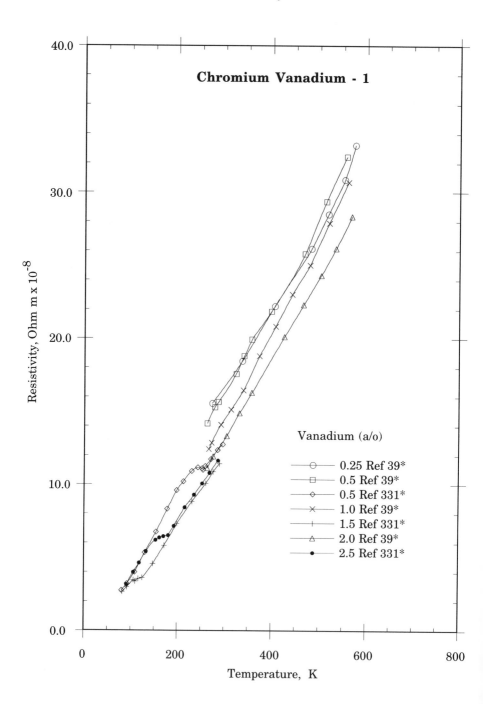

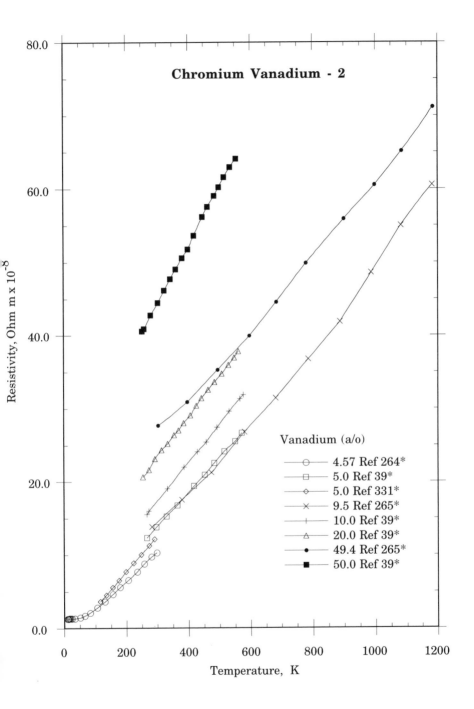

Chromium Vanadium - 2

Vanadium (a/o)

- ○ 4.57 Ref 264*
- □ 5.0 Ref 39*
- ◇ 5.0 Ref 331*
- ✕ 9.5 Ref 265*
- ＋ 10.0 Ref 39*
- △ 20.0 Ref 39*
- ● 49.4 Ref 265*
- ■ 50.0 Ref 39*

Resistivity, Ohm m x 10^{-8}

Temperature, K

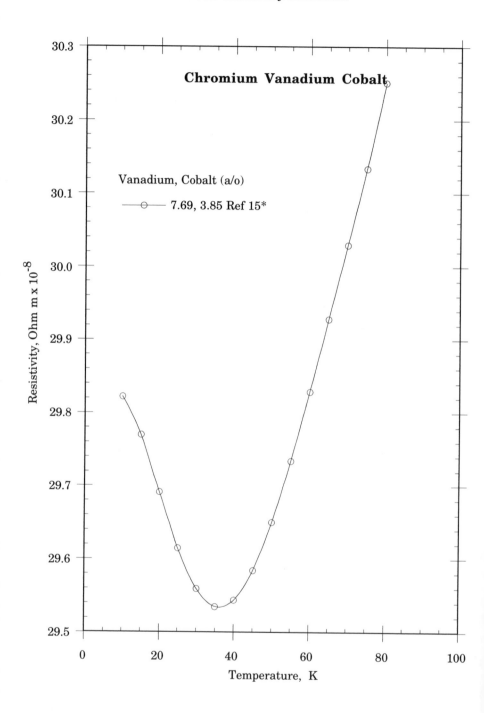

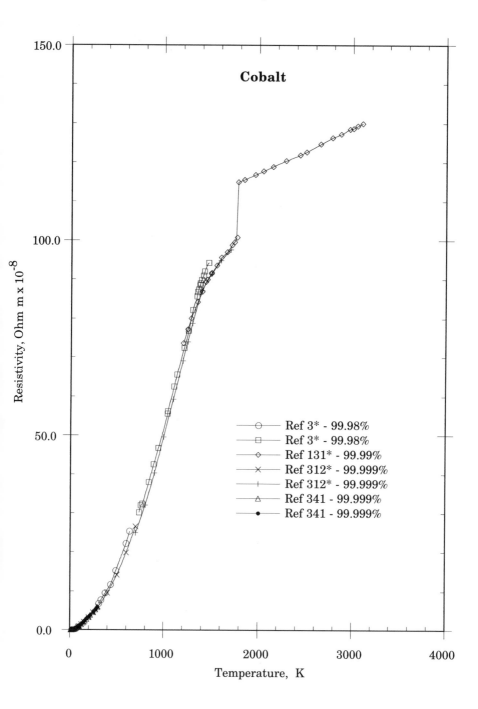

Cobalt

Resistivity, Ohm m x 10^{-8}

Temperature, K

Ref 3* - 99.98%
Ref 3* - 99.98%
Ref 131* - 99.99%
Ref 312* - 99.999%
Ref 312* - 99.999%
Ref 341 - 99.999%
Ref 341 - 99.999%

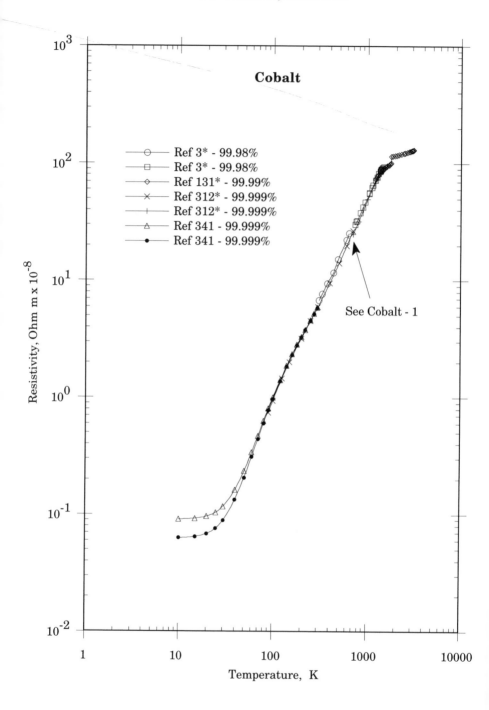

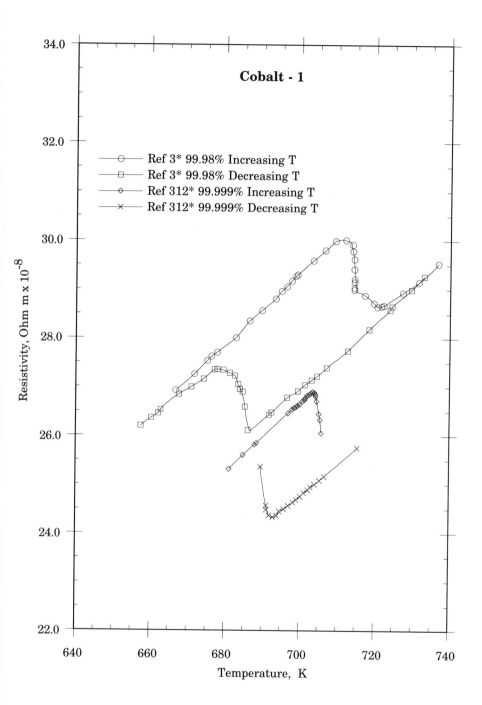

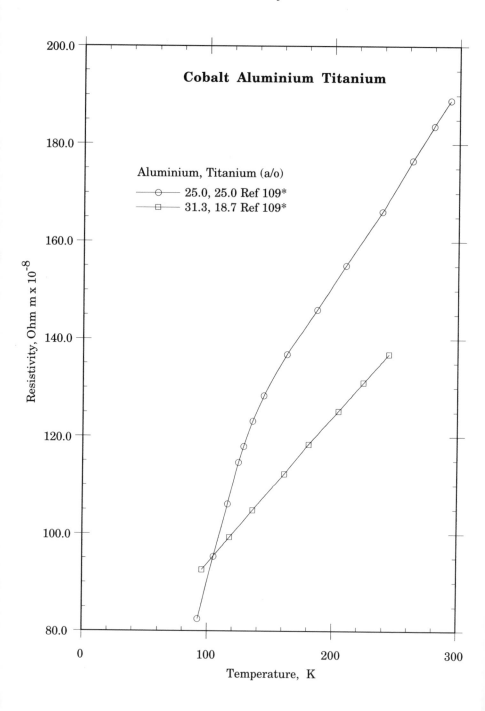

Cobalt Aluminium Titanium

Aluminium, Titanium (a/o)
—○— 25.0, 25.0 Ref 109*
—□— 31.3, 18.7 Ref 109*

Resistivity, Ohm m x 10⁻⁸

Temperature, K

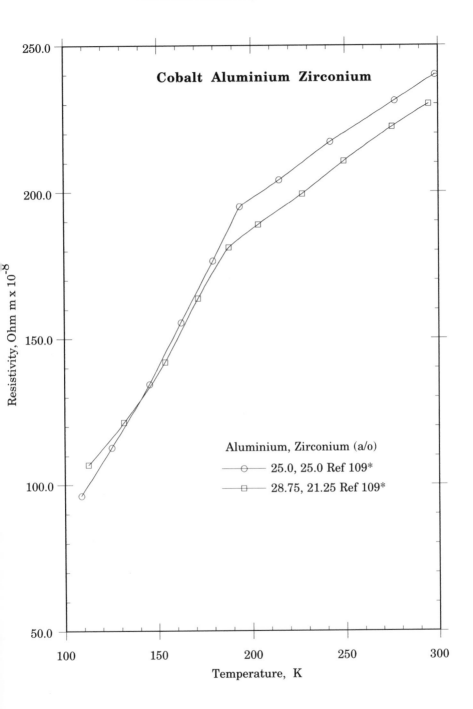

Cobalt Aluminium Zirconium

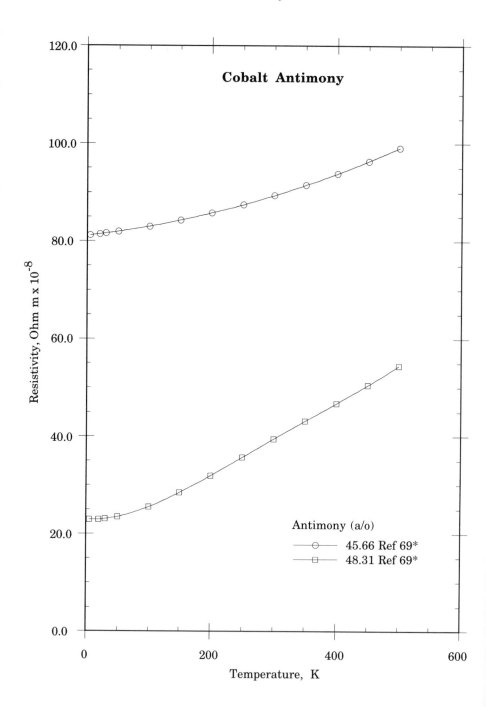

Cobalt Antimony

Resistivity, Ohm m x 10^{-8}

Temperature, K

Antimony (a/o)
45.66 Ref 69*
48.31 Ref 69*

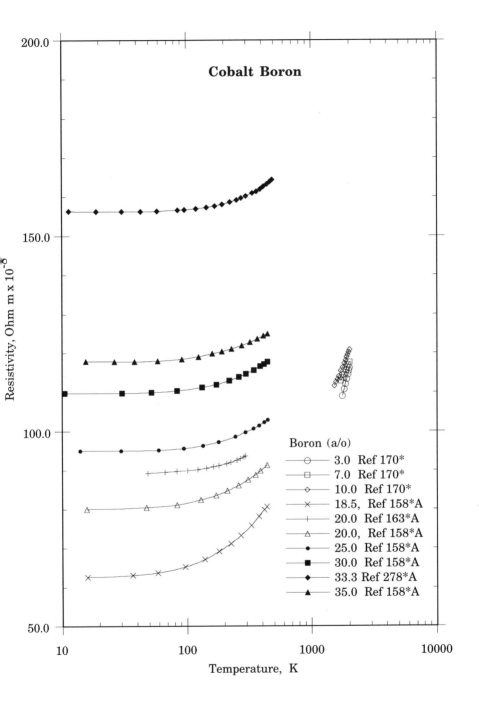

Cobalt Boron

Boron (a/o)
- ⊖— 3.0 Ref 170*
- ☐— 7.0 Ref 170*
- ◇— 10.0 Ref 170*
- ✕— 18.5, Ref 158*A
- +— 20.0 Ref 163*A
- △— 20.0, Ref 158*A
- ●— 25.0 Ref 158*A
- ■— 30.0 Ref 158*A
- ◆— 33.3 Ref 278*A
- ▲— 35.0 Ref 158*A

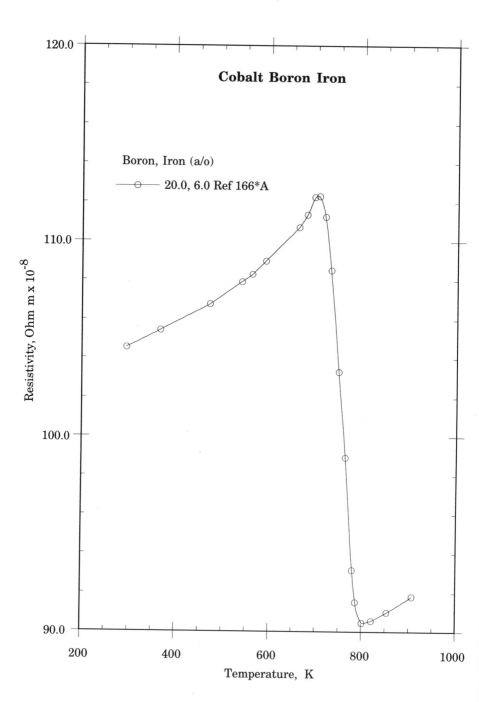

Cobalt Boron Iron

Boron, Iron (a/o)

──⊖── 20.0, 6.0 Ref 166*A

Resistivity, Ohm m x 10^{-8}

Temperature, K

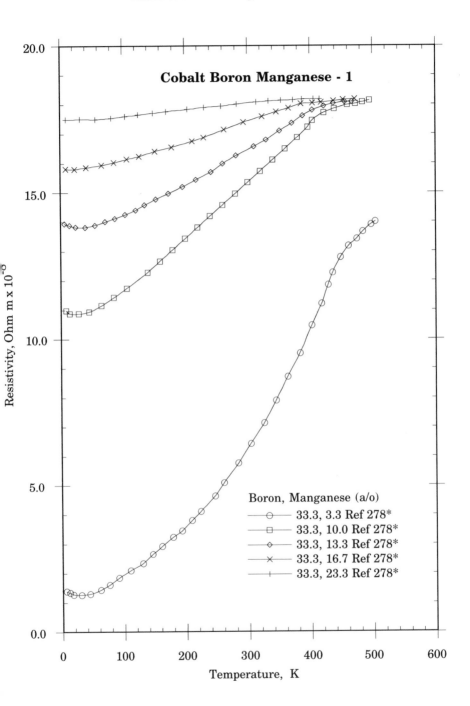

Cobalt Boron Manganese - 1

Boron, Manganese (a/o)
- —○— 33.3, 3.3 Ref 278*
- —□— 33.3, 10.0 Ref 278*
- —◇— 33.3, 13.3 Ref 278*
- —×— 33.3, 16.7 Ref 278*
- —+— 33.3, 23.3 Ref 278*

Resistivity, Ohm m x 10^{-8}

Temperature, K

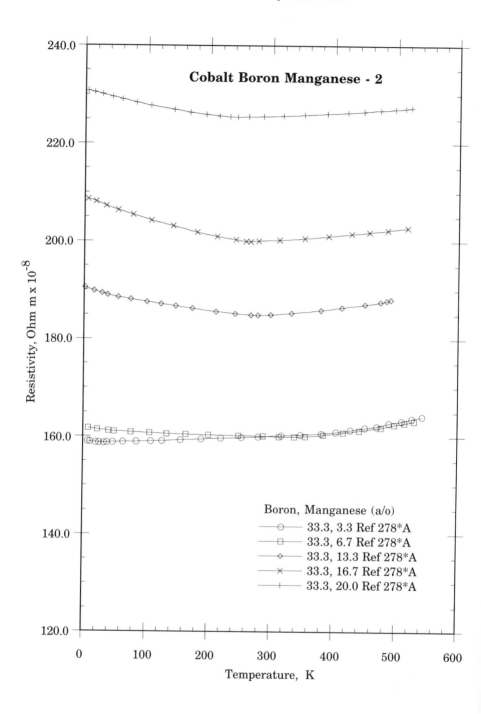

Cobalt Boron Manganese - 2

Boron, Manganese (a/o)
- —⊙— 33.3, 3.3 Ref 278*A
- —□— 33.3, 6.7 Ref 278*A
- —◇— 33.3, 13.3 Ref 278*A
- —×— 33.3, 16.7 Ref 278*A
- —+— 33.3, 20.0 Ref 278*A

Resistivity, Ohm m x 10^{-8}

Temperature, K

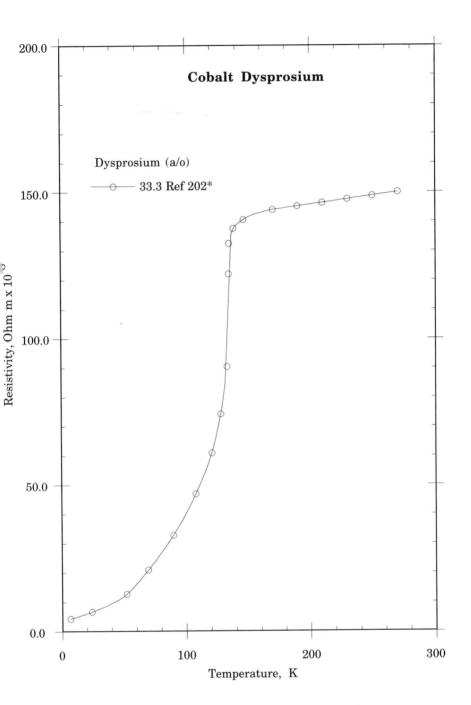

Cobalt Dysprosium

Dysprosium (a/o)

—⊙— 33.3 Ref 202*

Resistivity, Ohm m x 10

Temperature, K

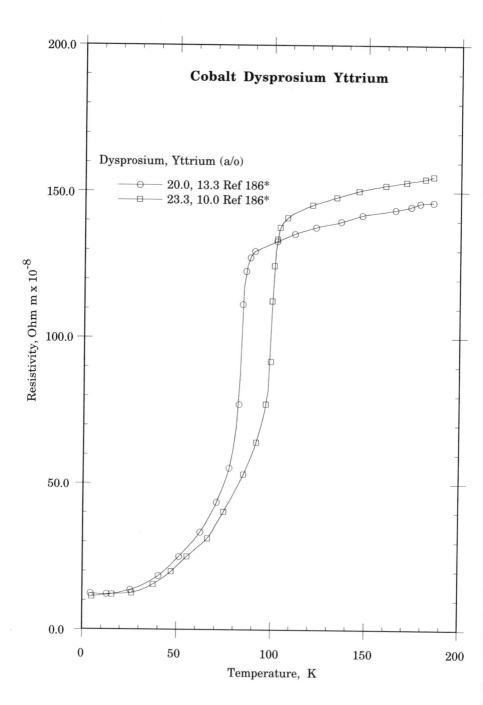

Cobalt Dysprosium Yttrium

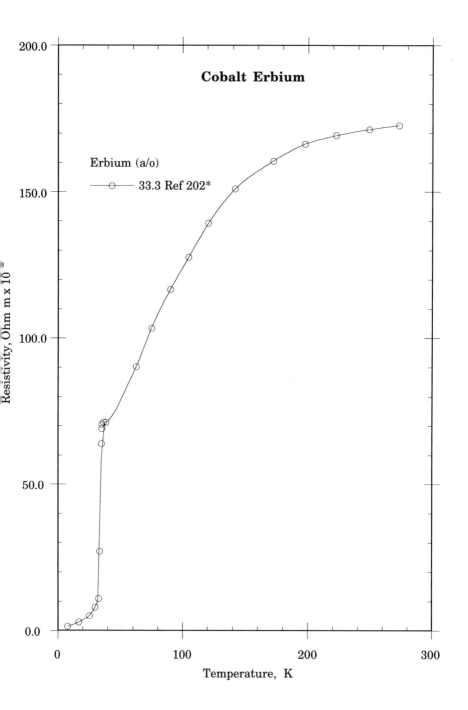

Cobalt Erbium

Resistivity, Ohm m x 10

Temperature, K

Erbium (a/o)
33.3 Ref 202*

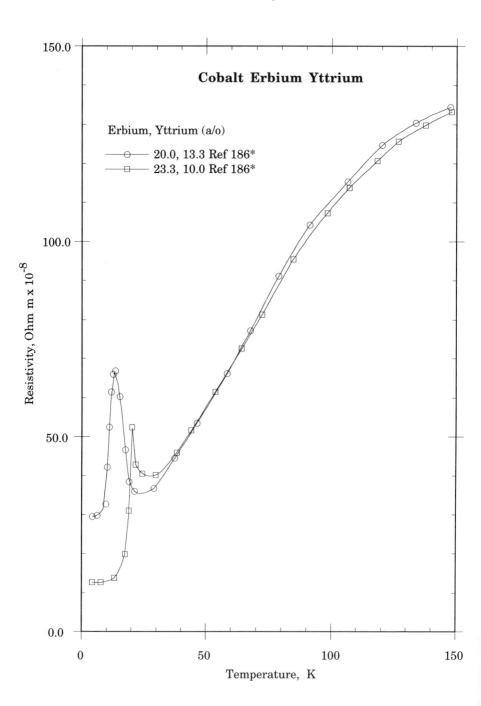

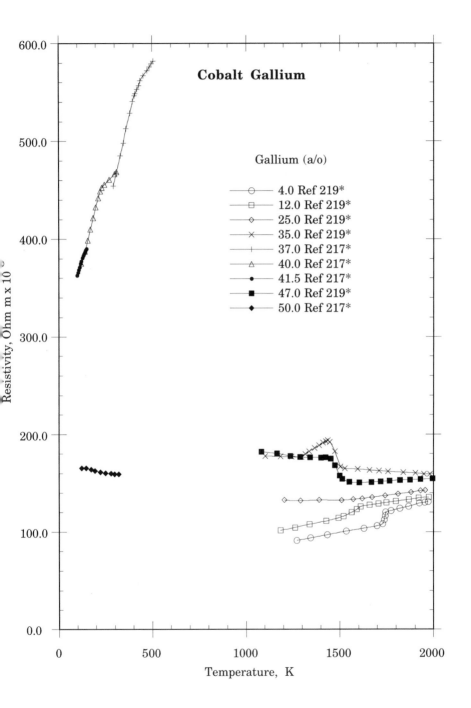

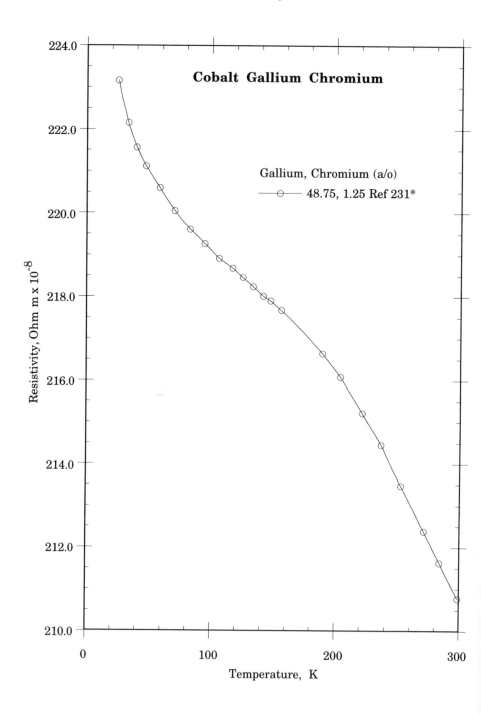

Cobalt Gallium Chromium

Gallium, Chromium (a/o)
—⊙— 48.75, 1.25 Ref 231*

Resistivity, Ohm m x 10^{-8}

Temperature, K

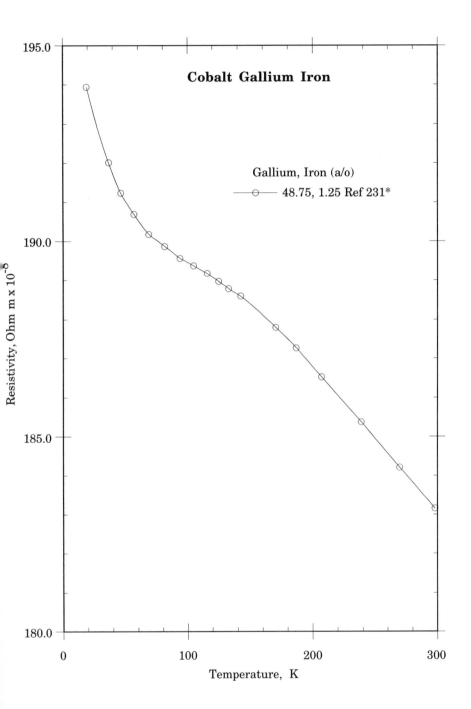

Cobalt Gallium Iron

Gallium, Iron (a/o)
───◌─── 48.75, 1.25 Ref 231*

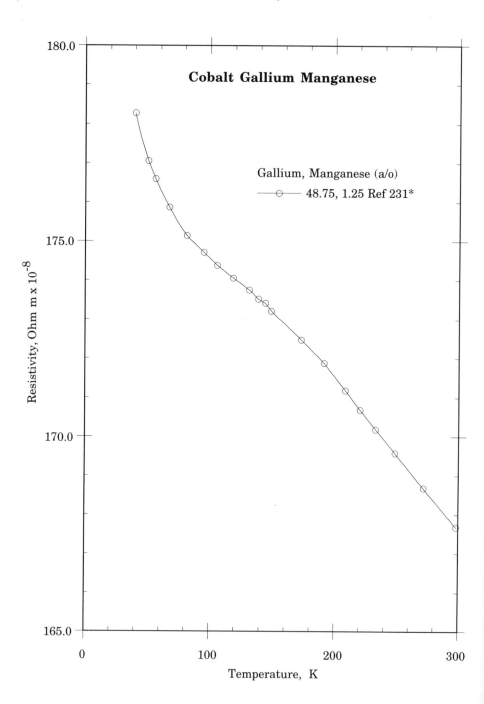

Cobalt Gallium Manganese

Gallium, Manganese (a/o)

48.75, 1.25 Ref 231*

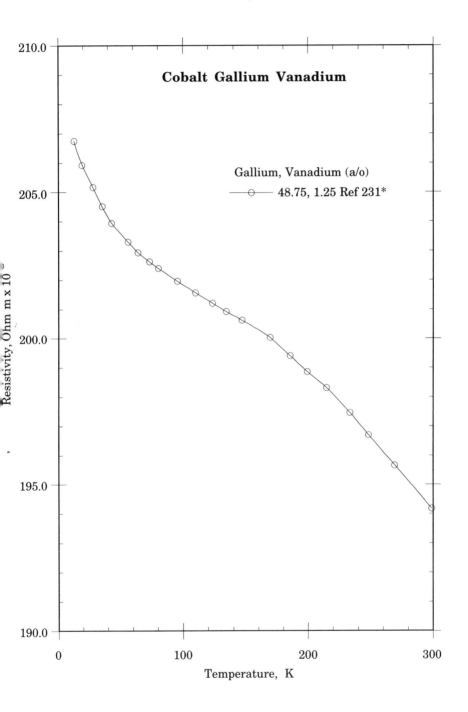

Cobalt Gallium Vanadium

Gallium, Vanadium (a/o)
— ⊙ — 48.75, 1.25 Ref 231*

Resistivity, Ohm m x 10⁻

Temperature, K

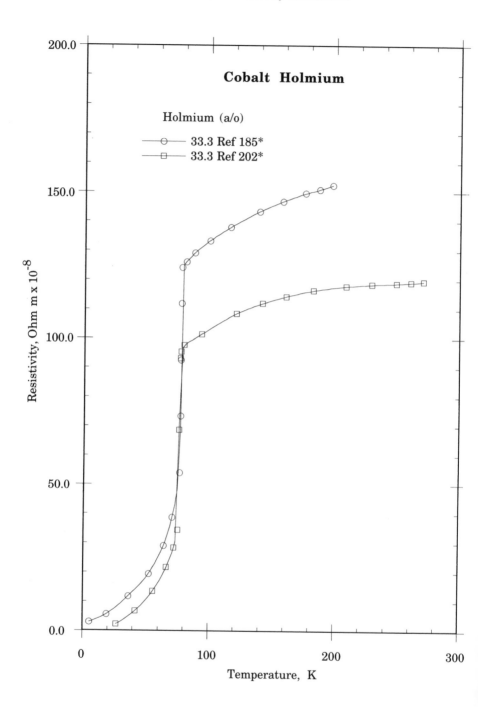

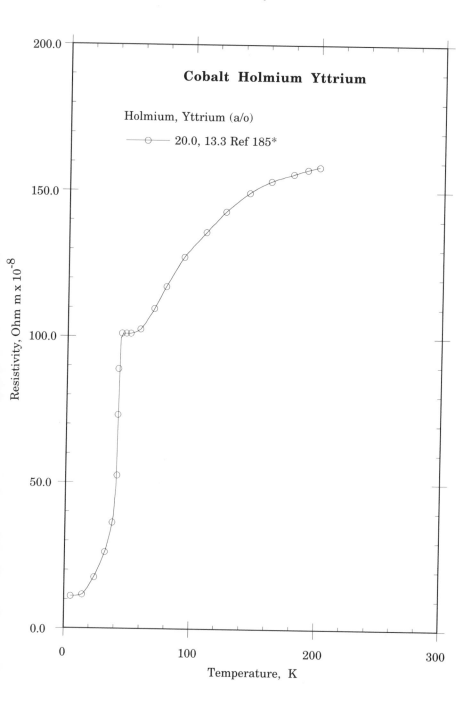

Cobalt Holmium Yttrium

Holmium, Yttrium (a/o)

───○─── 20.0, 13.3 Ref 185*

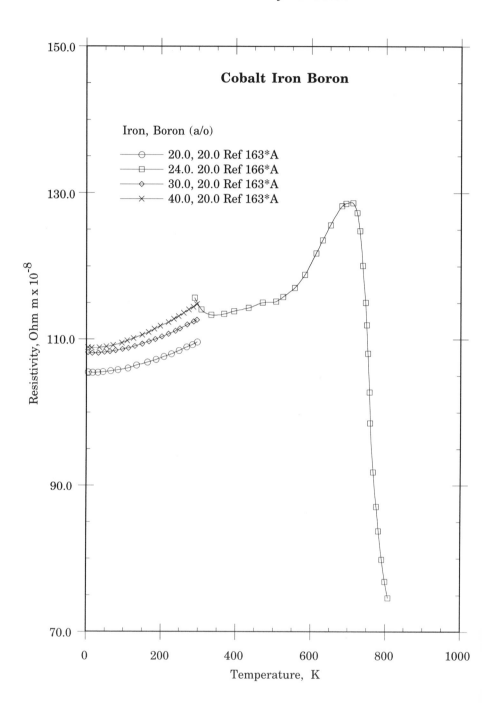

Cobalt Iron Boron

Iron, Boron (a/o)

- ─○─ 20.0, 20.0 Ref 163*A
- ─□─ 24.0. 20.0 Ref 166*A
- ─◇─ 30.0, 20.0 Ref 163*A
- ─✕─ 40.0, 20.0 Ref 163*A

Resistivity, Ohm m x 10^{-8}

Temperature, K

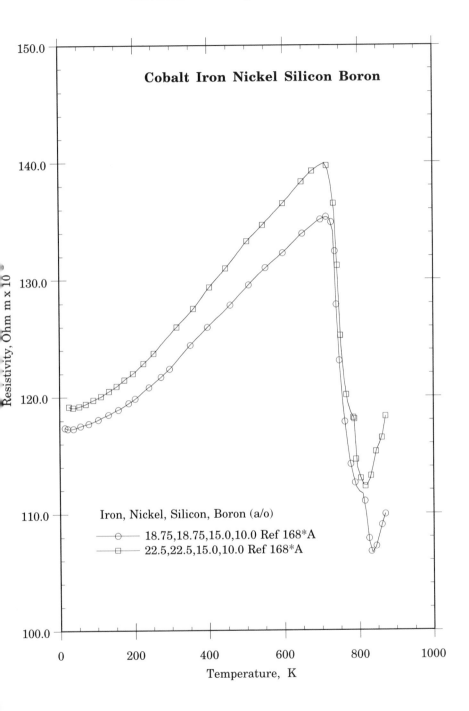

Cobalt Iron Nickel Silicon Boron

Iron, Nickel, Silicon, Boron (a/o)
- ○ 18.75,18.75,15.0,10.0 Ref 168*A
- □ 22.5,22.5,15.0,10.0 Ref 168*A

Resistivity, Ohm m x 10⁻⁸

Temperature, K

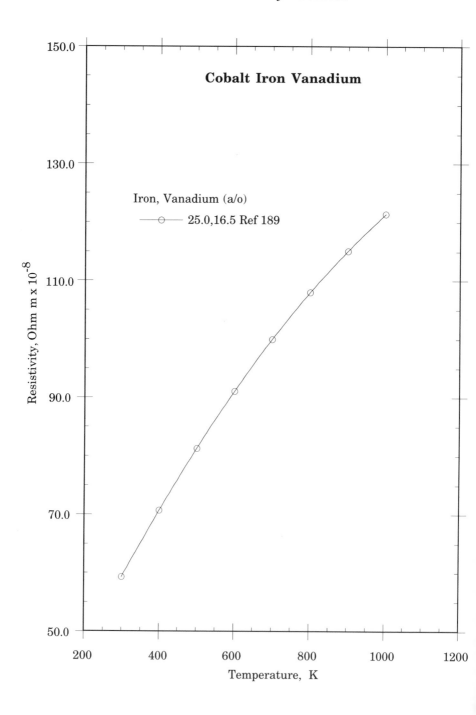

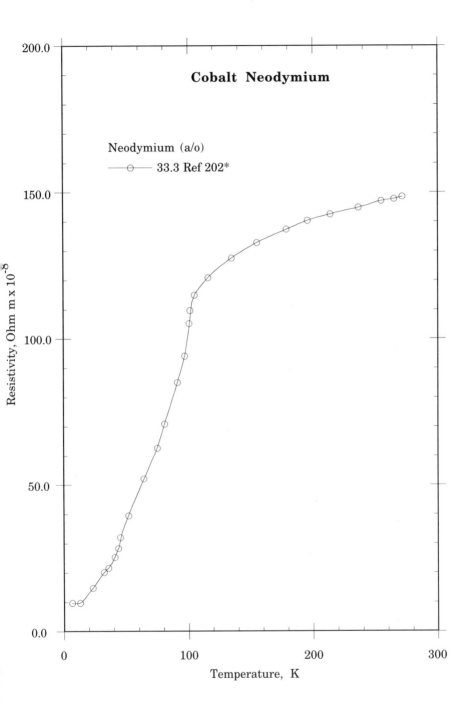

Cobalt Neodymium

Neodymium (a/o)
⊖ 33.3 Ref 202*

Resistivity, Ohm m x 10^{-8}

Temperature, K

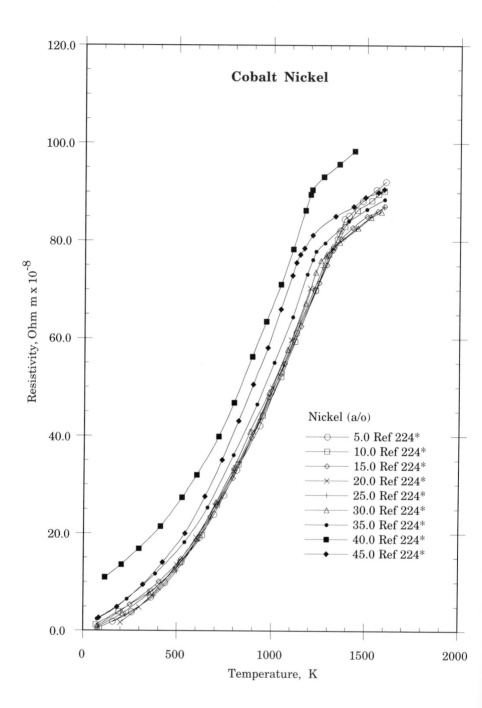

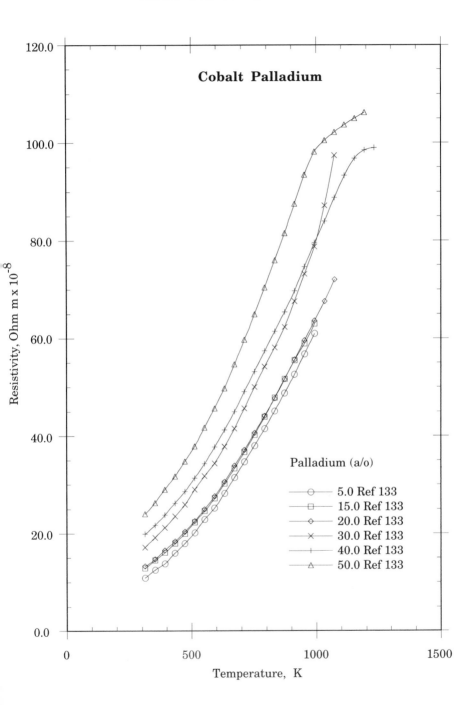

Cobalt Palladium

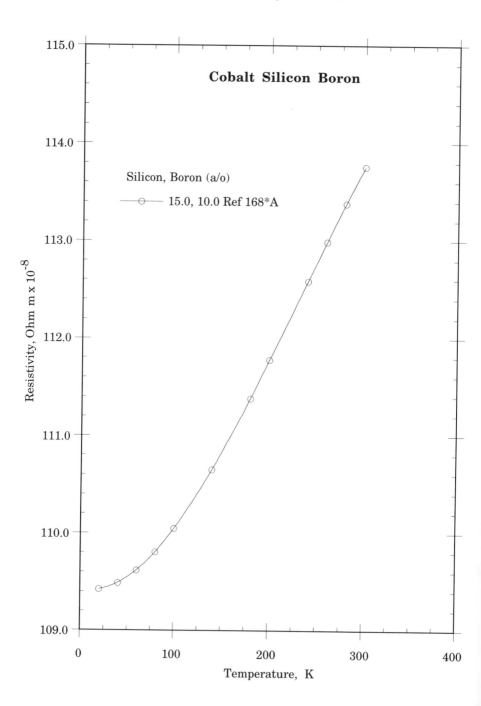

Cobalt Silicon Boron

Silicon, Boron (a/o)

———○——— 15.0, 10.0 Ref 168*A

Resistivity, Ohm m x 10⁻⁸

Temperature, K

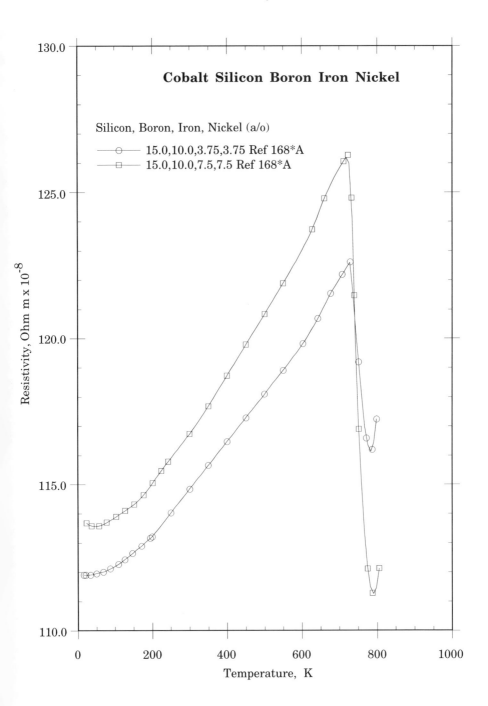

Cobalt Silicon Boron Iron Nickel

Silicon, Boron, Iron, Nickel (a/o)
- 15.0,10.0,3.75,3.75 Ref 168*A
- 15.0,10.0,7.5,7.5 Ref 168*A

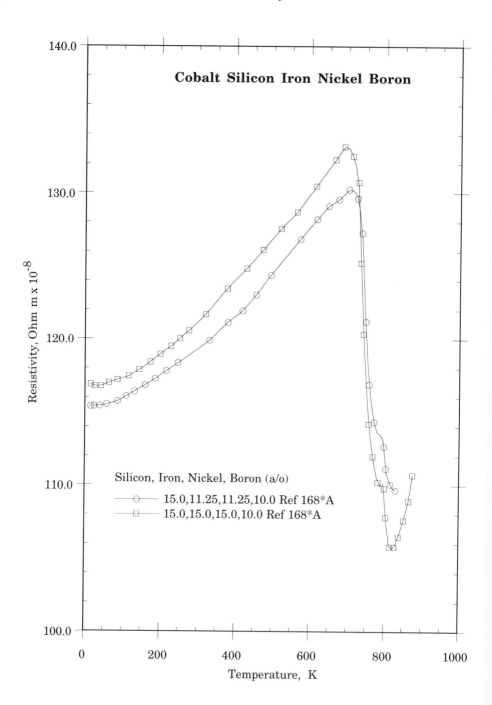

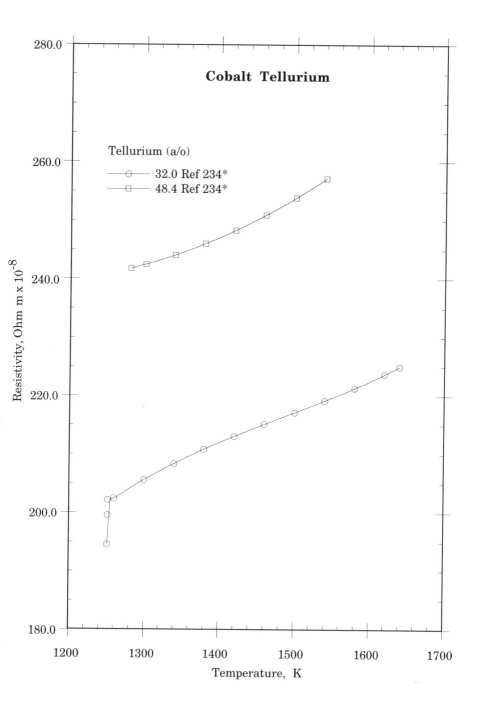

Cobalt Tellurium

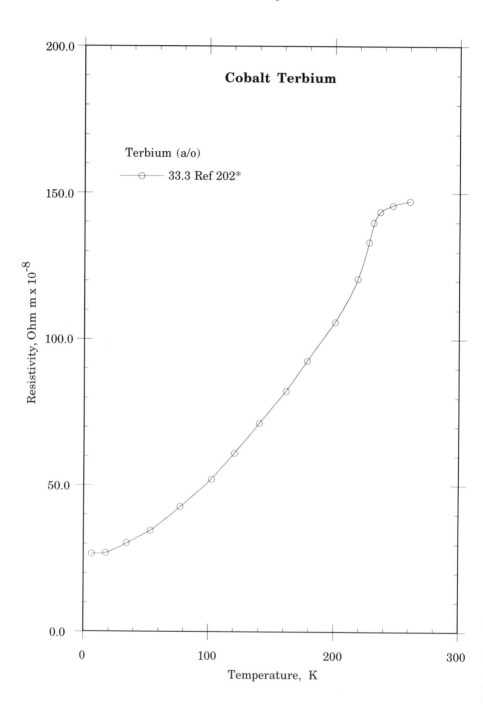

Cobalt Terbium

Terbium (a/o)
—o— 33.3 Ref 202*

Resistivity, Ohm m x 10^{-8}

Temperature, K

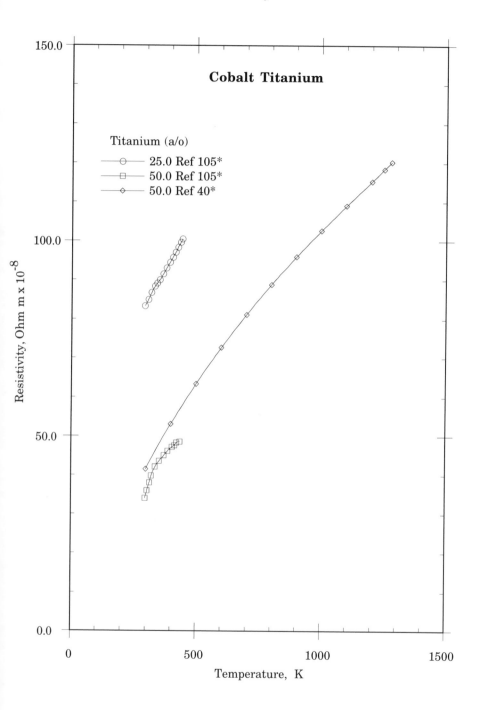

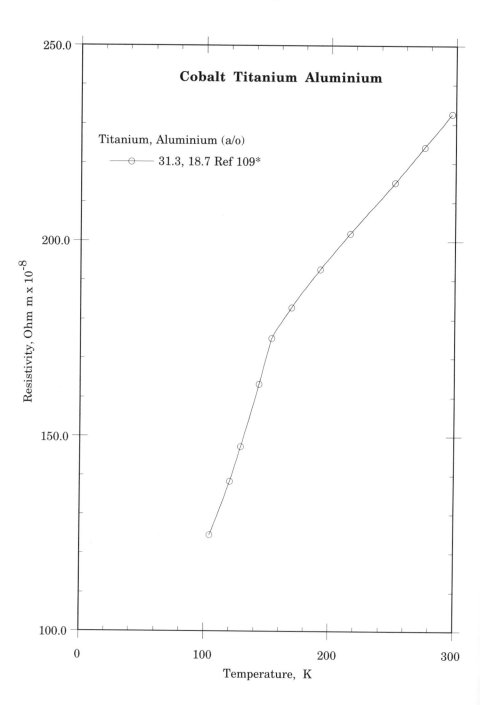

Cobalt Titanium Aluminium

Titanium, Aluminium (a/o)

———⊙——— 31.3, 18.7 Ref 109*

Resistivity, Ohm m x 10⁻⁸

Temperature, K

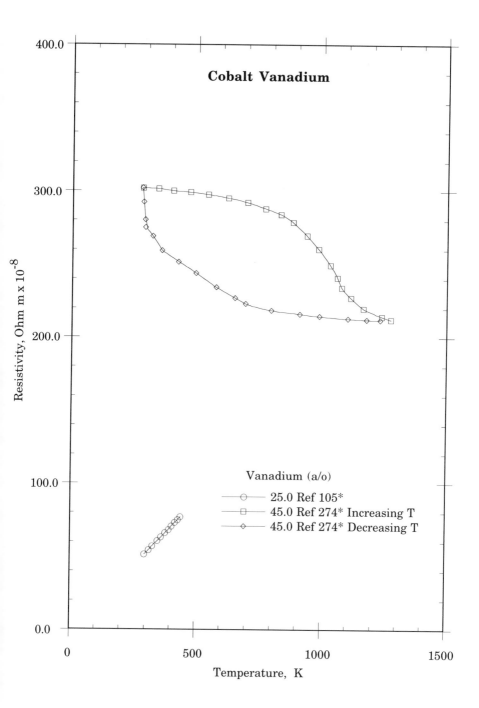

Cobalt Vanadium

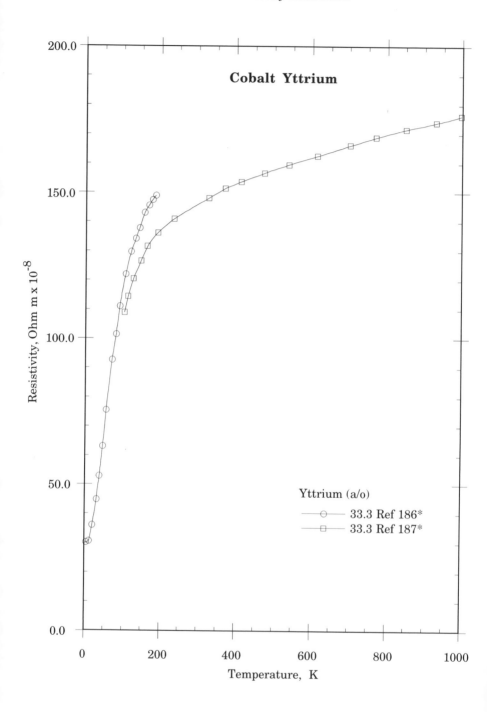

Cobalt Yttrium

Resistivity, Ohm m x 10⁻⁸

Temperature, K

Yttrium (a/o)
—⊙— 33.3 Ref 186*
—⊟— 33.3 Ref 187*

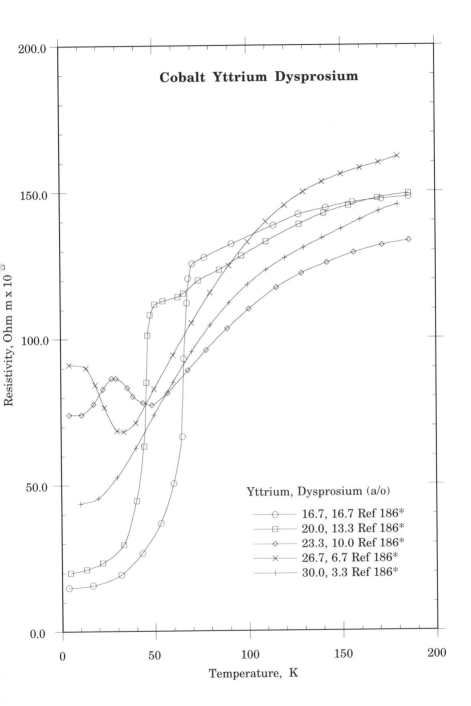

Cobalt Yttrium Dysprosium

Yttrium, Dysprosium (a/o)

—⊙— 16.7, 16.7 Ref 186*
—□— 20.0, 13.3 Ref 186*
—◇— 23.3, 10.0 Ref 186*
—×— 26.7, 6.7 Ref 186*
—+— 30.0, 3.3 Ref 186*

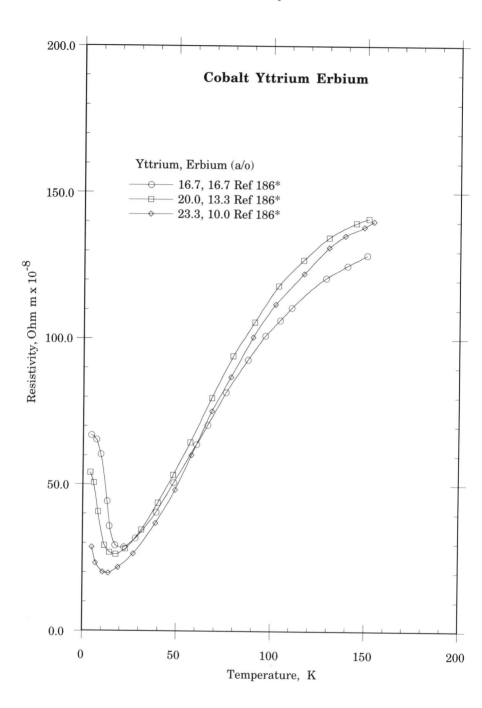

Cobalt Yttrium Erbium

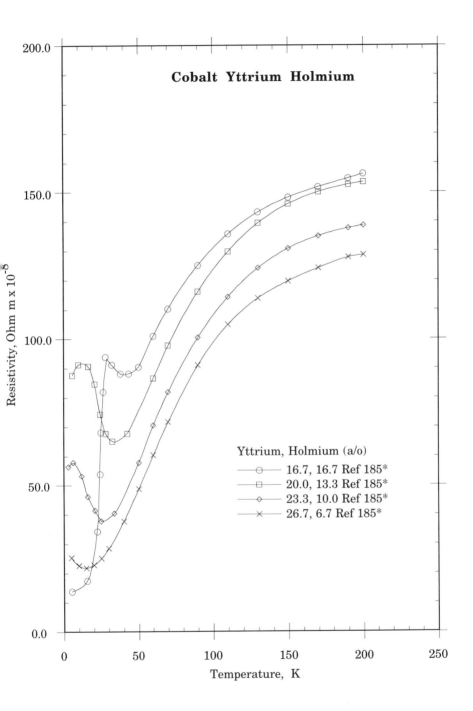

Cobalt Yttrium Holmium

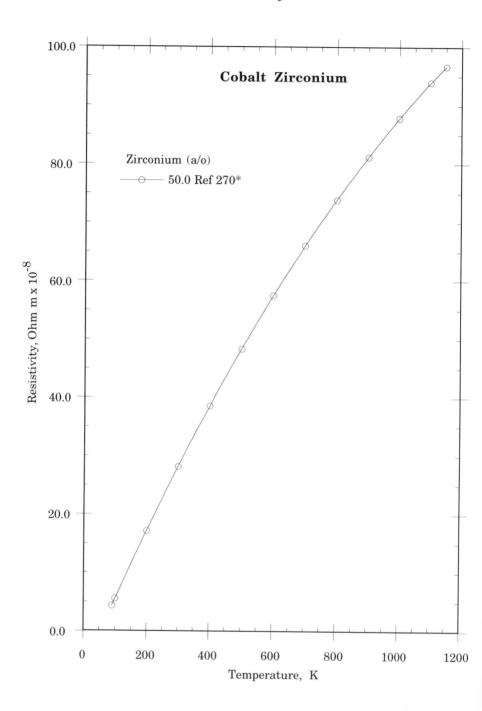

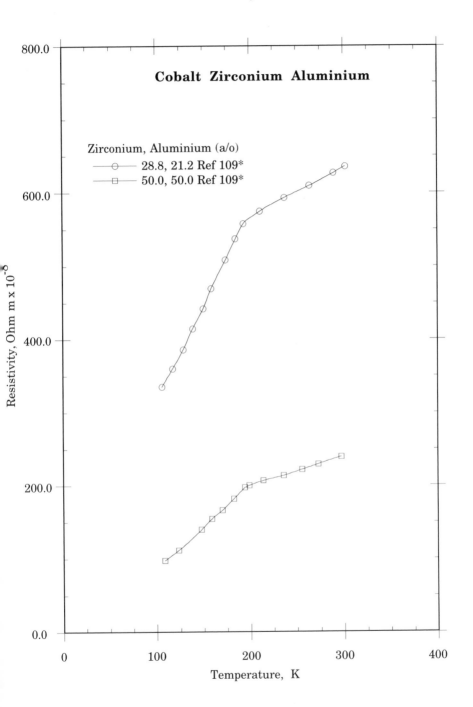

Cobalt Zirconium Aluminium

Zirconium, Aluminium (a/o)
- ○ 28.8, 21.2 Ref 109*
- □ 50.0, 50.0 Ref 109*

Resistivity, Ohm m x 10⁻⁸

Temperature, K

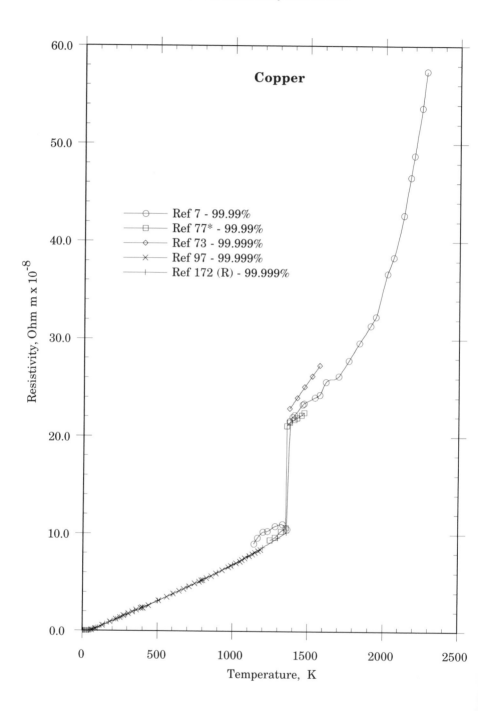

Copper

Ref 7 - 99.99%
Ref 77* - 99.99%
Ref 73 - 99.999%
Ref 97 - 99.999%
Ref 172 (R) - 99.999%

Resistivity, Ohm m x 10^{-8}

Temperature, K

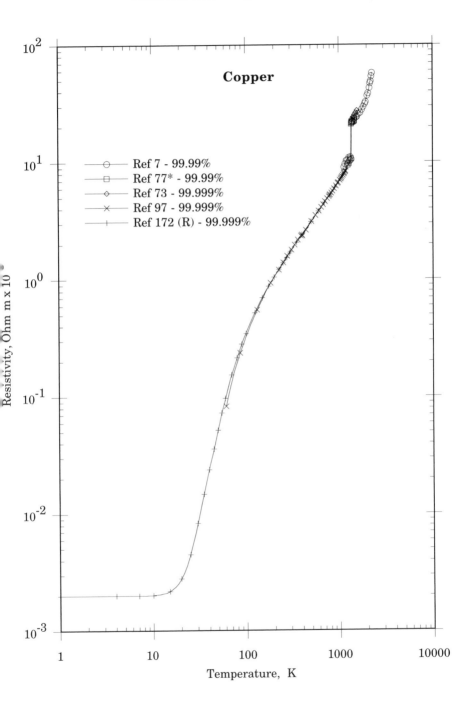

Copper

Ref 7 - 99.99%
Ref 77* - 99.99%
Ref 73 - 99.999%
Ref 97 - 99.999%
Ref 172 (R) - 99.999%

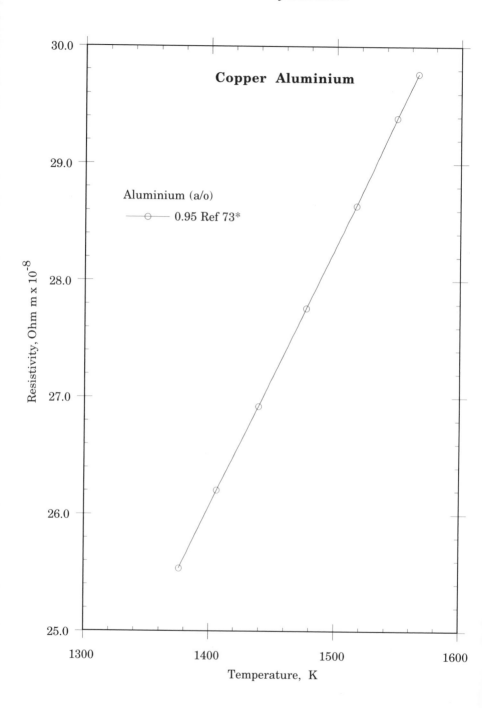

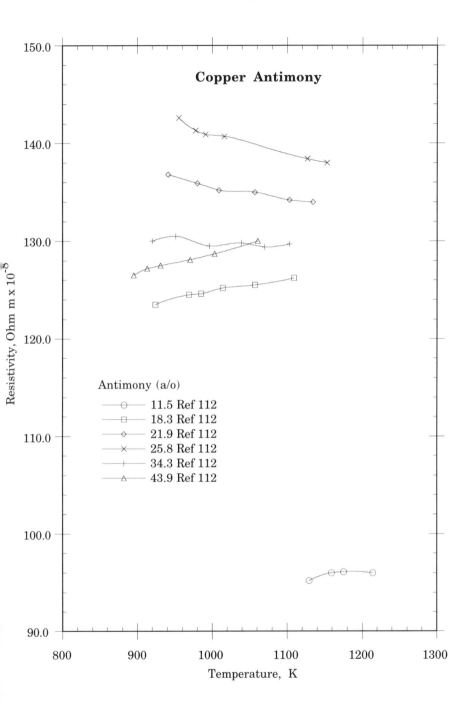

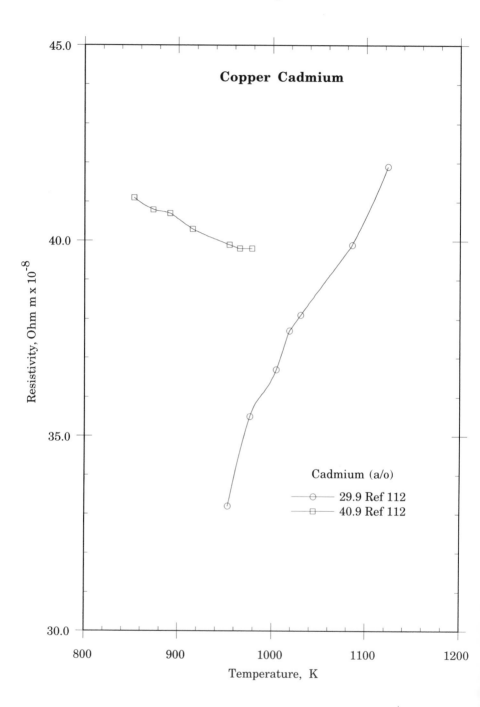

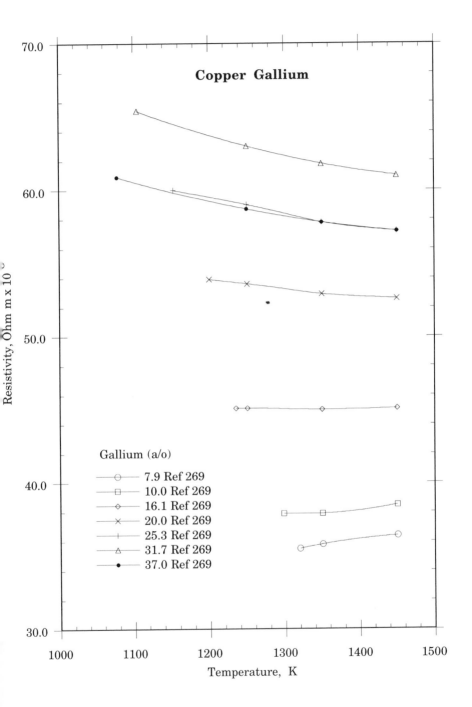

Copper Gallium

Resistivity, Ohm m x 10⁰

Temperature, K

Gallium (a/o)
- ○ 7.9 Ref 269
- □ 10.0 Ref 269
- ◇ 16.1 Ref 269
- × 20.0 Ref 269
- + 25.3 Ref 269
- △ 31.7 Ref 269
- ● 37.0 Ref 269

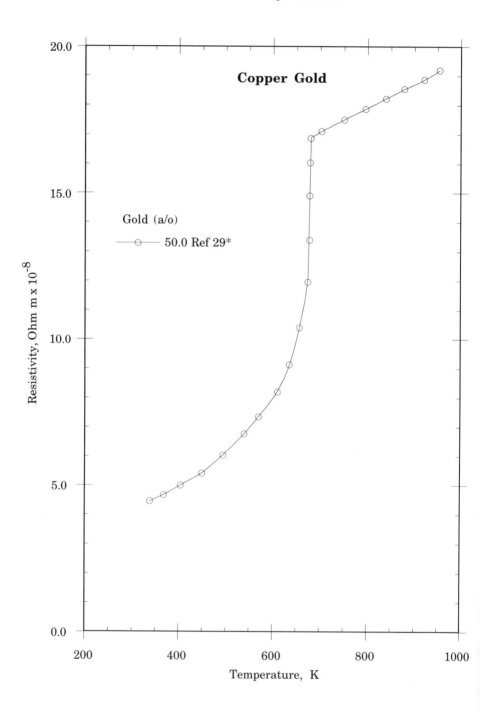

Copper Gold

Gold (a/o)

—⊙— 50.0 Ref 29*

Resistivity, Ohm m x 10^{-8}

Temperature, K

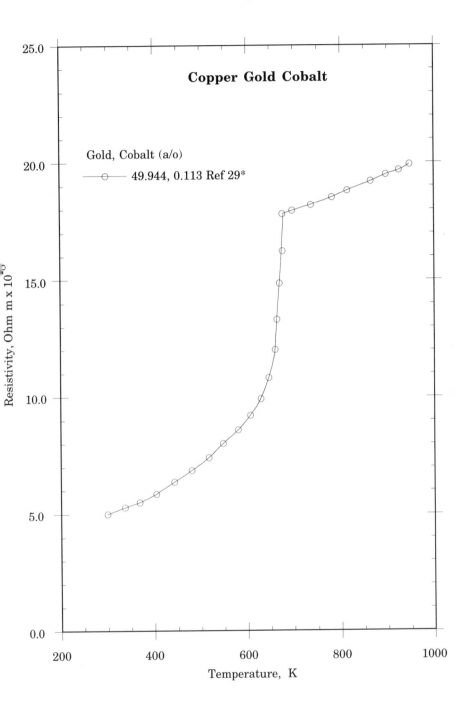

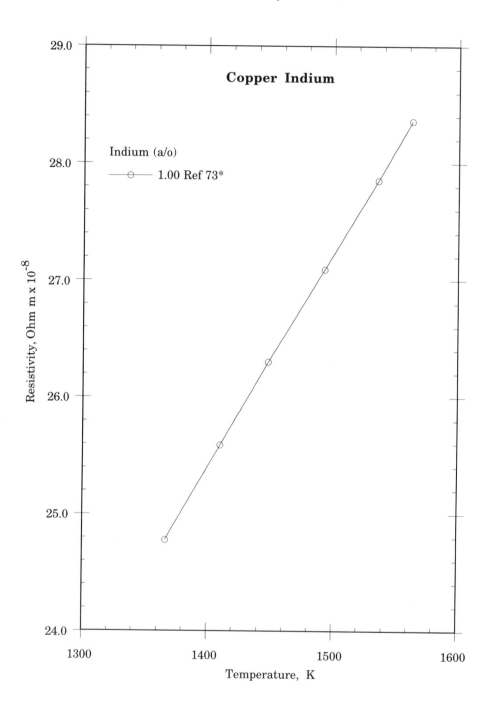

Copper Indium

Indium (a/o)

—⊖— 1.00 Ref 73*

Resistivity, Ohm m x 10^{-8}

Temperature, K

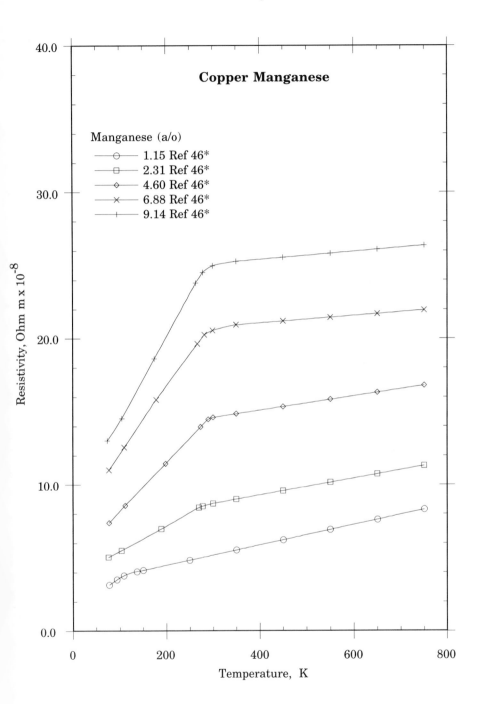

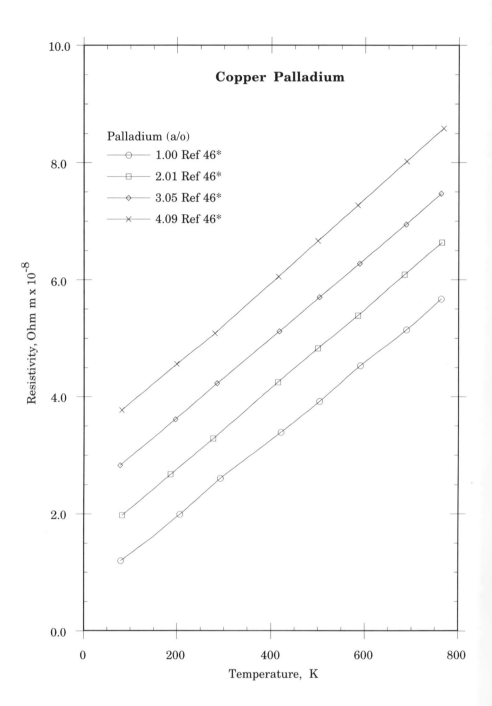

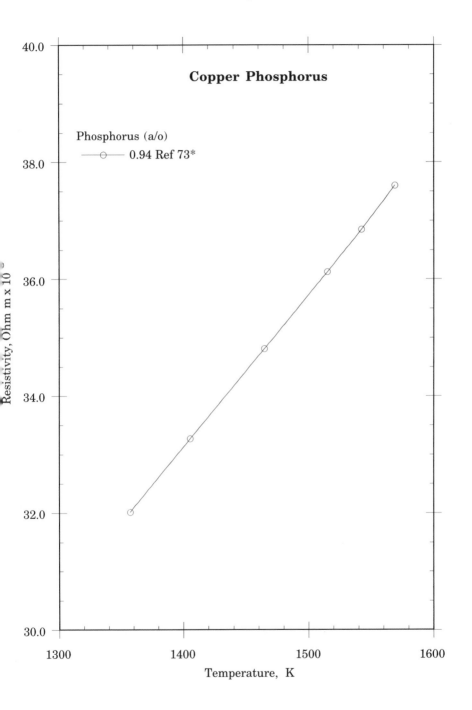

Copper Phosphorus

Phosphorus (a/o)
— ○ — 0.94 Ref 73*

Resistivity, Ohm m x 10⁻

Temperature, K

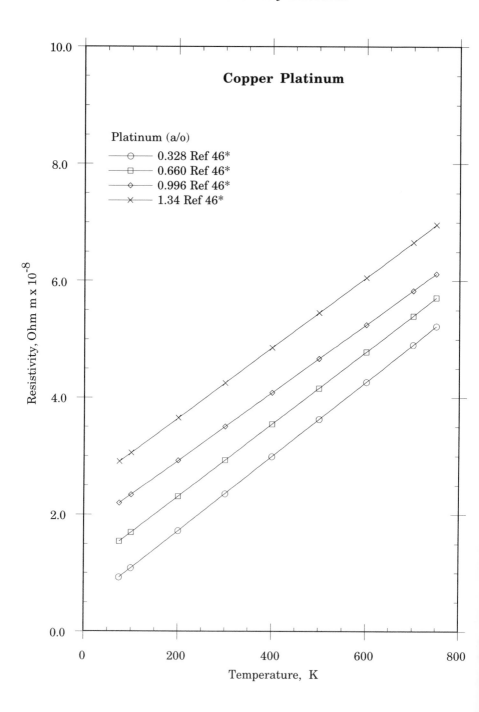

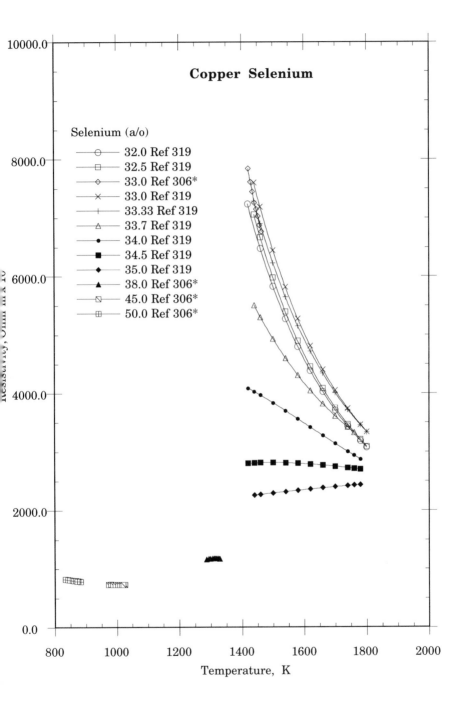

Copper Selenium

Selenium (a/o)

32.0 Ref 319
32.5 Ref 319
33.0 Ref 306*
33.0 Ref 319
33.33 Ref 319
33.7 Ref 319
34.0 Ref 319
34.5 Ref 319
35.0 Ref 319
38.0 Ref 306*
45.0 Ref 306*
50.0 Ref 306*

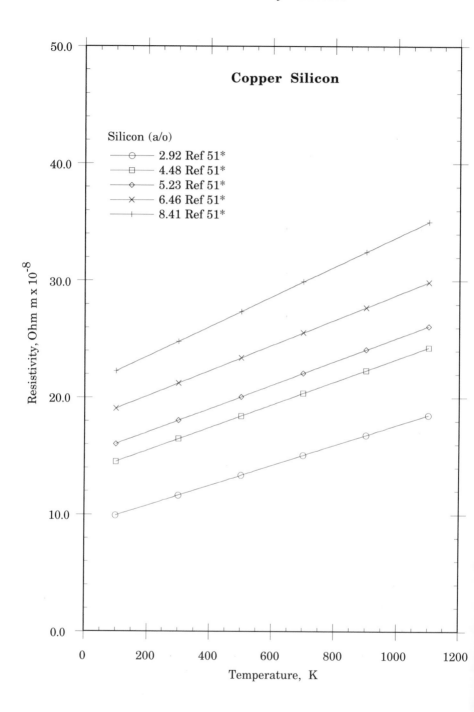

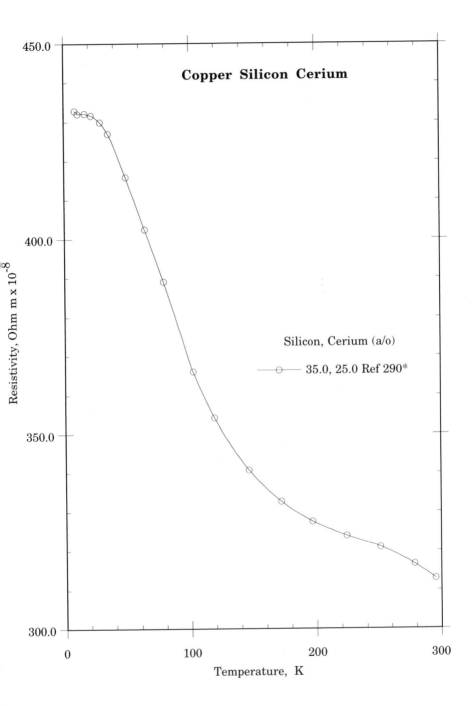

Copper Silicon Cerium

Silicon, Cerium (a/o)

⊸— 35.0, 25.0 Ref 290*

Resistivity, Ohm m x 10^{-8}

Temperature, K

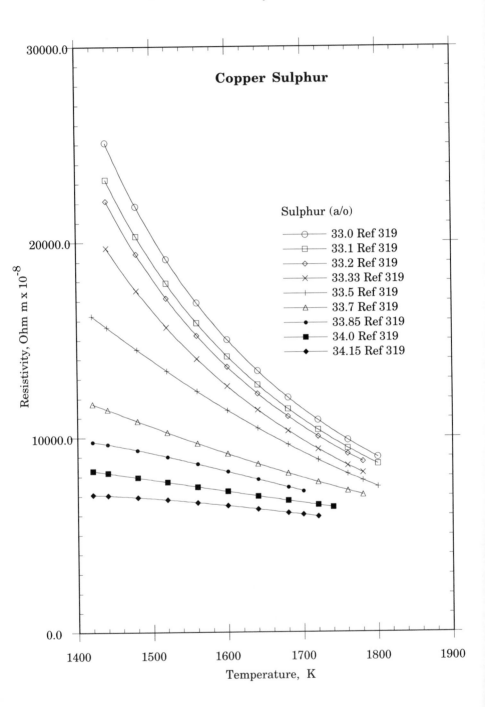

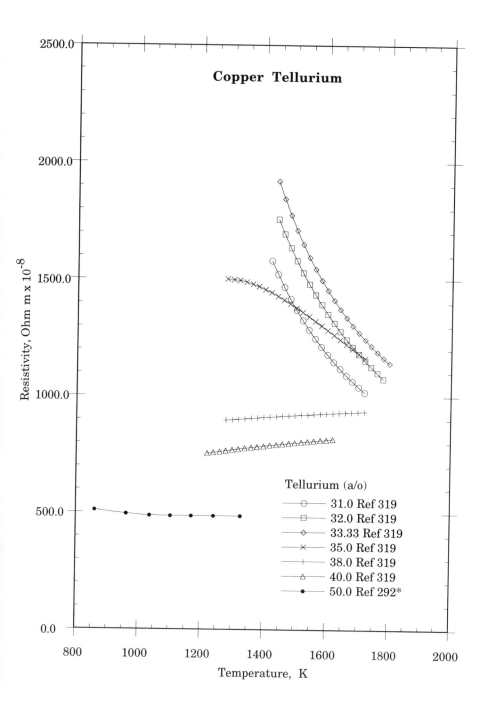

Copper Tellurium

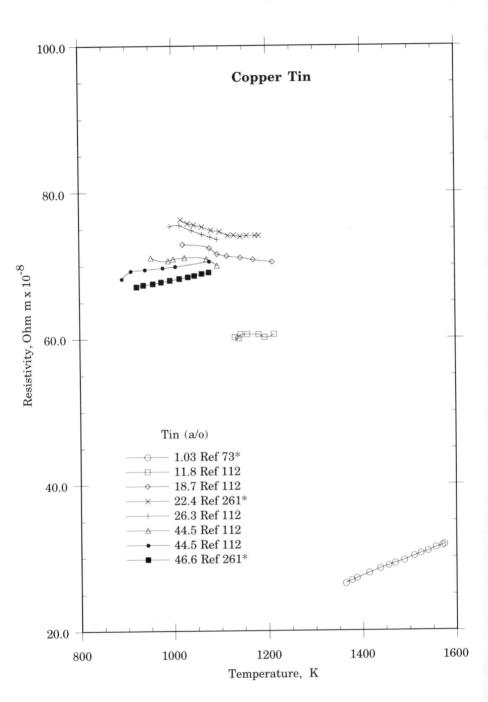

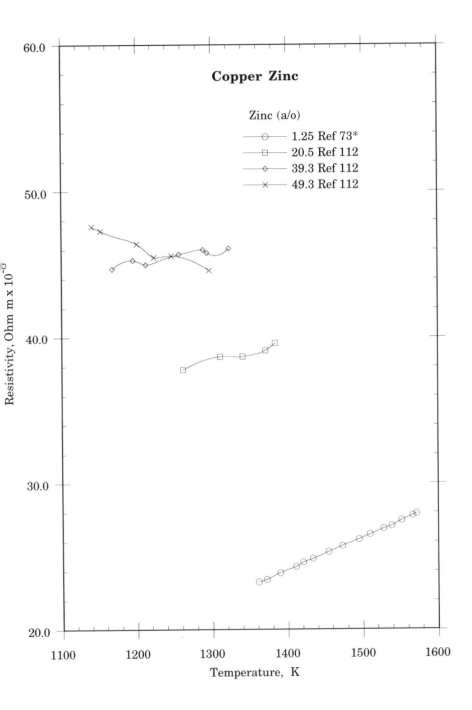

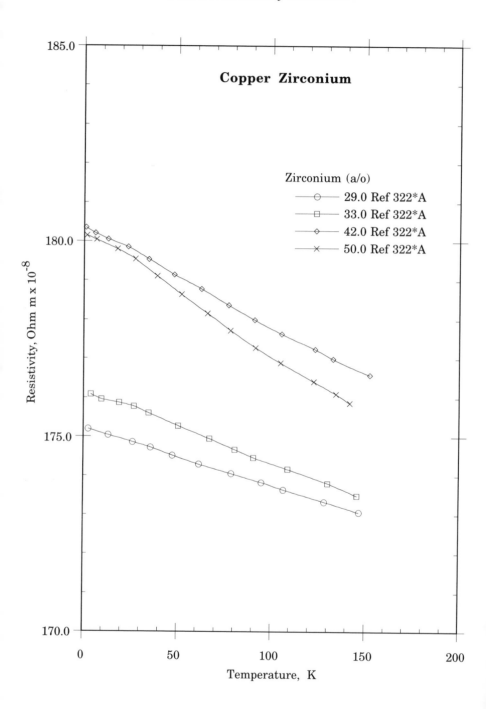

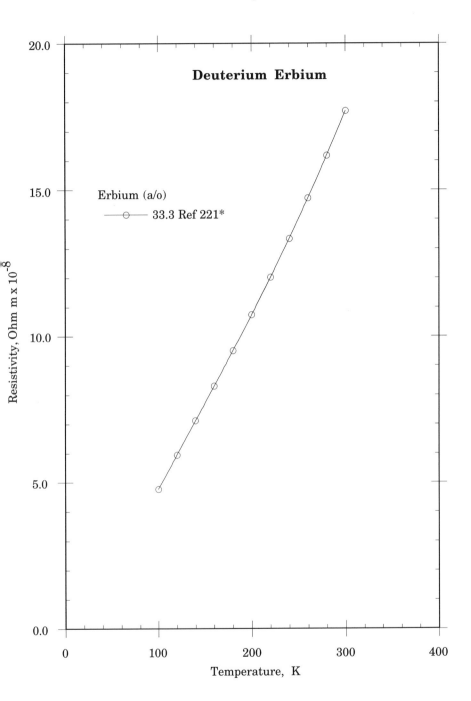

Deuterium Erbium

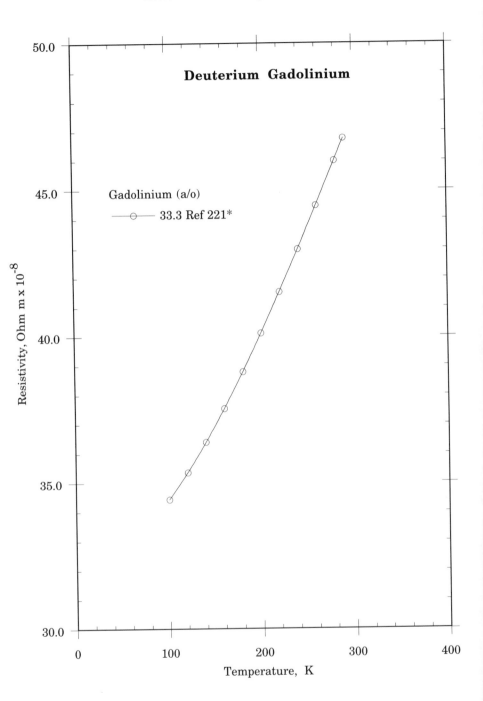

Deuterium Gadolinium

Gadolinium (a/o)
—◦— 33.3 Ref 221*

Resistivity, Ohm m x 10^{-8}

Temperature, K

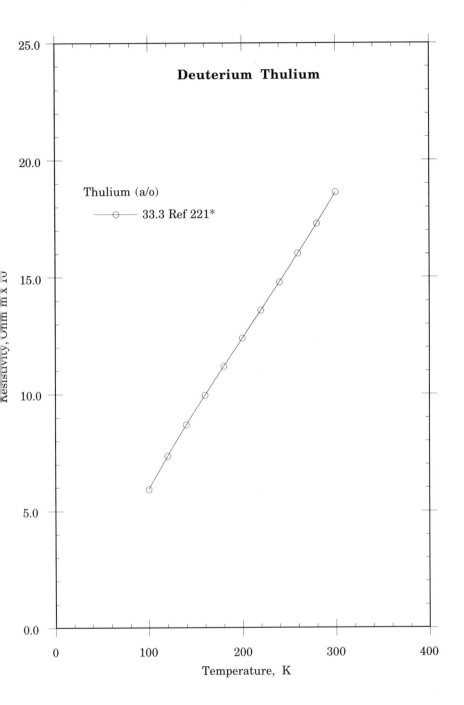

Deuterium Thulium

Thulium (a/o)

——⊖—— 33.3 Ref 221*

Temperature, K

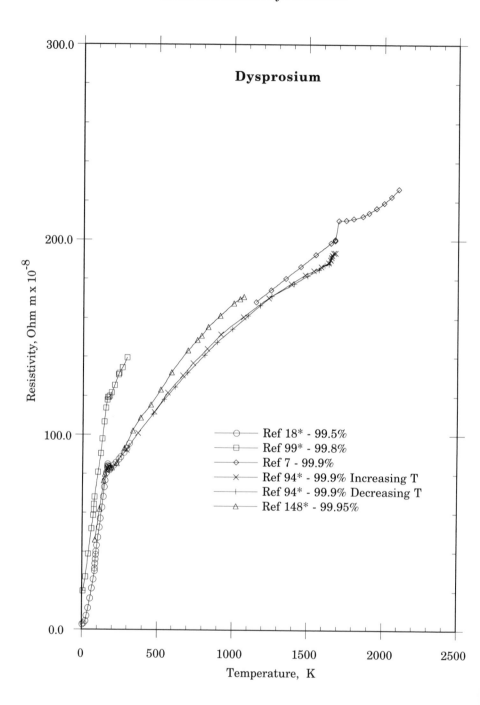

Dysprosium

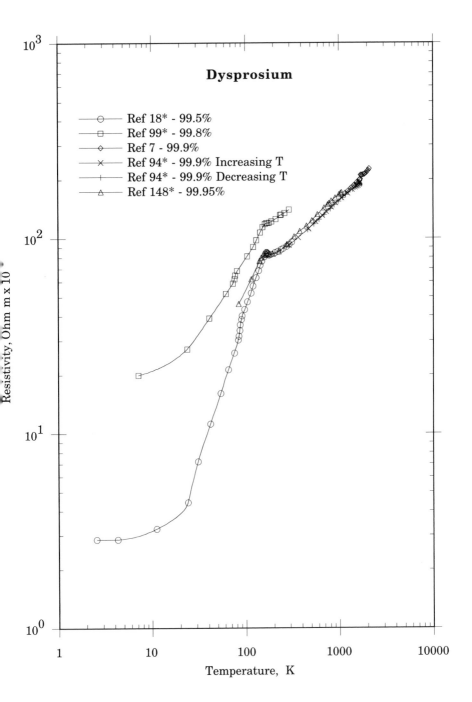

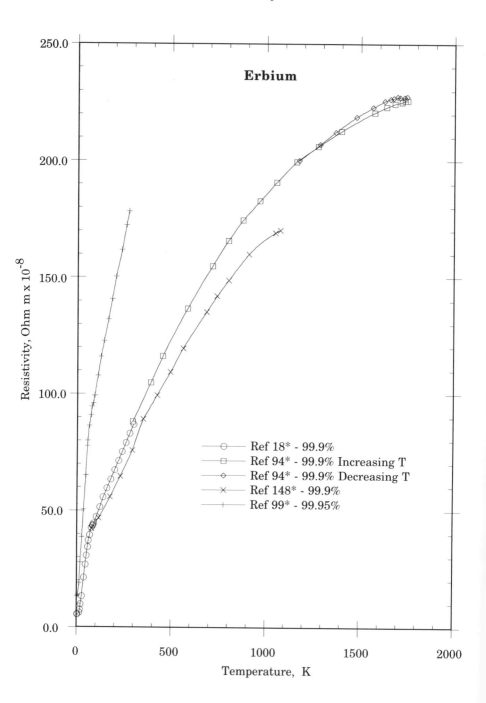

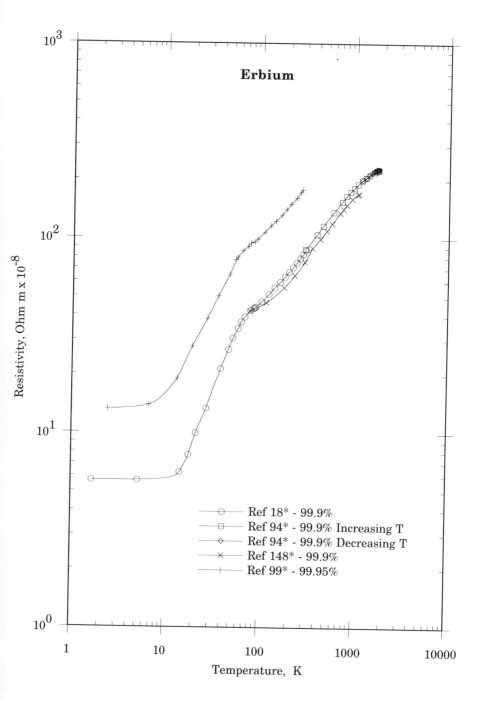

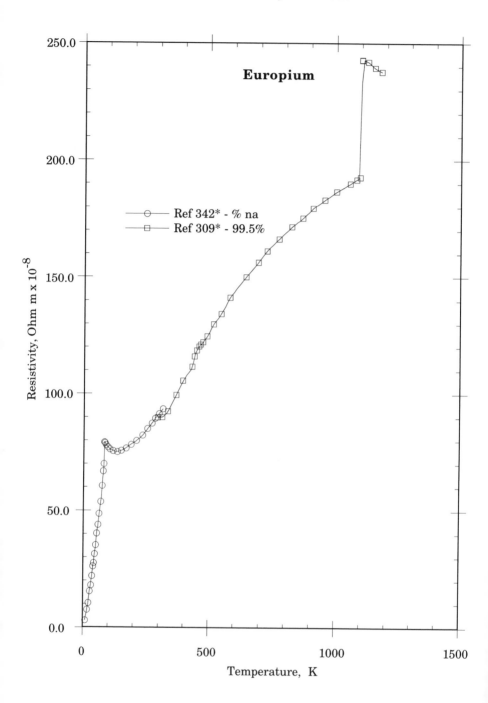

Europium

Ref 342* - % na
Ref 309* - 99.5%

Resistivity, Ohm m x 10⁻⁸

Temperature, K

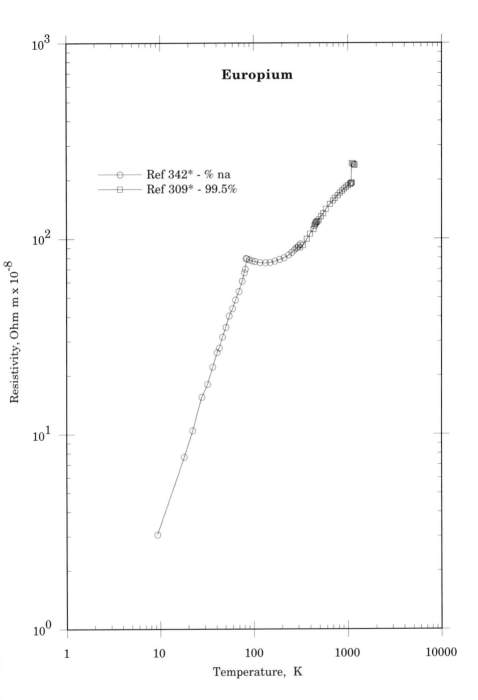

Europium

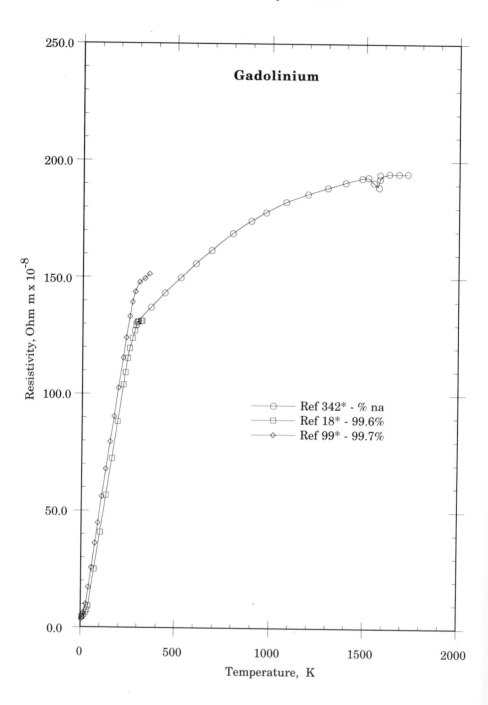

Gadolinium

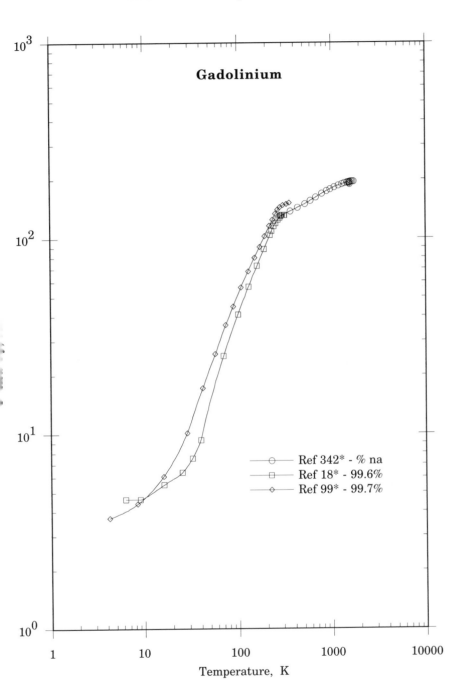

Gadolinium

Ref 342* - % na
Ref 18* - 99.6%
Ref 99* - 99.7%

Temperature, K

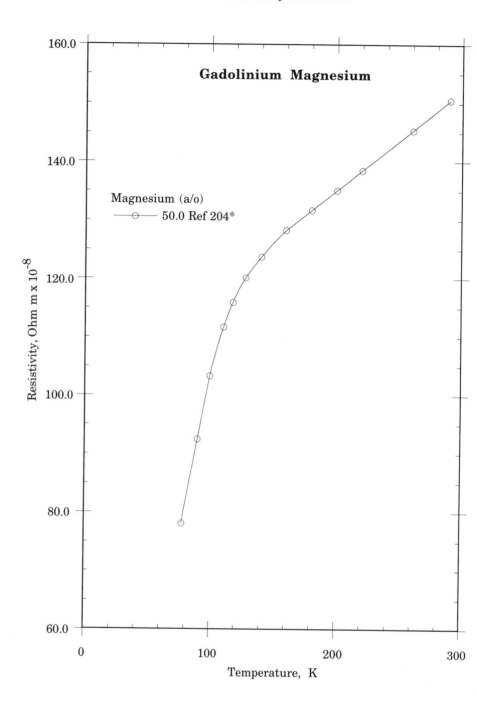

Gadolinium Magnesium

Magnesium (a/o)
—○— 50.0 Ref 204*

Resistivity, Ohm m x 10^{-8}

Temperature, K

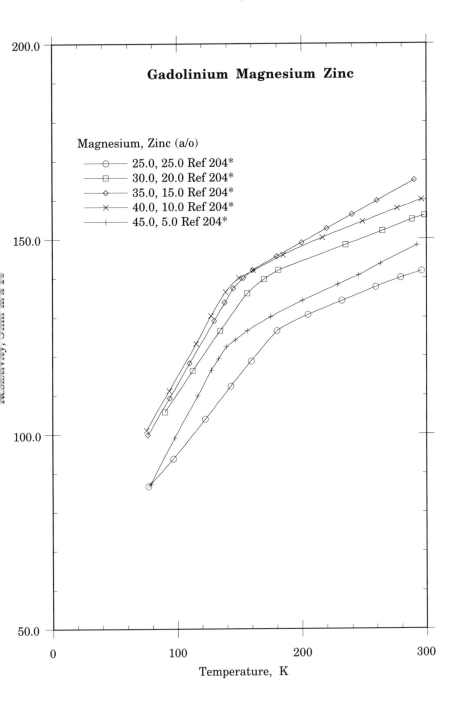

Gadolinium Magnesium Zinc

Magnesium, Zinc (a/o)
- ─⊖─ 25.0, 25.0 Ref 204*
- ─□─ 30.0, 20.0 Ref 204*
- ─◇─ 35.0, 15.0 Ref 204*
- ─✕─ 40.0, 10.0 Ref 204*
- ─┼─ 45.0, 5.0 Ref 204*

Temperature, K

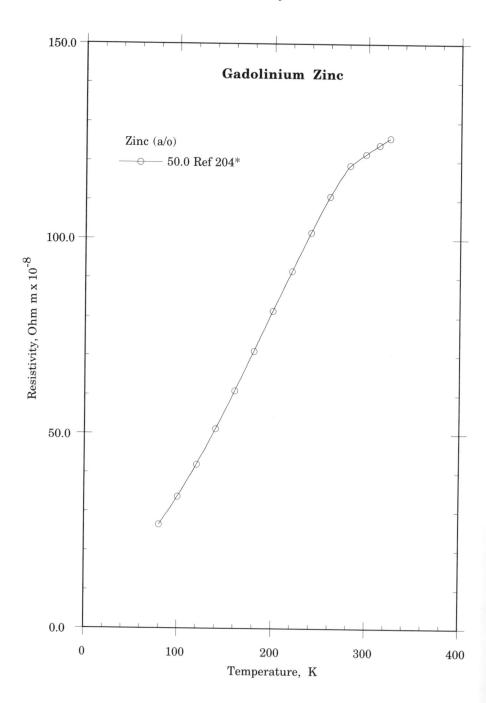

Gadolinium Zinc

Zinc (a/o)
—⊖— 50.0 Ref 204*

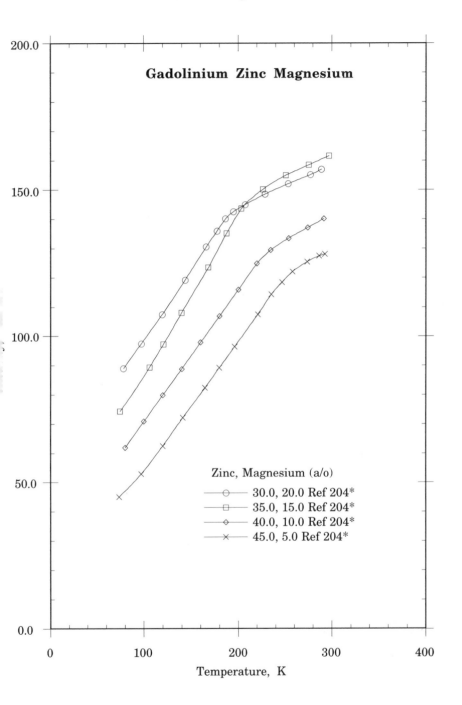

Gadolinium Zinc Magnesium

Zinc, Magnesium (a/o)
——⊖—— 30.0, 20.0 Ref 204*
——□—— 35.0, 15.0 Ref 204*
——◇—— 40.0, 10.0 Ref 204*
——✕—— 45.0, 5.0 Ref 204*

Temperature, K

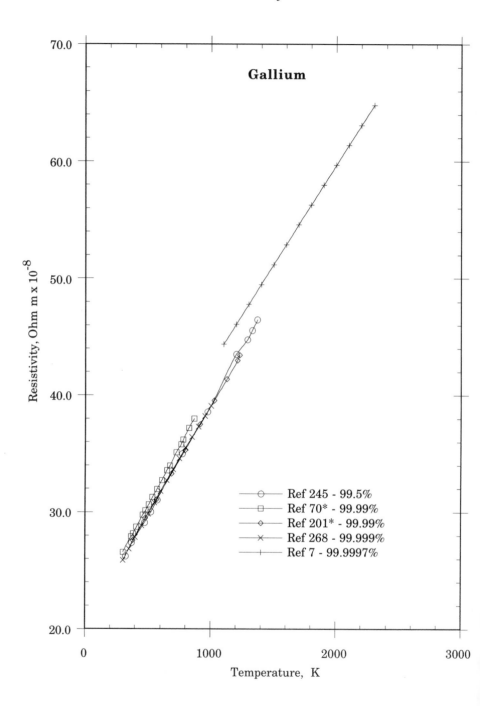

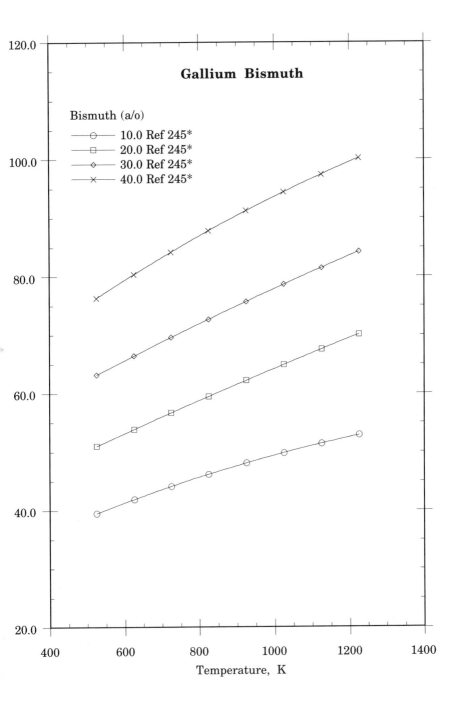

Gallium Bismuth

Bismuth (a/o)
- 10.0 Ref 245*
- 20.0 Ref 245*
- 30.0 Ref 245*
- 40.0 Ref 245*

Temperature, K

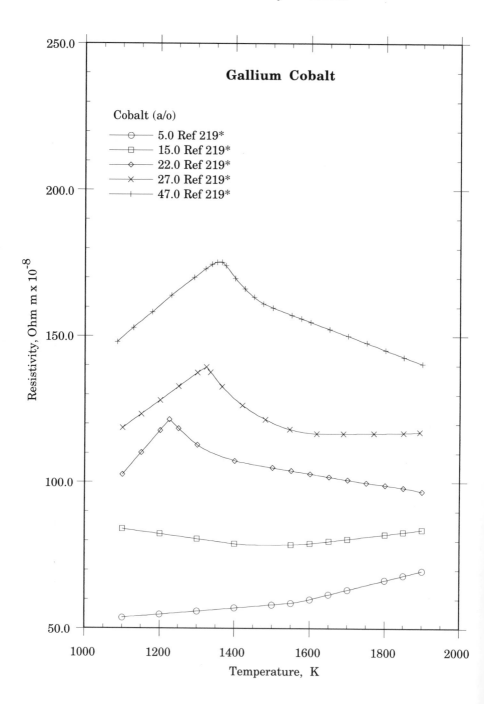

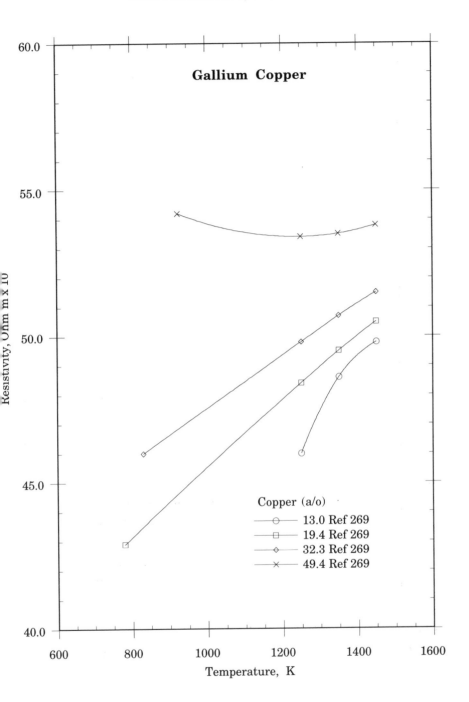

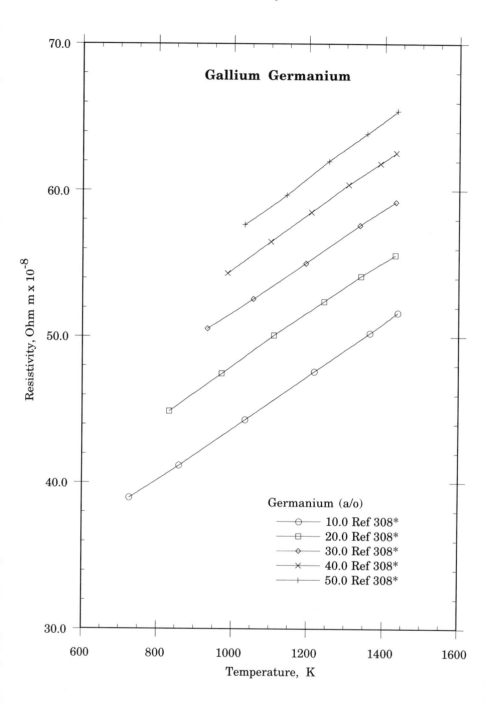

Gallium Germanium

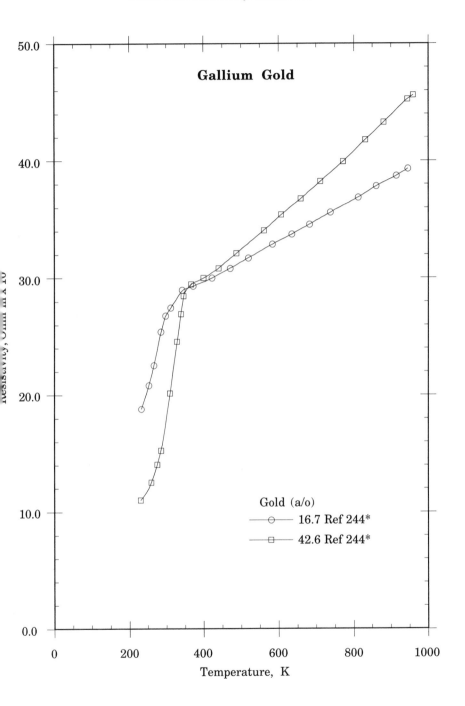

Gallium Gold

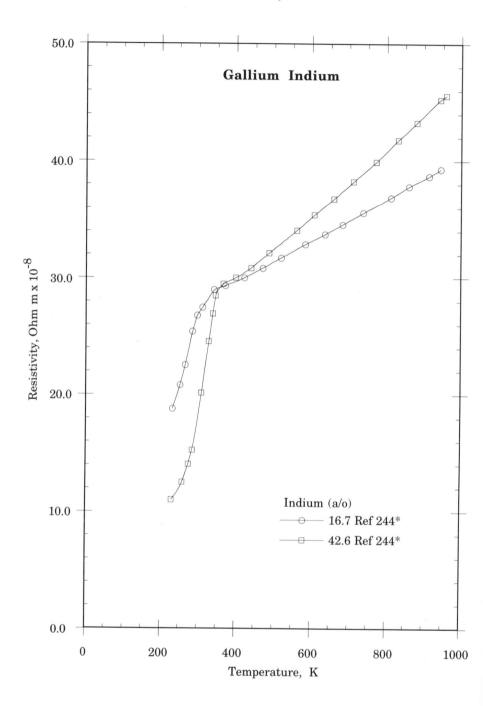

Gallium Indium

Resistivity, Ohm m x 10^{-8}

Temperature, K

Indium (a/o)
16.7 Ref 244*
42.6 Ref 244*

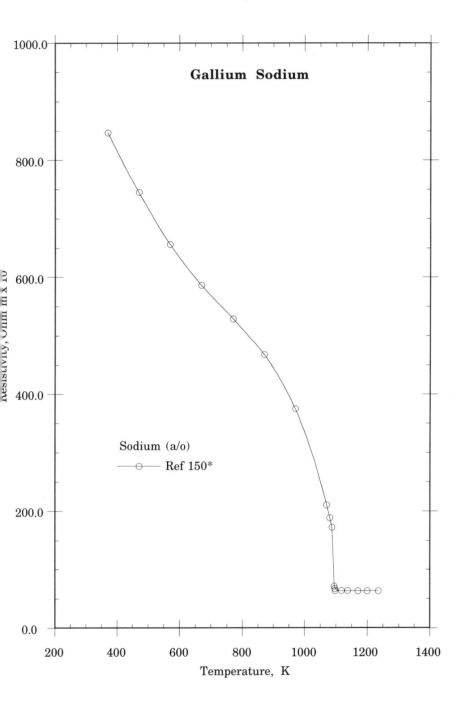

Gallium Sodium

Sodium (a/o)
Ref 150*

Temperature, K

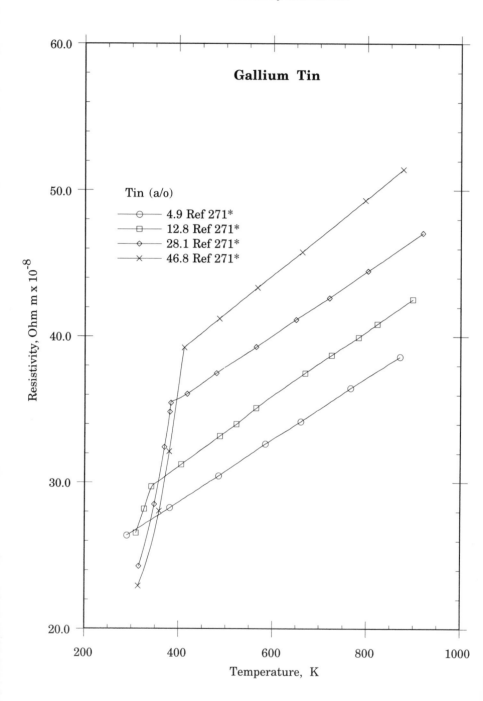

Gallium Tin

Tin (a/o)
- —○— 4.9 Ref 271*
- —□— 12.8 Ref 271*
- —◇— 28.1 Ref 271*
- —✕— 46.8 Ref 271*

Resistivity, Ohm m x 10^{-8}

Temperature, K

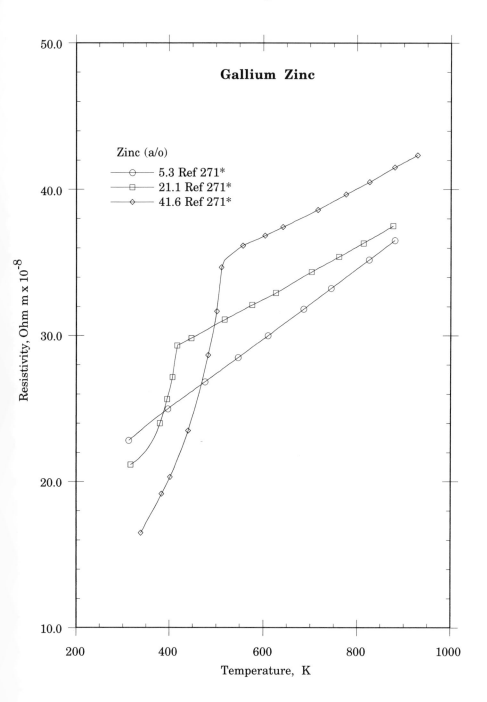

Gallium Zinc

Zinc (a/o)
- ○ 5.3 Ref 271*
- □ 21.1 Ref 271*
- ◇ 41.6 Ref 271*

Resistivity, Ohm m x 10^{-8}

Temperature, K

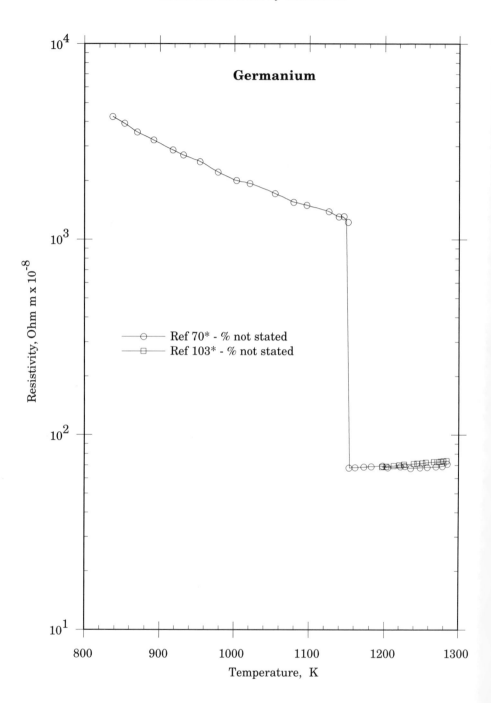

Germanium

Resistivity, Ohm m x 10^{-8}

Temperature, K

Ref 70* - % not stated
Ref 103* - % not stated

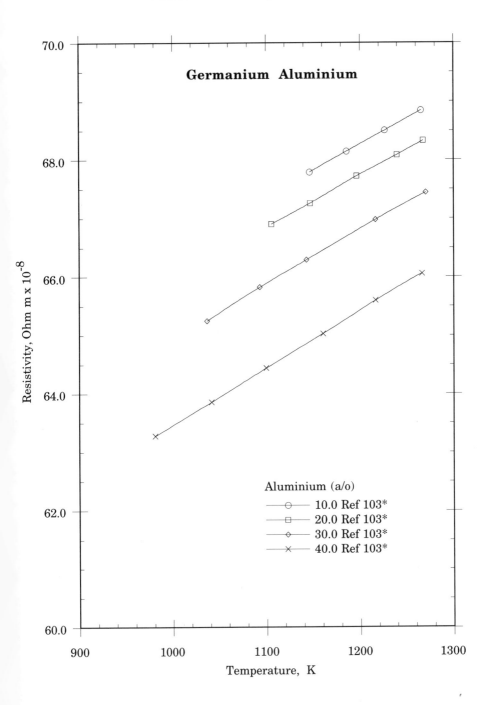

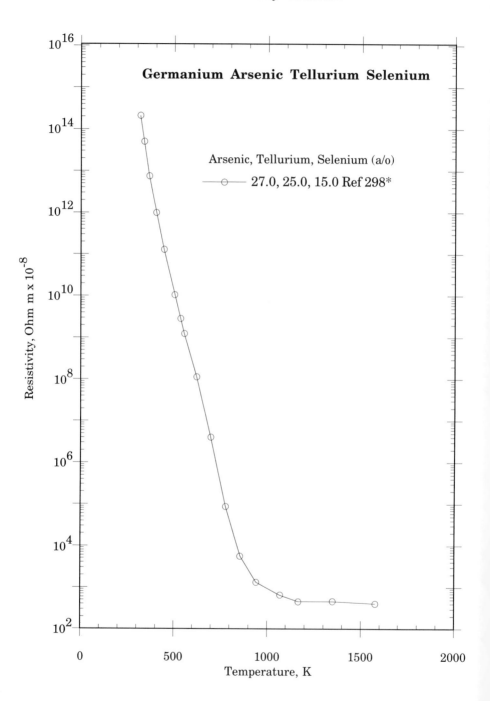

Germanium Arsenic Tellurium Selenium

Arsenic, Tellurium, Selenium (a/o)
27.0, 25.0, 15.0 Ref 298*

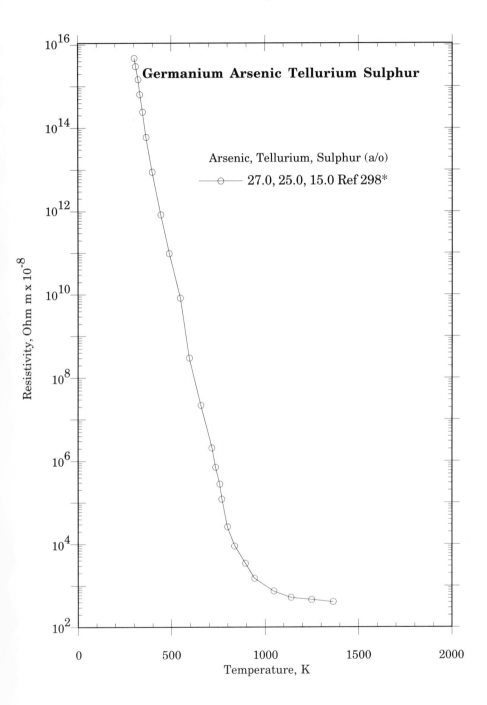

Germanium Arsenic Tellurium Sulphur

Arsenic, Tellurium, Sulphur (a/o)

27.0, 25.0, 15.0 Ref 298*

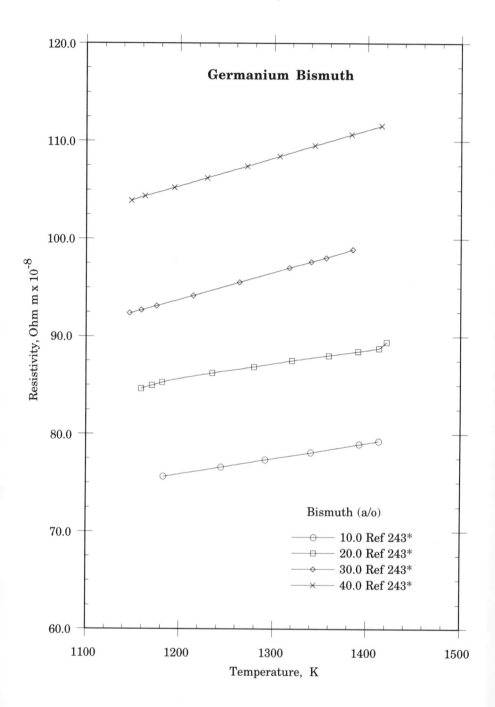

Germanium Bismuth

Bismuth (a/o)

─○─	10.0 Ref 243*
─□─	20.0 Ref 243*
─◇─	30.0 Ref 243*
─✕─	40.0 Ref 243*

Resistivity, Ohm m x 10⁻⁸

Temperature, K

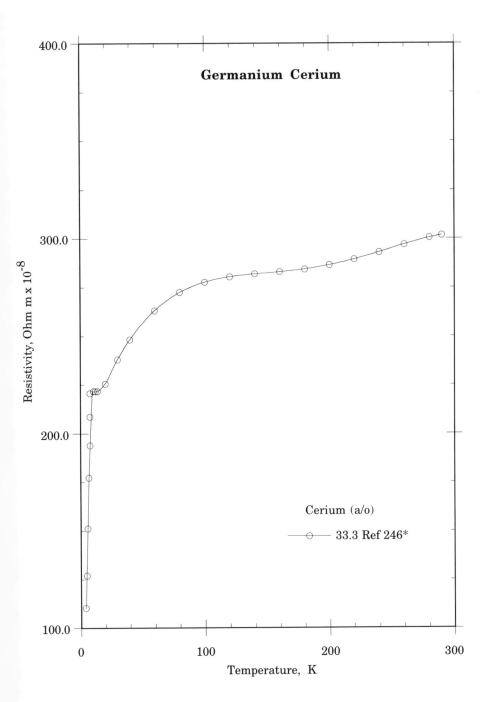

Germanium Cerium

Resistivity, Ohm m x 10⁻⁸

Temperature, K

Cerium (a/o)

33.3 Ref 246*

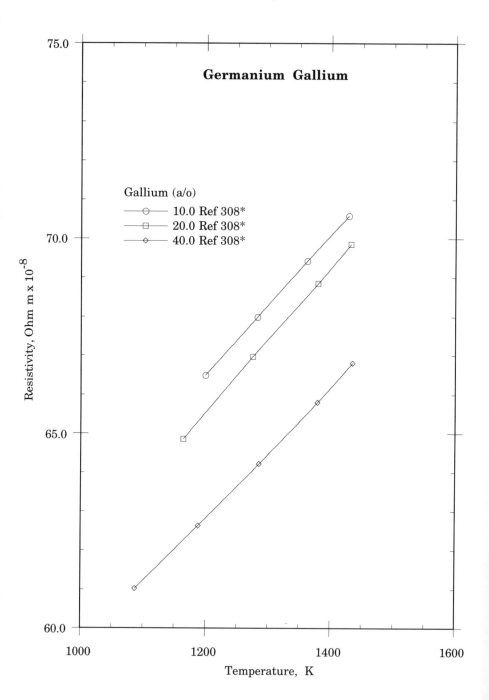

Germanium Gallium

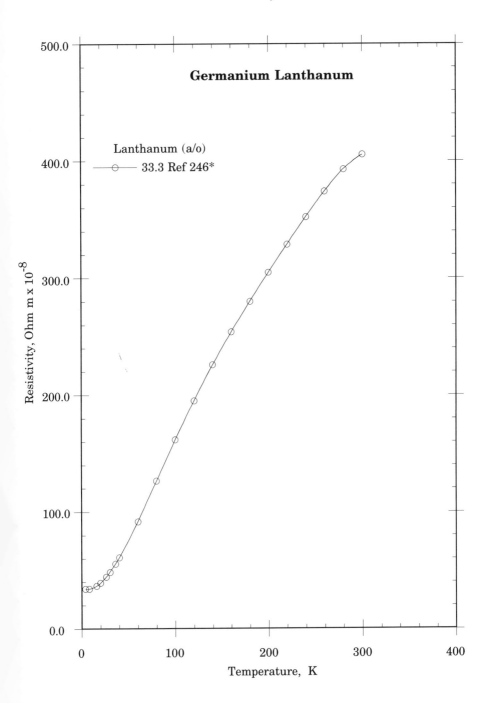

Germanium Lanthanum

Lanthanum (a/o)
⊖ 33.3 Ref 246*

Resistivity, Ohm m x 10^{-8}

Temperature, K

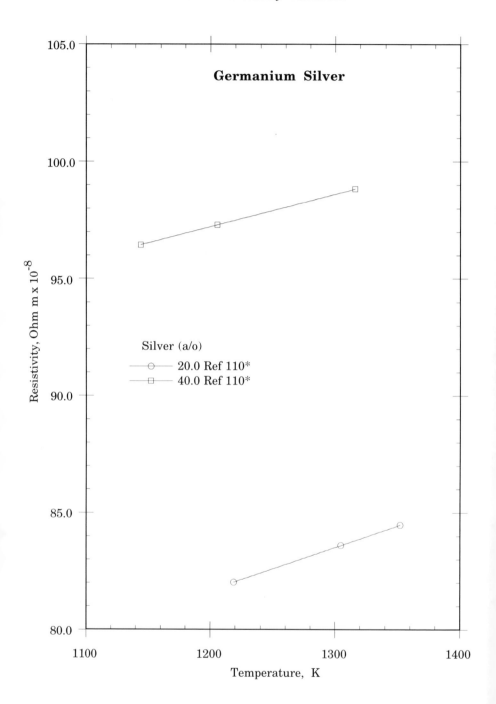

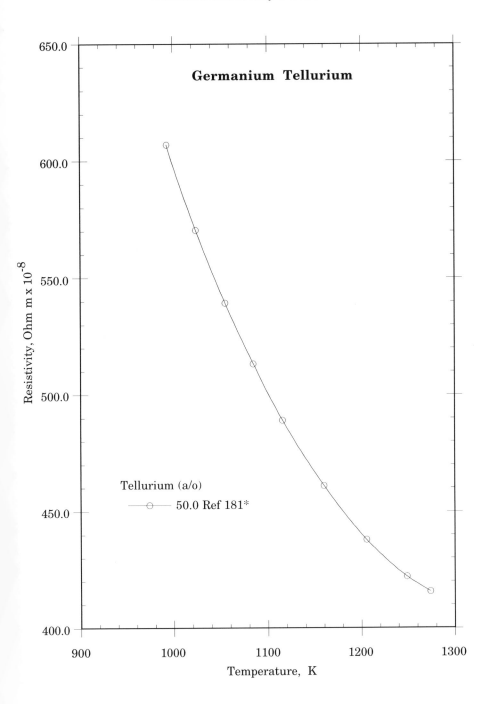

Germanium Tellurium

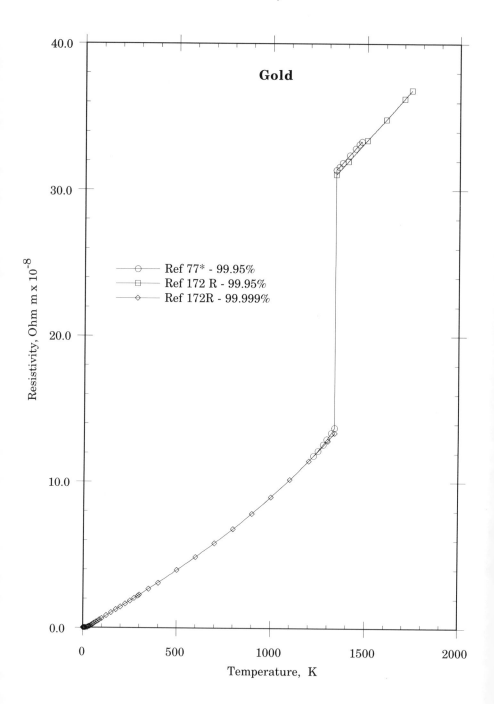

Gold

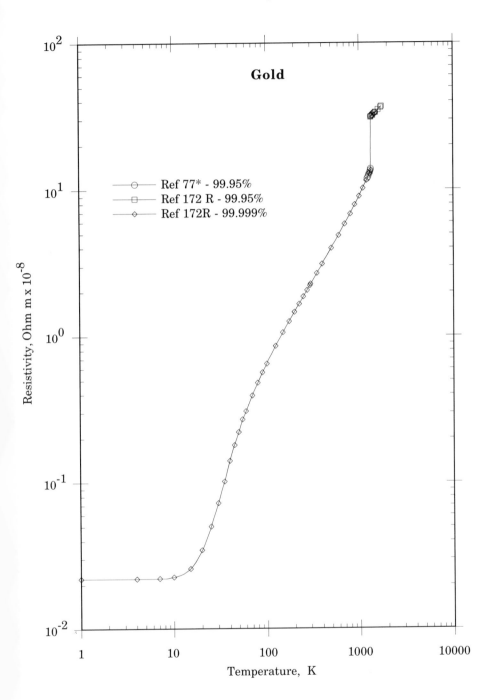

Gold

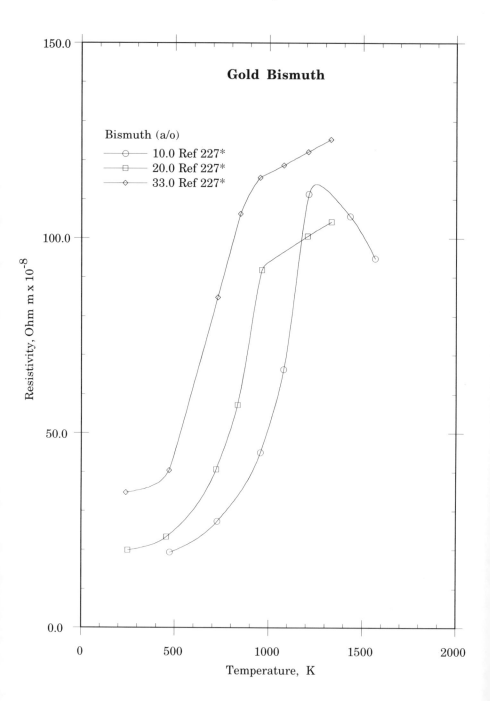

Gold Bismuth

Bismuth (a/o)
- ─○─ 10.0 Ref 227*
- ─□─ 20.0 Ref 227*
- ─◇─ 33.0 Ref 227*

Resistivity, Ohm m x 10^{-8}

Temperature, K

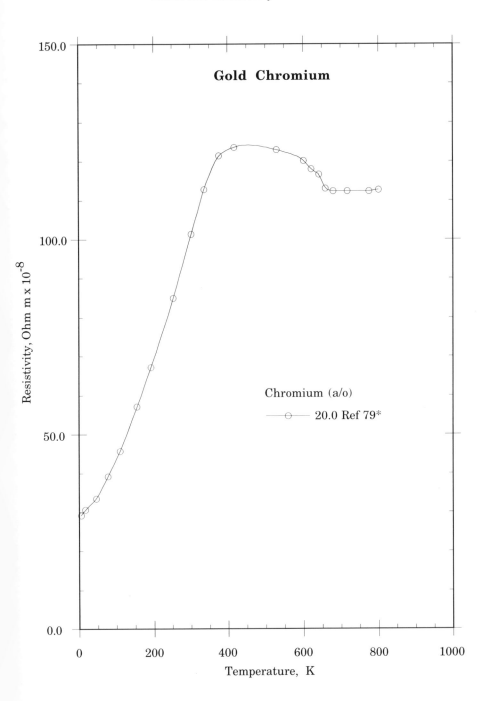

Gold Chromium

Chromium (a/o)

—⊖— 20.0 Ref 79*

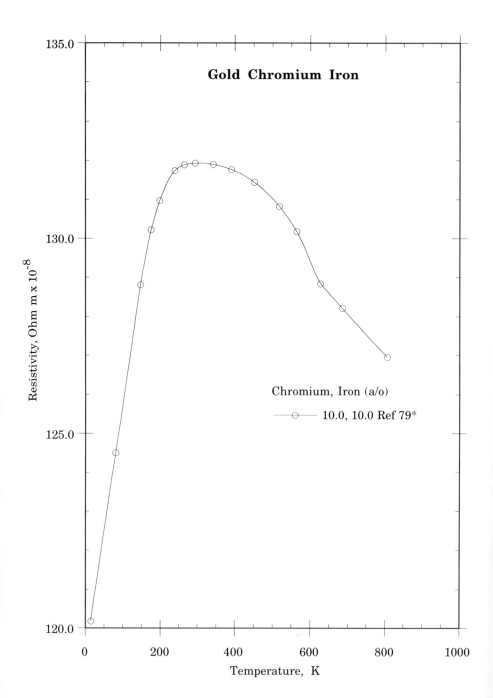

Gold Chromium Iron

Resistivity, Ohm m x 10⁻⁸

Temperature, K

Chromium, Iron (a/o)

10.0, 10.0 Ref 79*

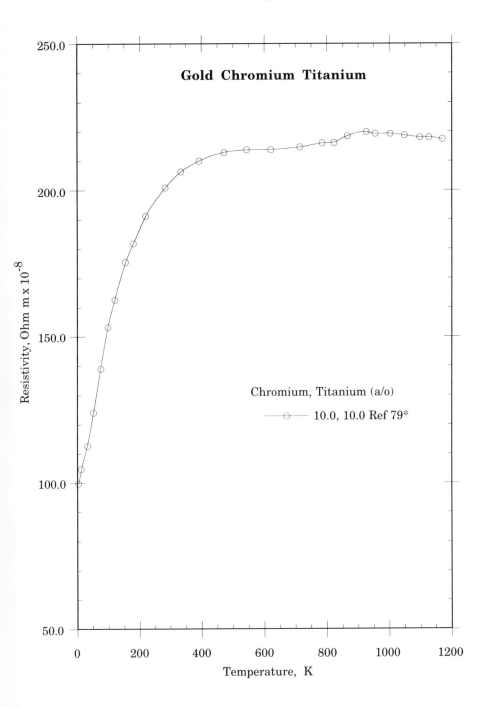

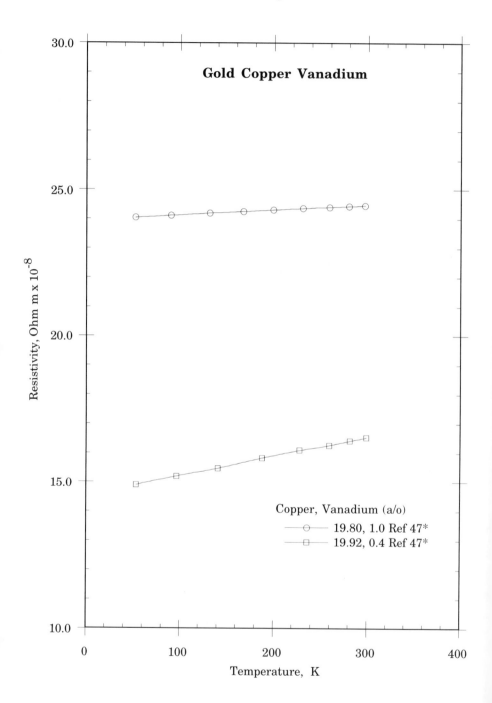

Gold Copper Vanadium

Copper, Vanadium (a/o)
19.80, 1.0 Ref 47*
19.92, 0.4 Ref 47*

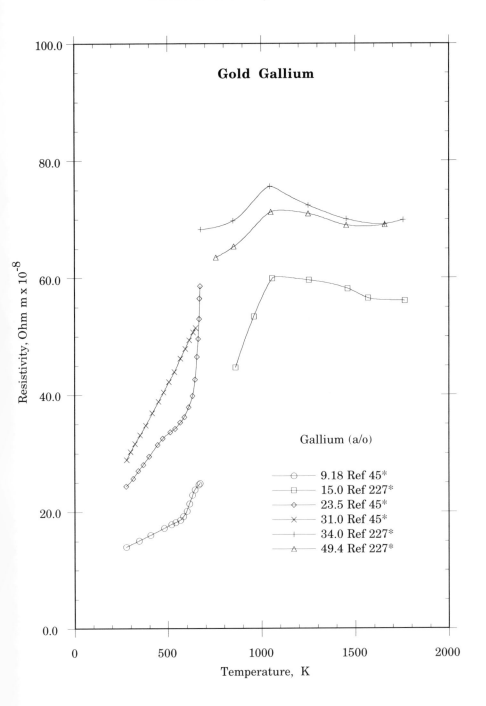

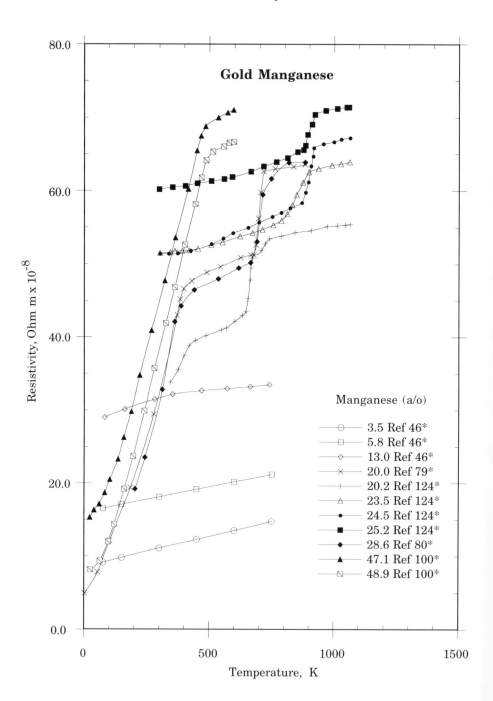

Gold Manganese

Resistivity, Ohm m x 10⁻⁸

Temperature, K

Manganese (a/o)

—⊖— 3.5 Ref 46*
—◻— 5.8 Ref 46*
—◇— 13.0 Ref 46*
—✕— 20.0 Ref 79*
—+— 20.2 Ref 124*
—△— 23.5 Ref 124*
—●— 24.5 Ref 124*
—■— 25.2 Ref 124*
—◆— 28.6 Ref 80*
—▲— 47.1 Ref 100*
—◩— 48.9 Ref 100*

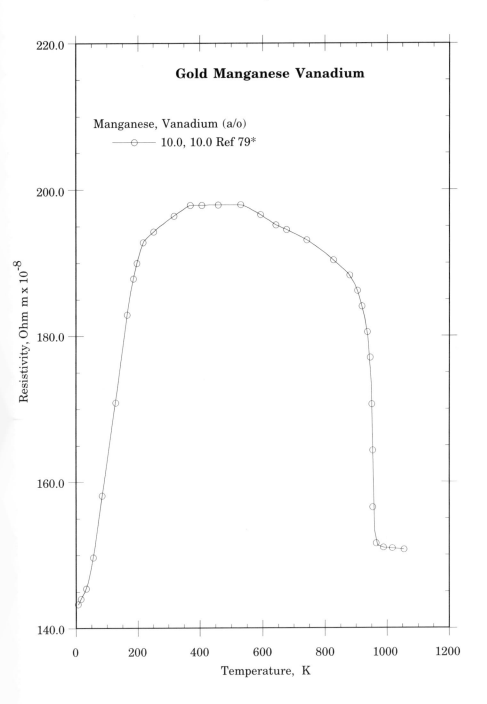

Gold Manganese Vanadium

Manganese, Vanadium (a/o)
10.0, 10.0 Ref 79*

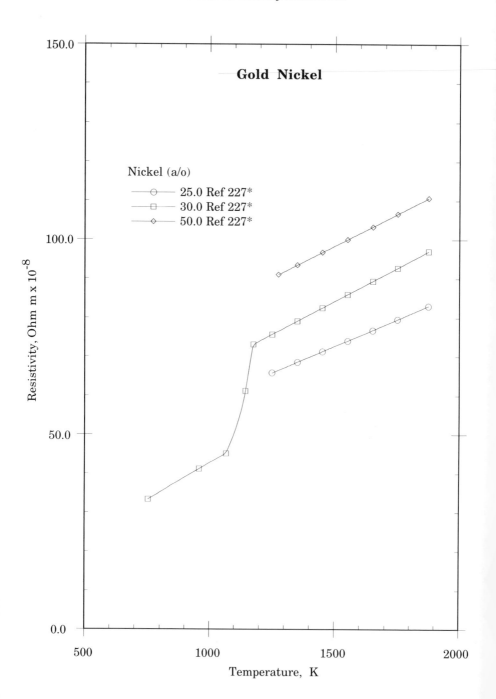

Gold Nickel

Nickel (a/o)
- 25.0 Ref 227*
- 30.0 Ref 227*
- 50.0 Ref 227*

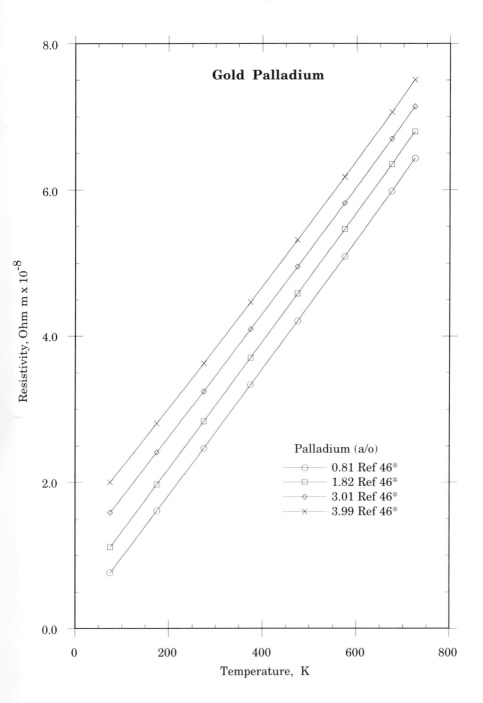

Gold Palladium

Resistivity, Ohm m x 10^{-8}

Temperature, K

Palladium (a/o)
- ○ 0.81 Ref 46*
- □ 1.82 Ref 46*
- ◇ 3.01 Ref 46*
- × 3.99 Ref 46*

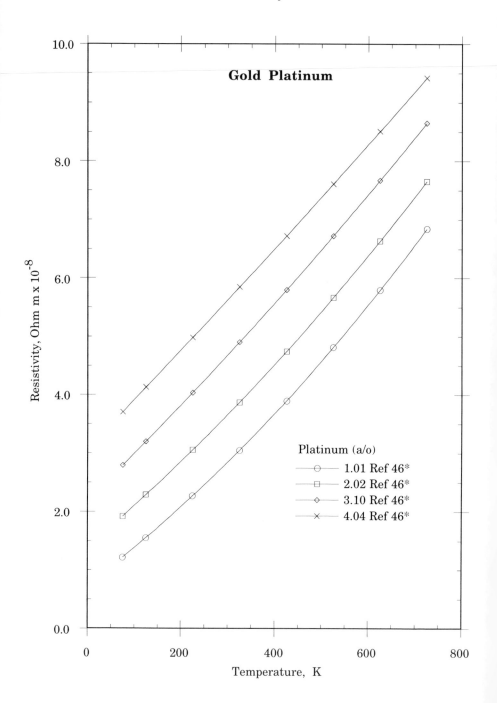

Gold Platinum

Platinum (a/o)
- ──○── 1.01 Ref 46*
- ──□── 2.02 Ref 46*
- ──◇── 3.10 Ref 46*
- ──×── 4.04 Ref 46*

Resistivity, Ohm m x 10^{-8}

Temperature, K

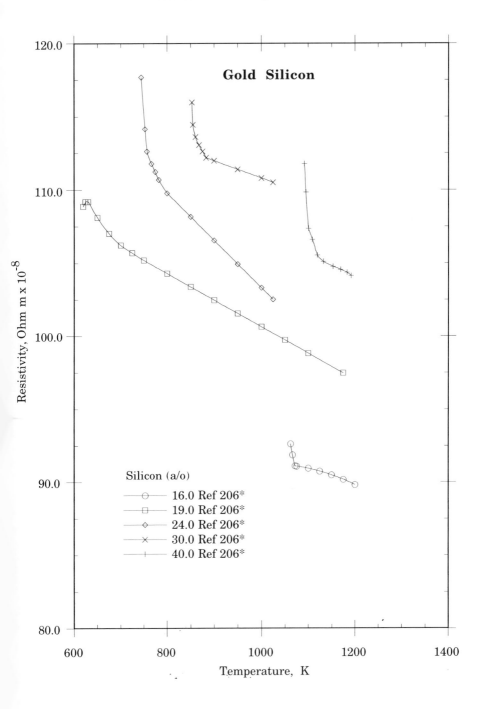

Gold Silicon

Silicon (a/o)
○ 16.0 Ref 206*
□ 19.0 Ref 206*
◇ 24.0 Ref 206*
× 30.0 Ref 206*
+ 40.0 Ref 206*

Resistivity, Ohm m x 10^{-8}

Temperature, K

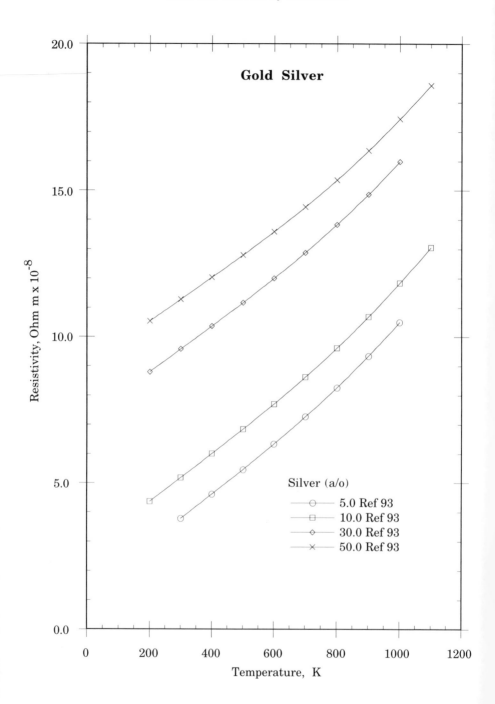

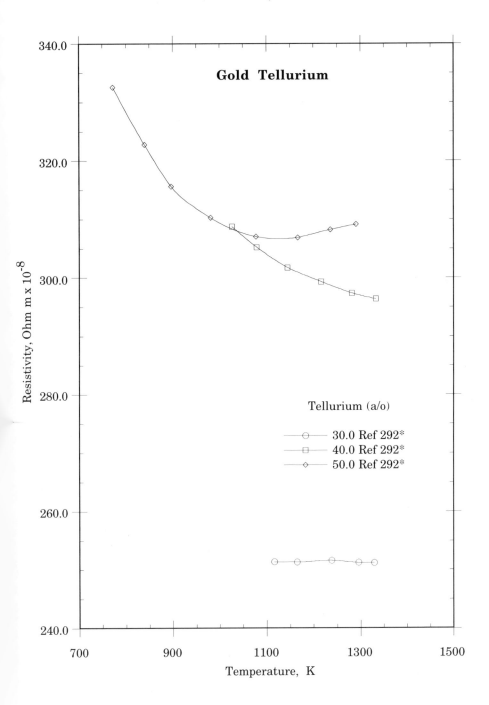

Gold Tellurium

Resistivity, Ohm m x 10^{-8}

Temperature, K

Tellurium (a/o)

30.0 Ref 292*
40.0 Ref 292*
50.0 Ref 292*

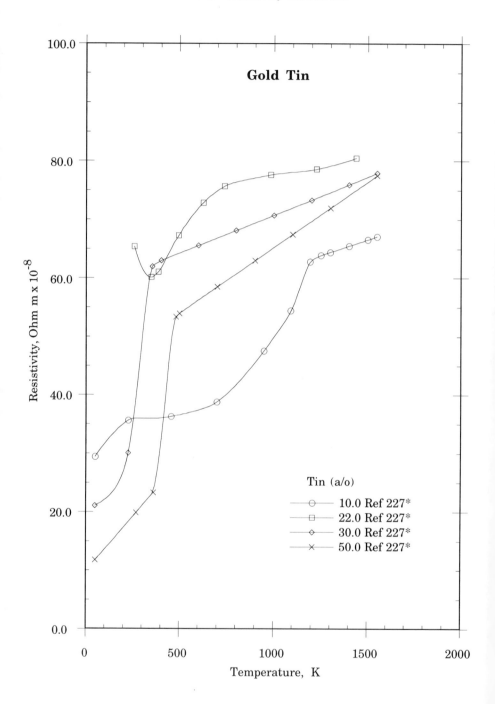

Gold Tin

Resistivity, Ohm m x 10^{-8}

Temperature, K

Tin (a/o)

- 10.0 Ref 227*
- 22.0 Ref 227*
- 30.0 Ref 227*
- 50.0 Ref 227*

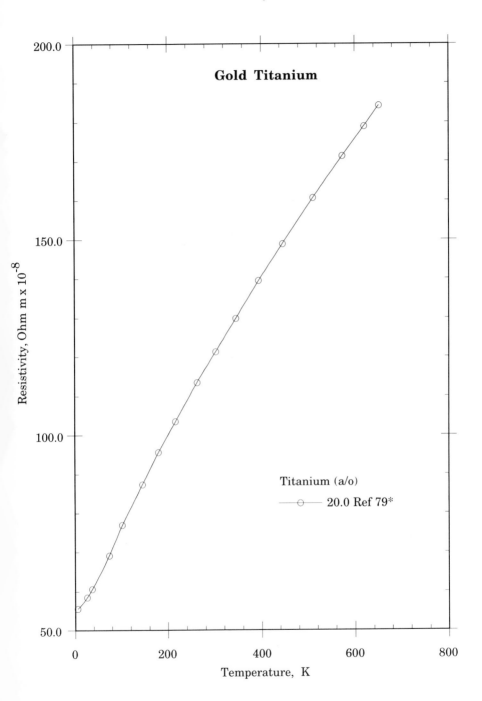

Gold Titanium

Resistivity, Ohm m x 10^{-8}

Temperature, K

Titanium (a/o)
20.0 Ref 79*

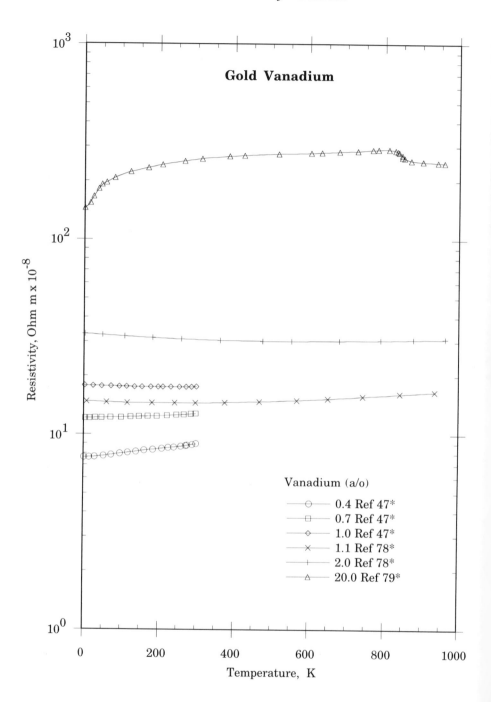

Gold Vanadium

Resistivity, Ohm m x 10^{-8}

Temperature, K

Vanadium (a/o)

○— 0.4 Ref 47*
□— 0.7 Ref 47*
◇— 1.0 Ref 47*
×— 1.1 Ref 78*
+— 2.0 Ref 78*
△— 20.0 Ref 79*

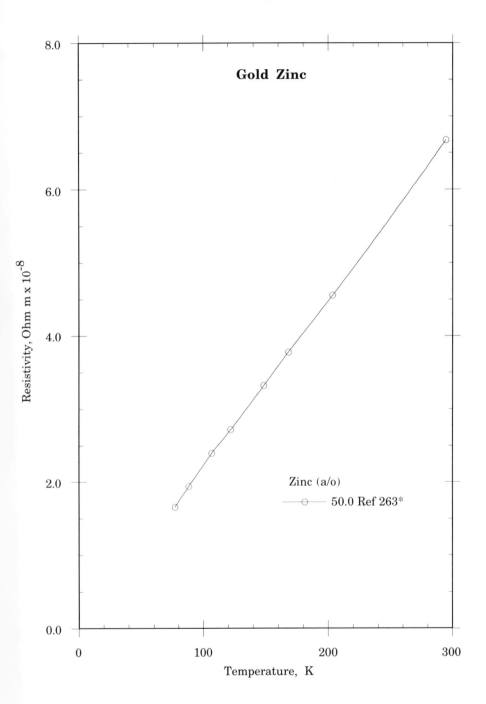

Gold Zinc

Zinc (a/o)
50.0 Ref 263*

Resistivity, Ohm m x 10⁻⁸

Temperature, K

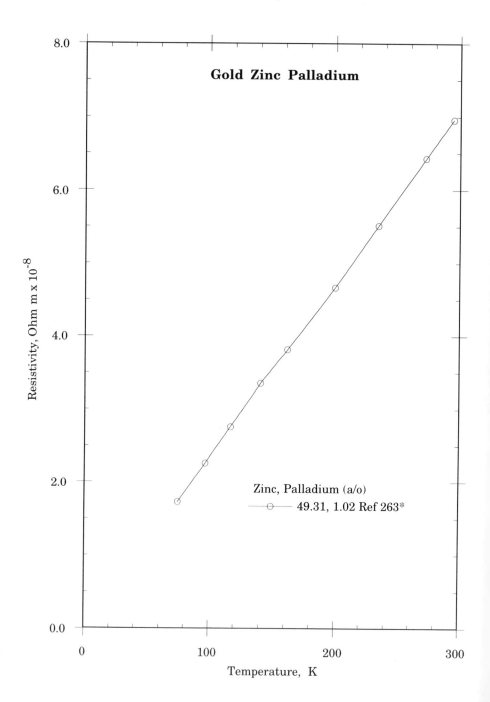

Gold Zinc Palladium

Zinc, Palladium (a/o)
49.31, 1.02 Ref 263*

Resistivity, Ohm m x 10^{-8}

Temperature, K

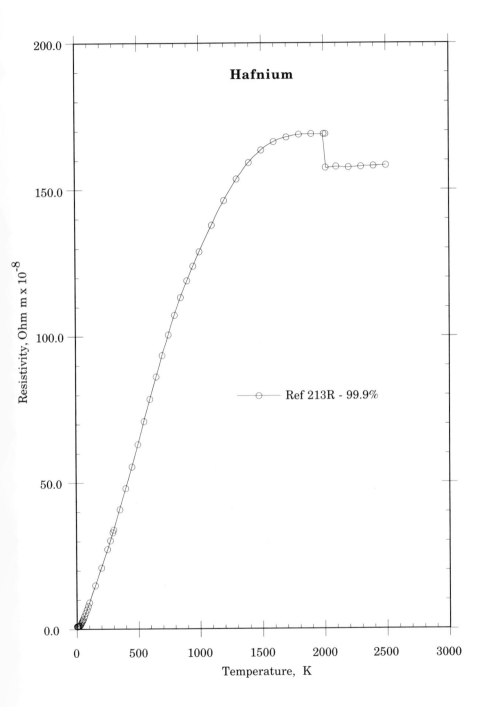

Hafnium

Resistivity, Ohm m x 10^{-8}

Temperature, K

Ref 213R - 99.9%

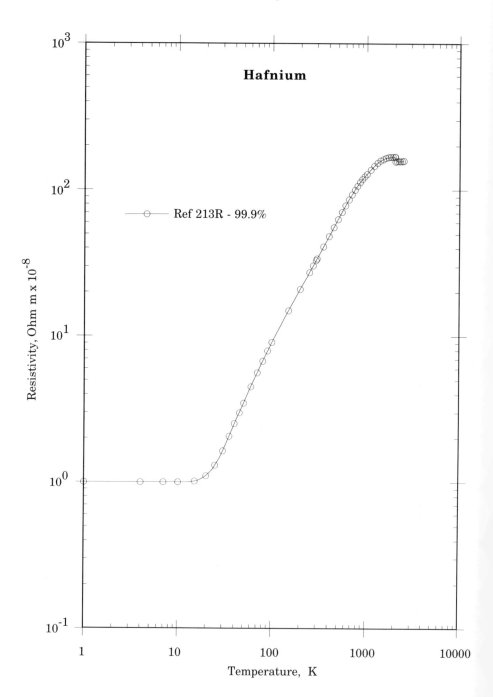

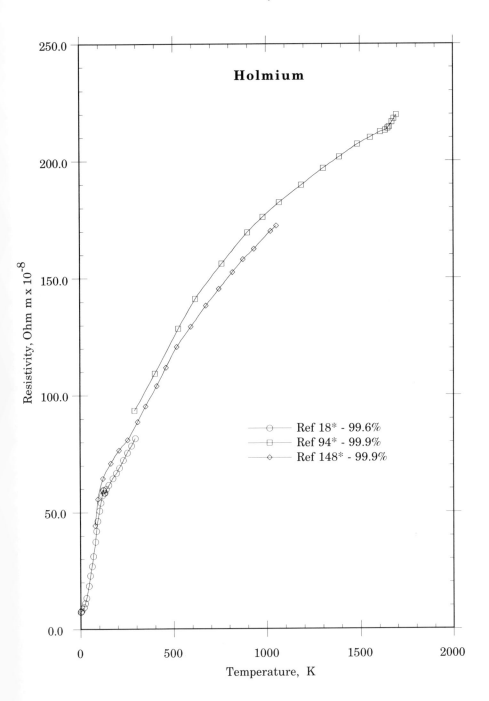

Holmium

Ref 18* - 99.6%
Ref 94* - 99.9%
Ref 148* - 99.9%

Resistivity, Ohm m x 10^{-8}

Temperature, K

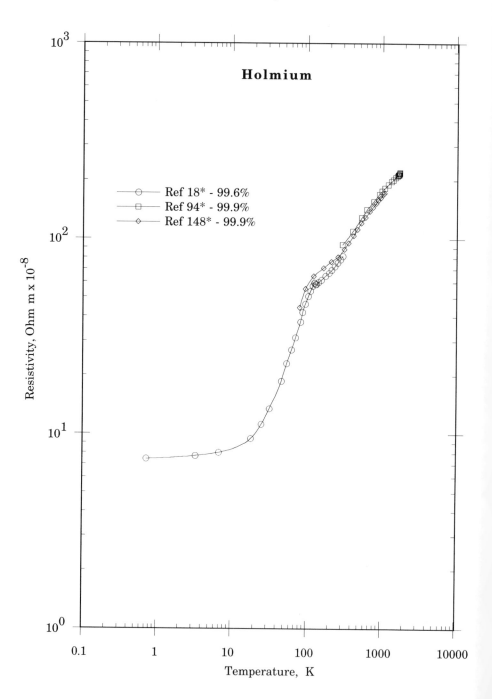

Holmium

Ref 18* - 99.6%
Ref 94* - 99.9%
Ref 148* - 99.9%

Resistivity, Ohm m x 10^{-8}

Temperature, K

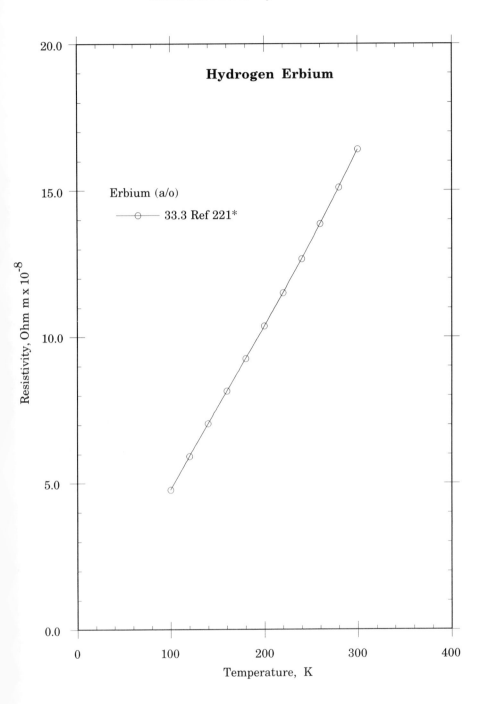

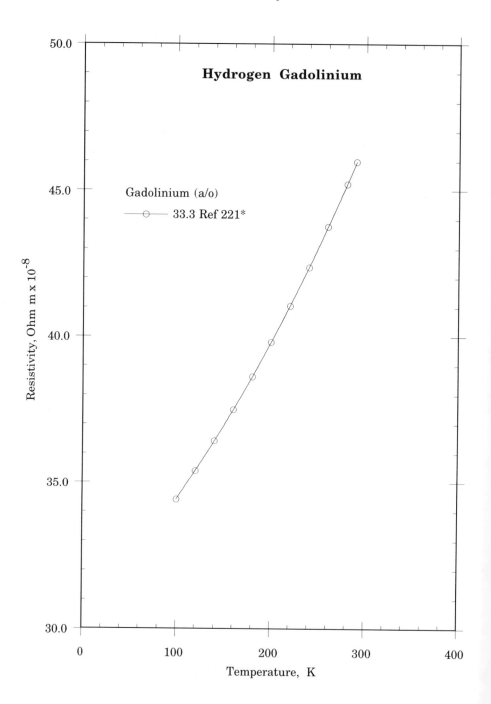

Hydrogen Gadolinium

Gadolinium (a/o)
—⊙— 33.3 Ref 221*

Resistivity, Ohm m x 10^{-8} (y-axis)

Temperature, K (x-axis)

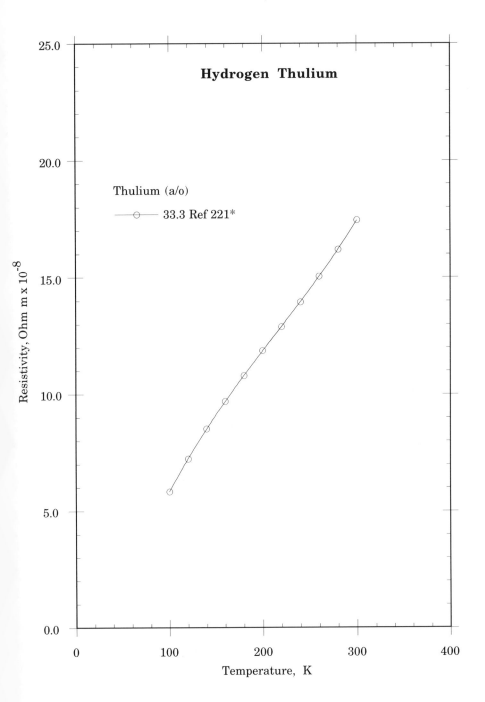

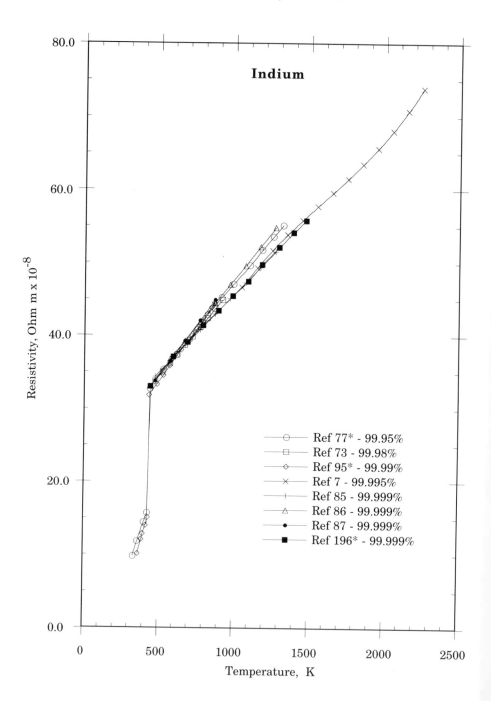

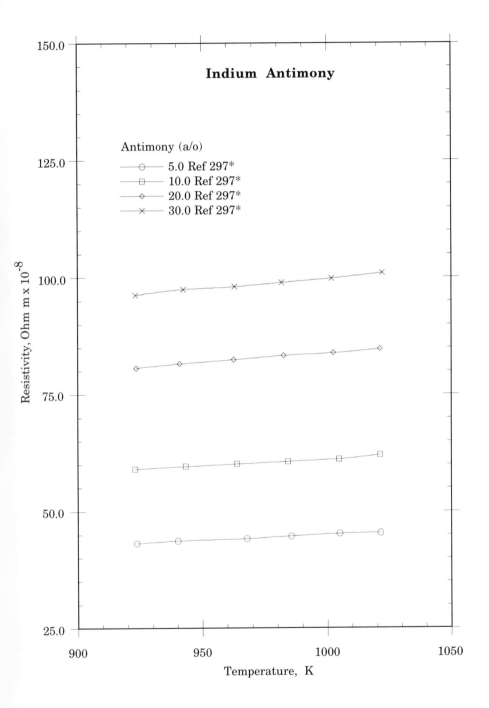

Indium Antimony

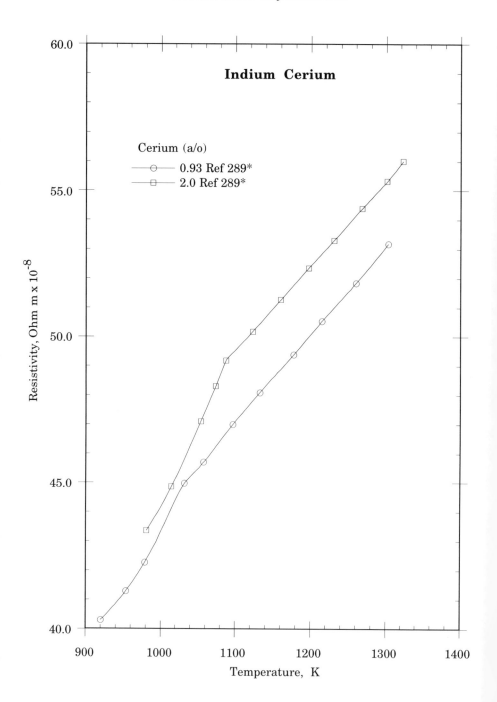

Indium Cerium

Cerium (a/o)
- ⊖ 0.93 Ref 289*
- ⊟ 2.0 Ref 289*

Resistivity, Ohm m x 10^{-8}

Temperature, K

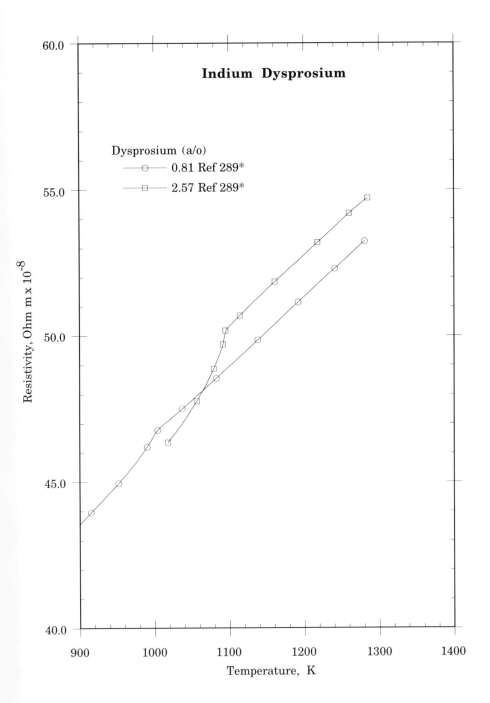

Indium Dysprosium

Dysprosium (a/o)
— ○ — 0.81 Ref 289*
— □ — 2.57 Ref 289*

Resistivity, Ohm m x 10⁻⁸

Temperature, K

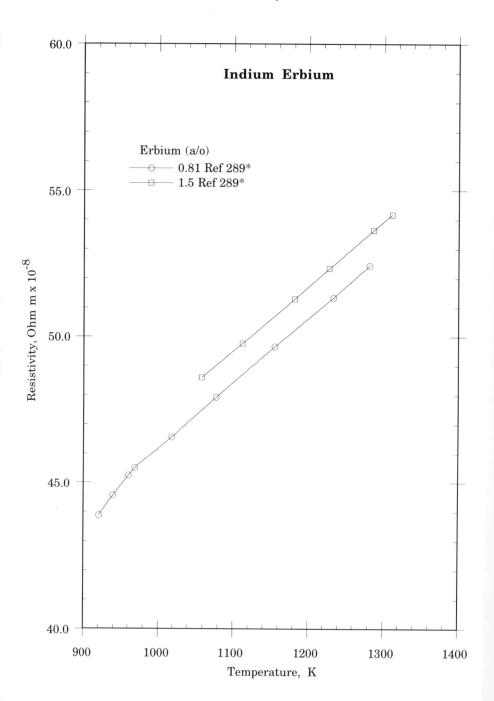

Indium Erbium

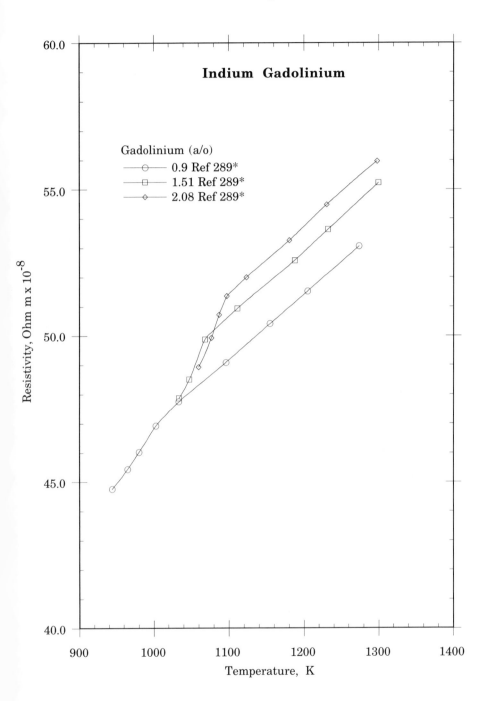

Indium Gadolinium

Gadolinium (a/o)
- 0.9 Ref 289*
- 1.51 Ref 289*
- 2.08 Ref 289*

Resistivity, Ohm m x 10^{-8}

Temperature, K

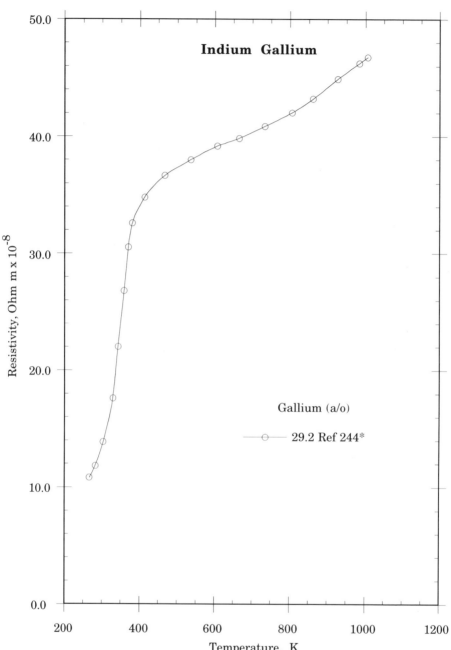

Indium Gallium

Resistivity, Ohm m x 10^{-8}

Temperature, K

Gallium (a/o)

29.2 Ref 244*

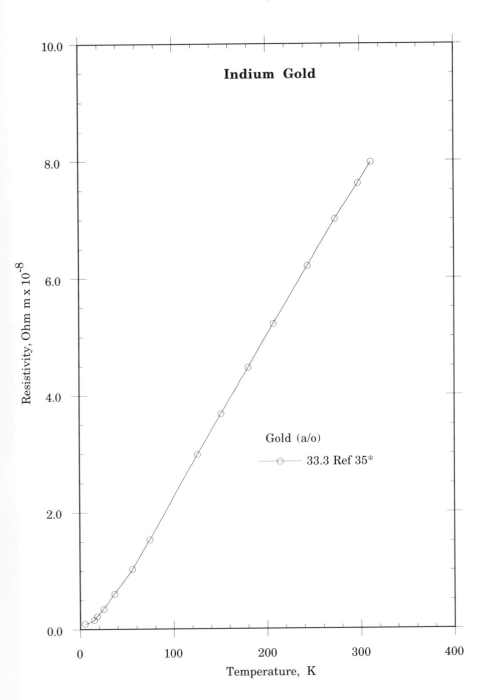

Indium Gold

Gold (a/o)

—⊙— 33.3 Ref 35*

Resistivity, Ohm m x 10^{-8}

Temperature, K

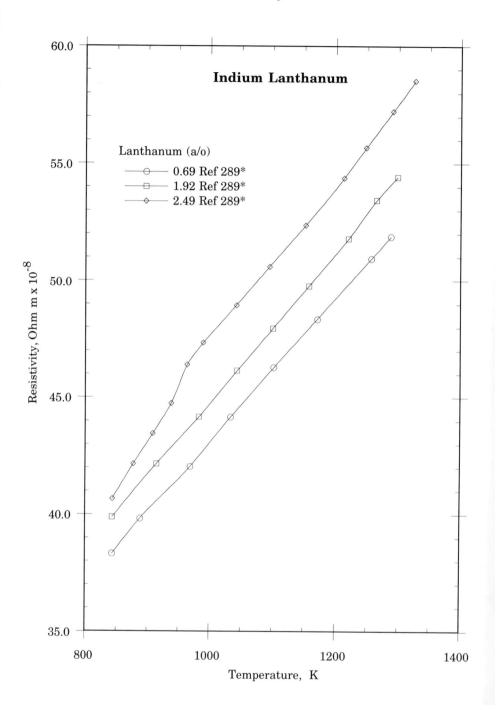

Indium Lanthanum

Lanthanum (a/o)
- ──⊖── 0.69 Ref 289*
- ──☐── 1.92 Ref 289*
- ──◇── 2.49 Ref 289*

Resistivity, Ohm m x 10^{-8}

Temperature, K

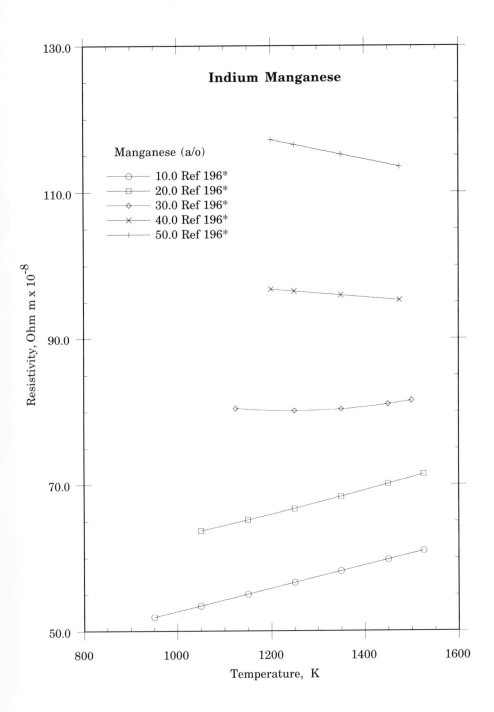

Indium Manganese

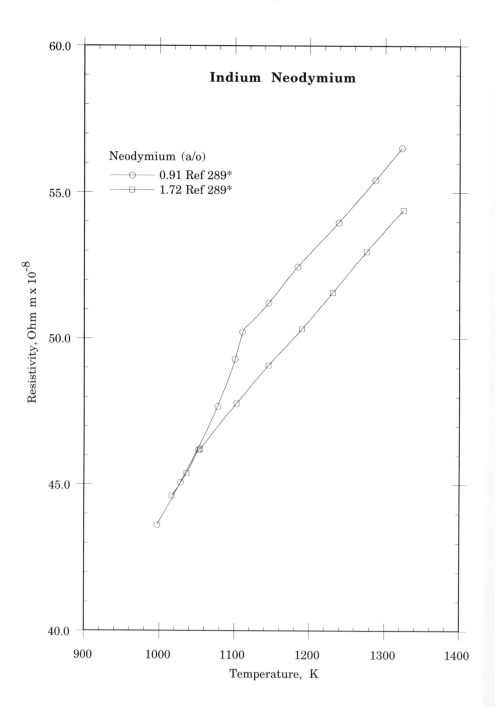

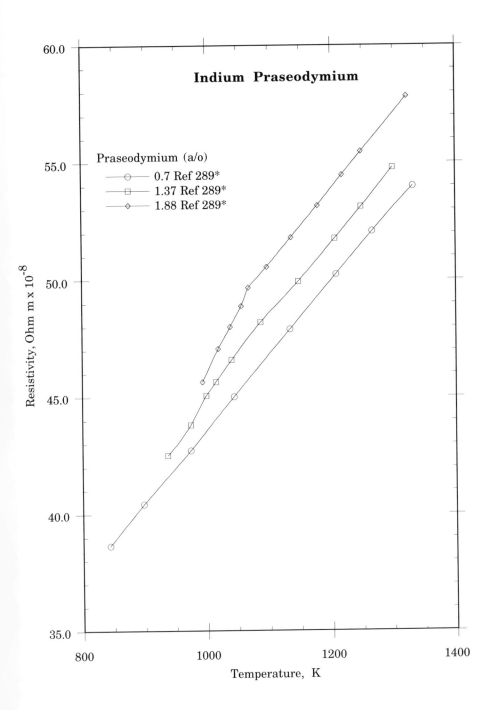

Indium Praseodymium

Praseodymium (a/o)
- ⊖— 0.7 Ref 289*
- □— 1.37 Ref 289*
- ◇— 1.88 Ref 289*

Resistivity, Ohm m x 10^{-8}

Temperature, K

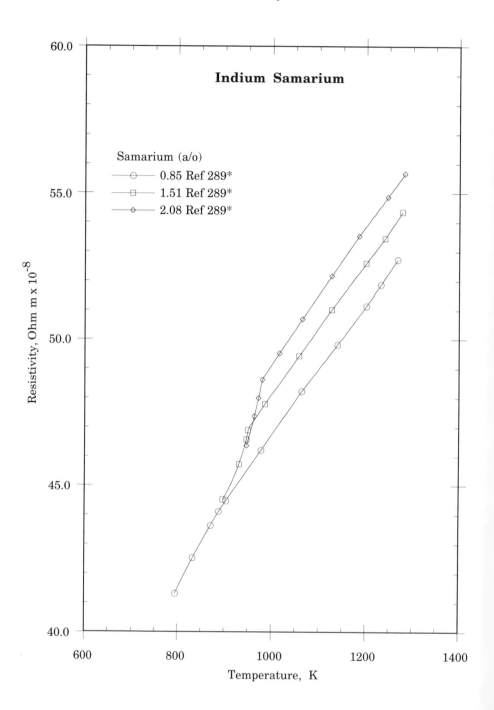

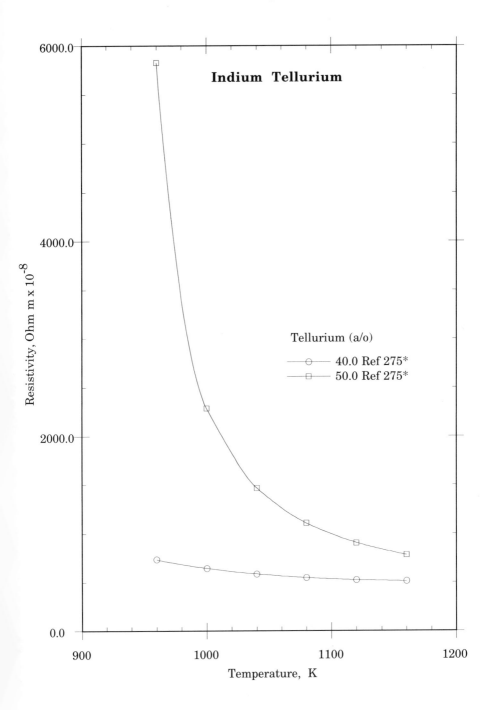

Indium Tellurium

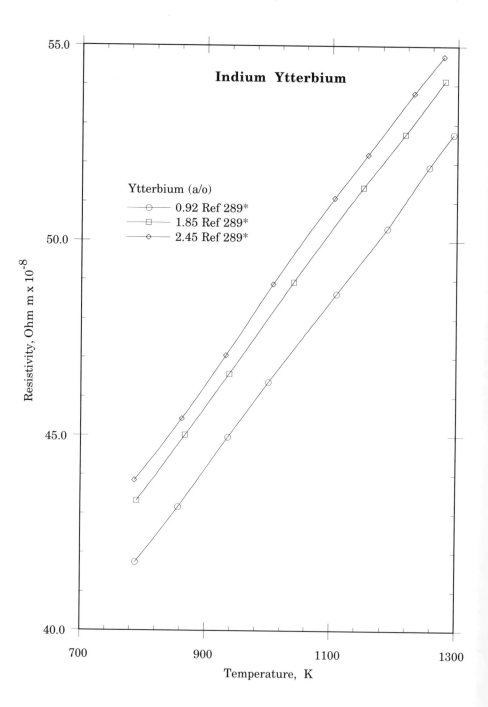

Indium Ytterbium

Ytterbium (a/o)
- ——○—— 0.92 Ref 289*
- ——□—— 1.85 Ref 289*
- ——◇—— 2.45 Ref 289*

Resistivity, Ohm m x 10^{-8}

Temperature, K

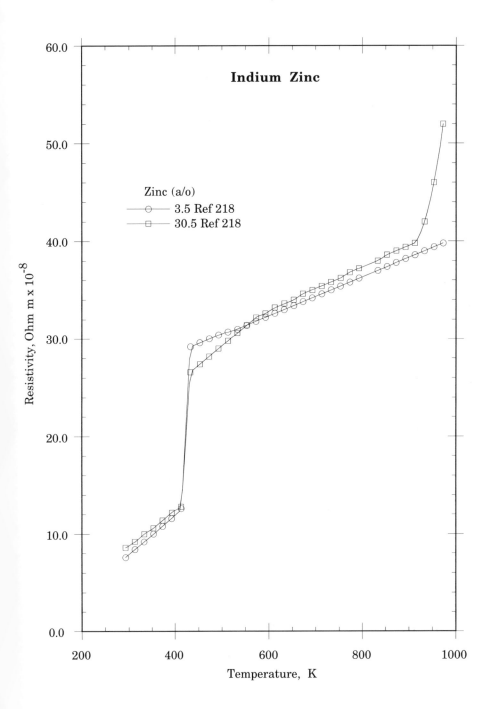

Indium Zinc

Zinc (a/o)
— ⊙ — 3.5 Ref 218
— □ — 30.5 Ref 218

Resistivity, Ohm m x 10^{-8}

Temperature, K

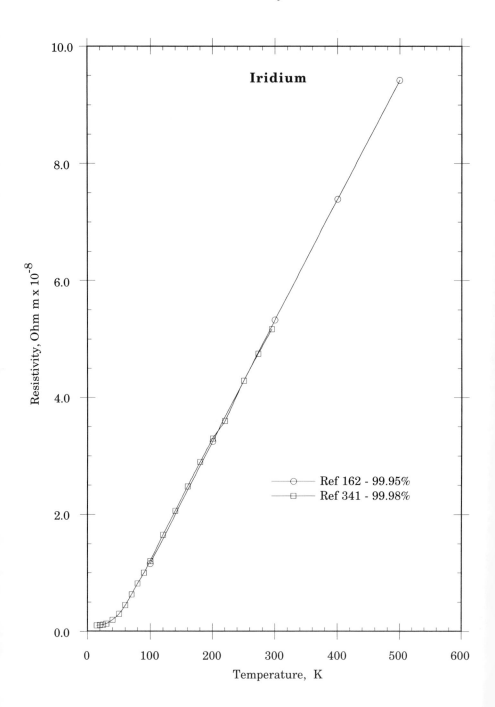

Iridium

Resistivity, Ohm m x 10^{-8}

Temperature, K

Ref 162 - 99.95%
Ref 341 - 99.98%

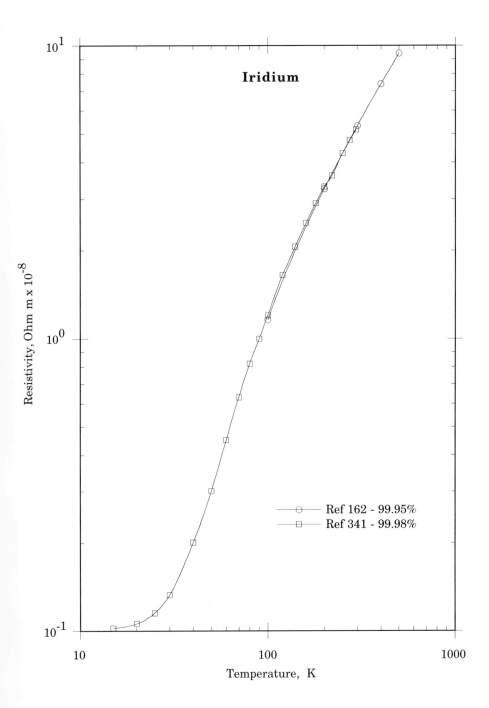

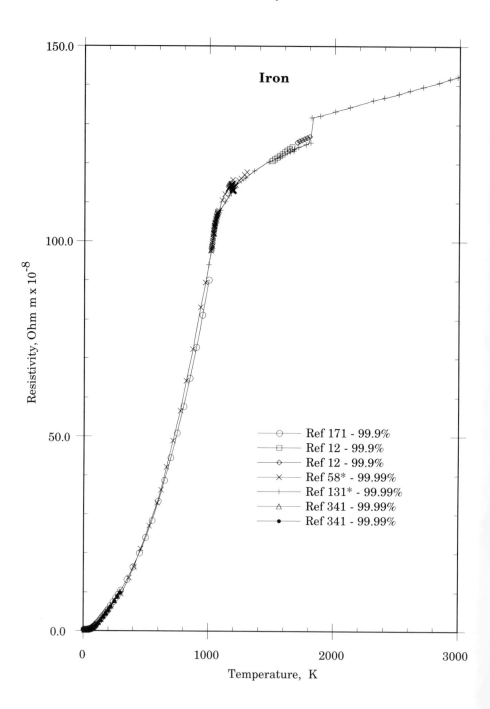

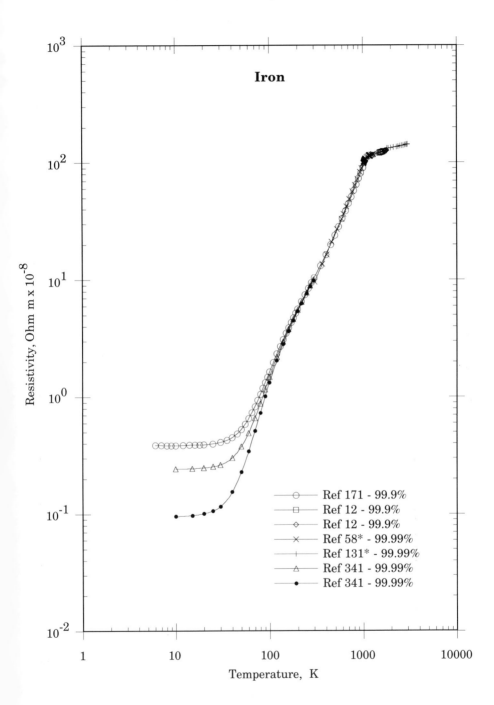

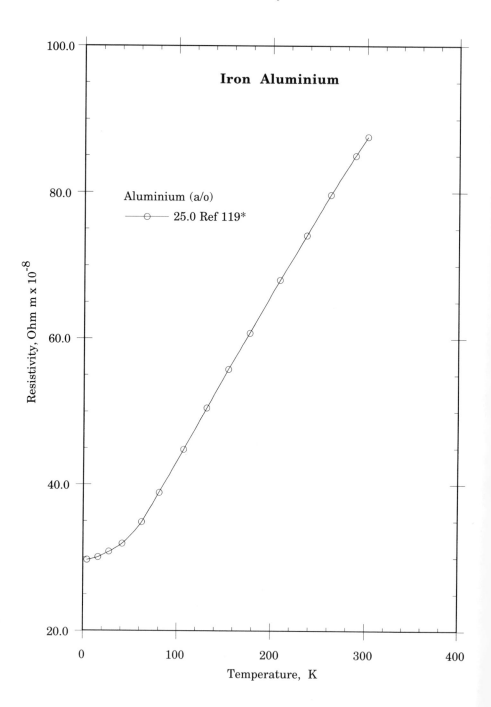

Iron Aluminium

Aluminium (a/o)
—o— 25.0 Ref 119*

Resistivity, Ohm m x 10⁻⁸ (y-axis)

Temperature, K (x-axis)

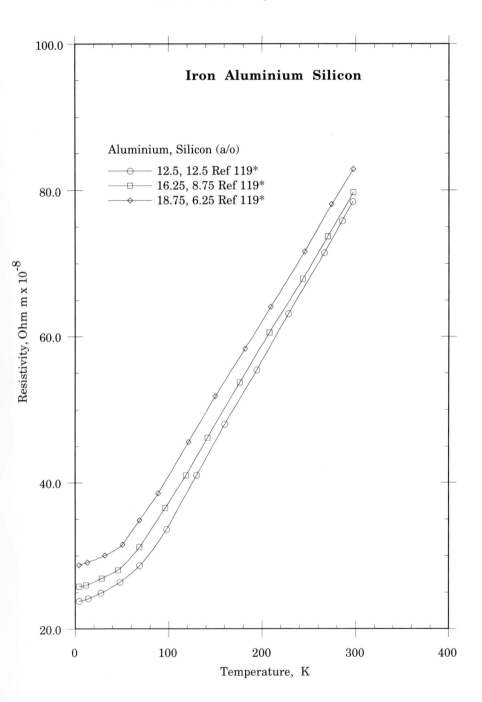

Iron Aluminium Silicon

Aluminium, Silicon (a/o)
- 12.5, 12.5 Ref 119*
- 16.25, 8.75 Ref 119*
- 18.75, 6.25 Ref 119*

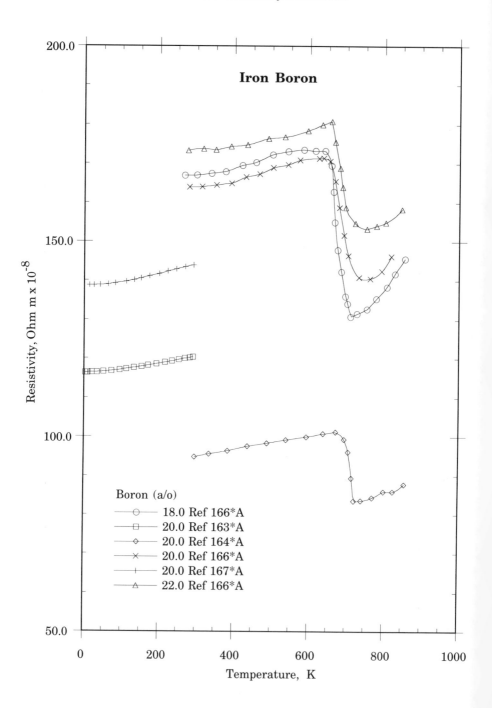

Iron Boron

Resistivity, Ohm m x 10^{-8}

Temperature, K

Boron (a/o)
- ⊖— 18.0 Ref 166*A
- ⊟— 20.0 Ref 163*A
- ◇— 20.0 Ref 164*A
- ✕— 20.0 Ref 166*A
- +— 20.0 Ref 167*A
- △— 22.0 Ref 166*A

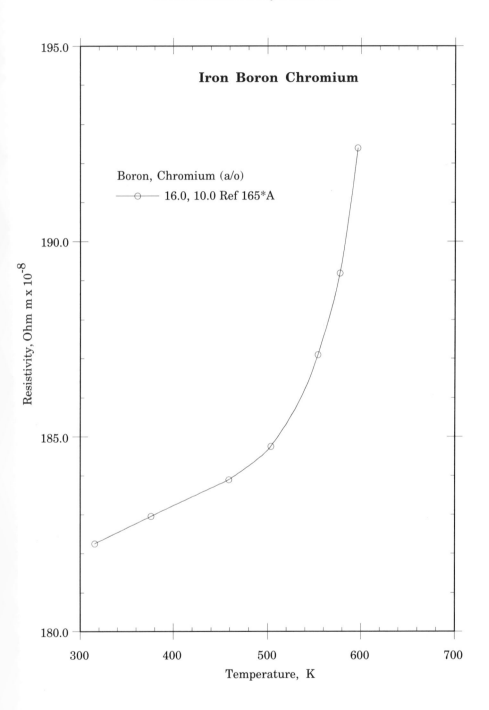

Iron Boron Chromium

Boron, Chromium (a/o)

16.0, 10.0 Ref 165*A

Resistivity, Ohm m x 10^{-8}

Temperature, K

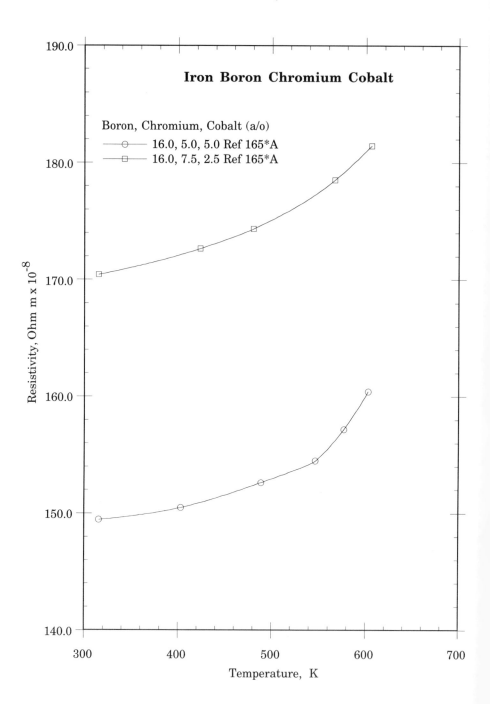

Iron Boron Chromium Cobalt

Boron, Chromium, Cobalt (a/o)
— ⊖ — 16.0, 5.0, 5.0 Ref 165*A
— ⊟ — 16.0, 7.5, 2.5 Ref 165*A

Resistivity, Ohm m x 10^{-8}

Temperature, K

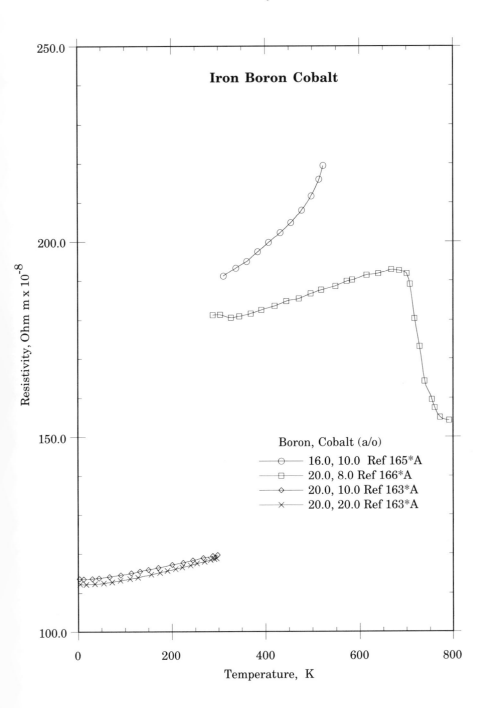

Iron Boron Cobalt

Boron, Cobalt (a/o)
16.0, 10.0 Ref 165*A
20.0, 8.0 Ref 166*A
20.0, 10.0 Ref 163*A
20.0, 20.0 Ref 163*A

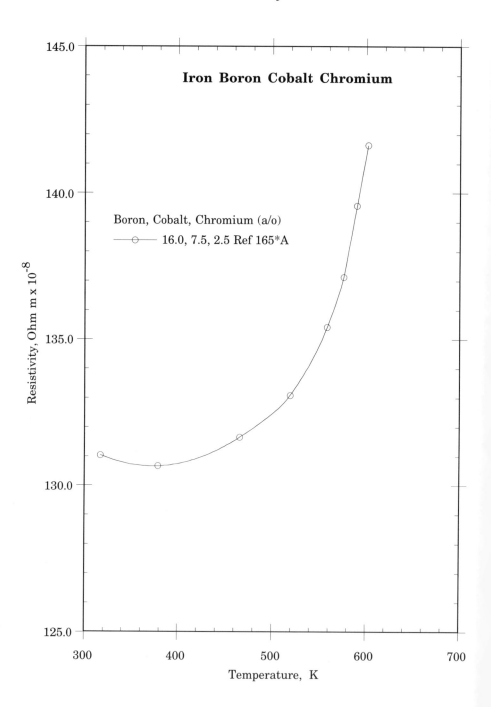

Iron Boron Cobalt Chromium

Boron, Cobalt, Chromium (a/o)

16.0, 7.5, 2.5 Ref 165*A

Resistivity, Ohm m x 10^{-8}

Temperature, K

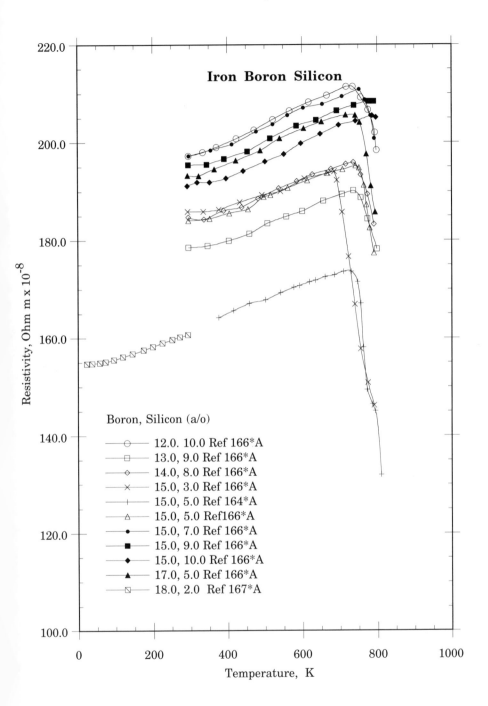

Iron Boron Silicon

Boron, Silicon (a/o)

○ 12.0. 10.0 Ref 166*A
□ 13.0, 9.0 Ref 166*A
◇ 14.0, 8.0 Ref 166*A
× 15.0, 3.0 Ref 166*A
＋ 15.0, 5.0 Ref 164*A
△ 15.0, 5.0 Ref166*A
● 15.0, 7.0 Ref 166*A
■ 15.0, 9.0 Ref 166*A
◆ 15.0, 10.0 Ref 166*A
▲ 17.0, 5.0 Ref 166*A
⬔ 18.0, 2.0 Ref 167*A

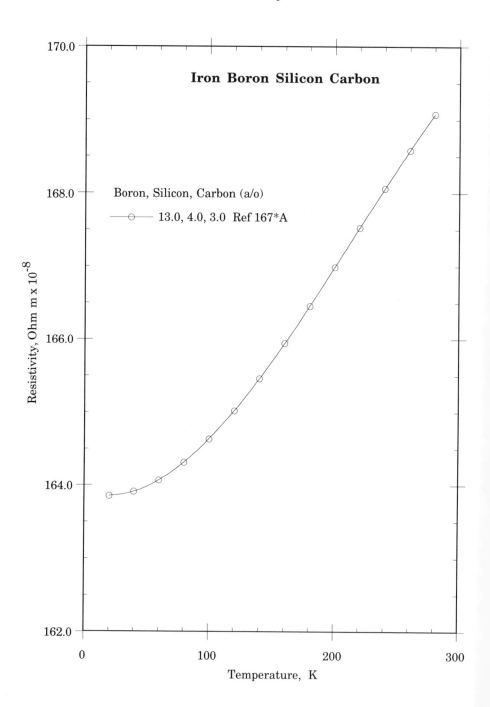

Iron Boron Silicon Carbon

Boron, Silicon, Carbon (a/o)

—○— 13.0, 4.0, 3.0 Ref 167*A

Resistivity, Ohm m x 10⁻⁸

Temperature, K

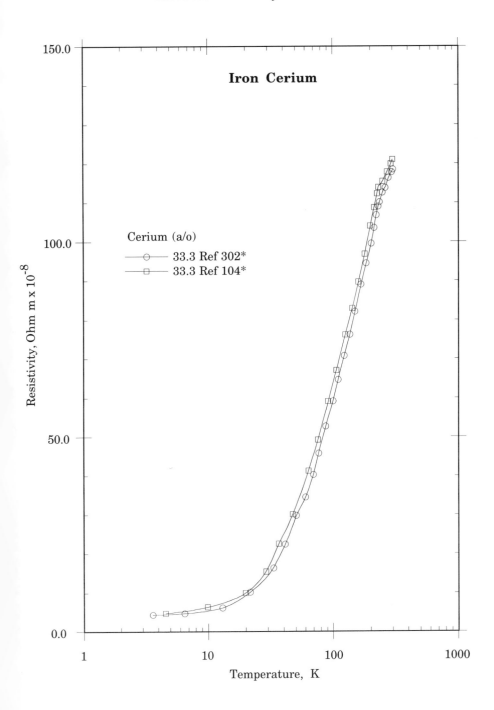

Iron Cerium

Cerium (a/o)

⊙ 33.3 Ref 302*
□ 33.3 Ref 104*

Resistivity, Ohm m x 10^{-8}

Temperature, K

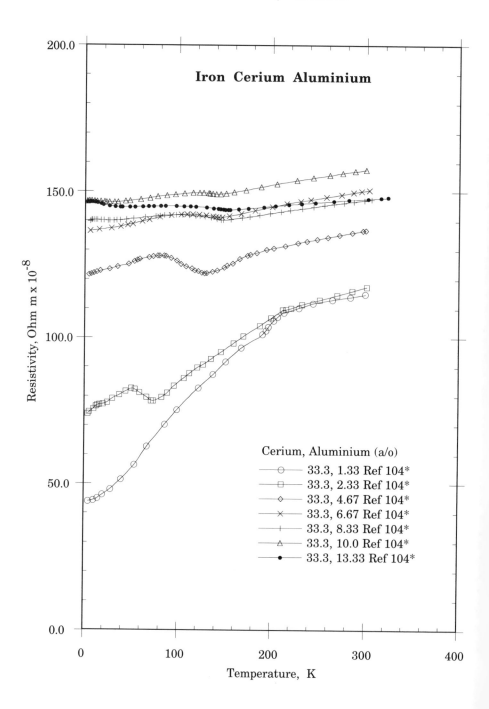

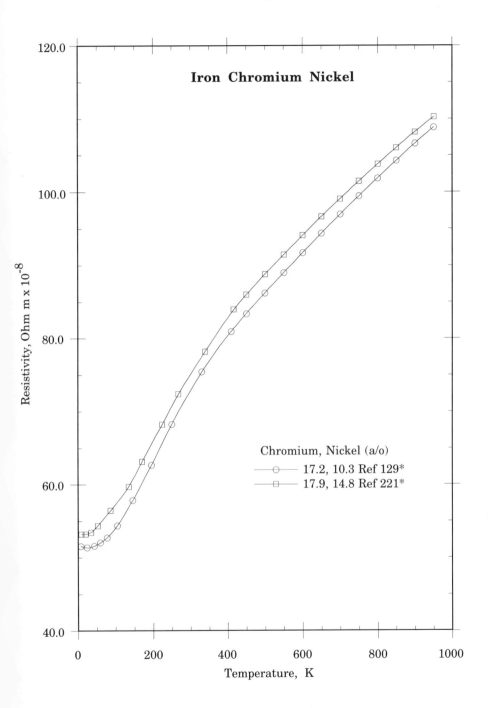

Iron Chromium Nickel

Chromium, Nickel (a/o)

- ─○─ 17.2, 10.3 Ref 129*
- ─□─ 17.9, 14.8 Ref 221*

Resistivity, Ohm m x 10^{-8}

Temperature, K

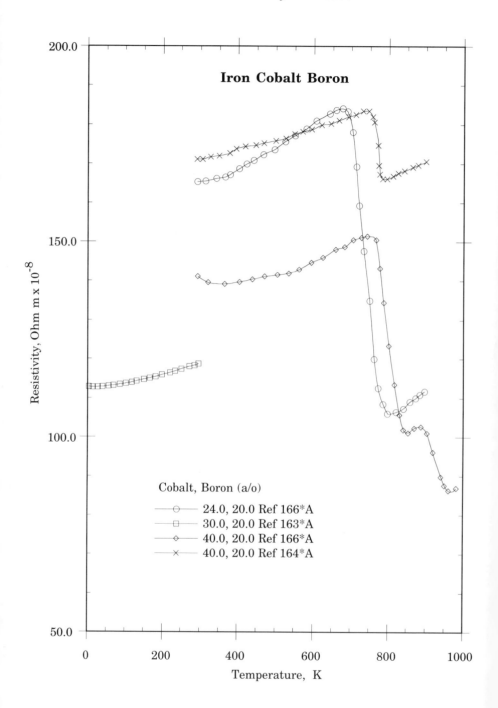

Iron Cobalt Boron

Cobalt, Boron (a/o)

- 24.0, 20.0 Ref 166*A
- 30.0, 20.0 Ref 163*A
- 40.0, 20.0 Ref 166*A
- 40.0, 20.0 Ref 164*A

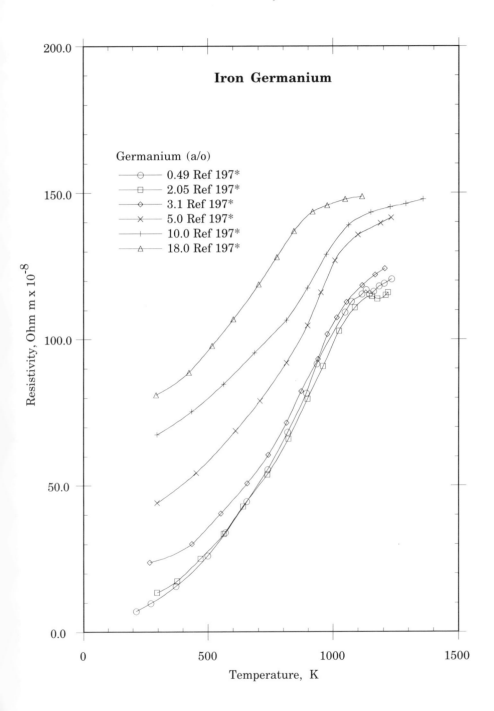

Iron Germanium

Germanium (a/o)
- ⊖ 0.49 Ref 197*
- ⊟ 2.05 Ref 197*
- ◇ 3.1 Ref 197*
- ✕ 5.0 Ref 197*
- ─┼─ 10.0 Ref 197*
- △ 18.0 Ref 197*

Resistivity, Ohm m x 10^{-8}

Temperature, K

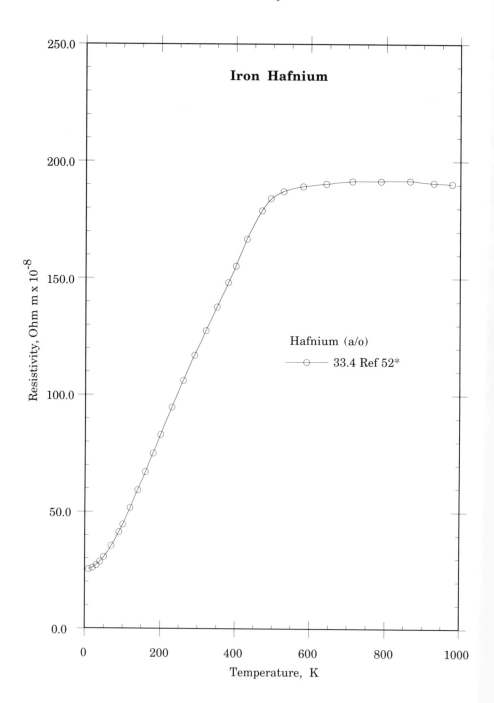

Iron Hafnium

Resistivity, Ohm m x 10^{-8}

Hafnium (a/o)

⊸⊶ 33.4 Ref 52*

Temperature, K

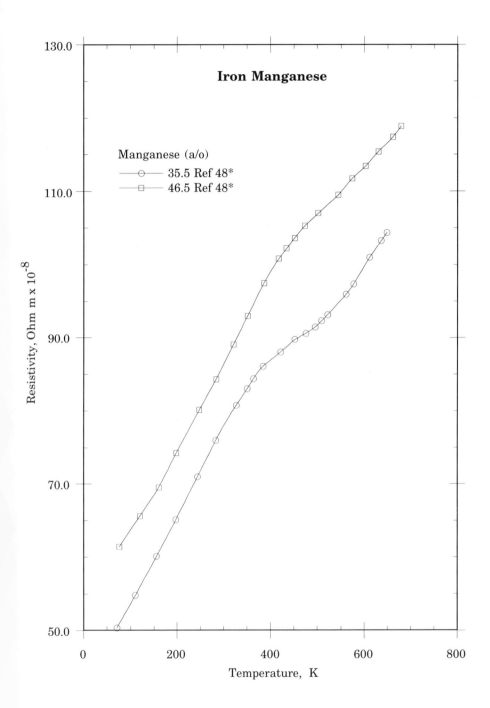

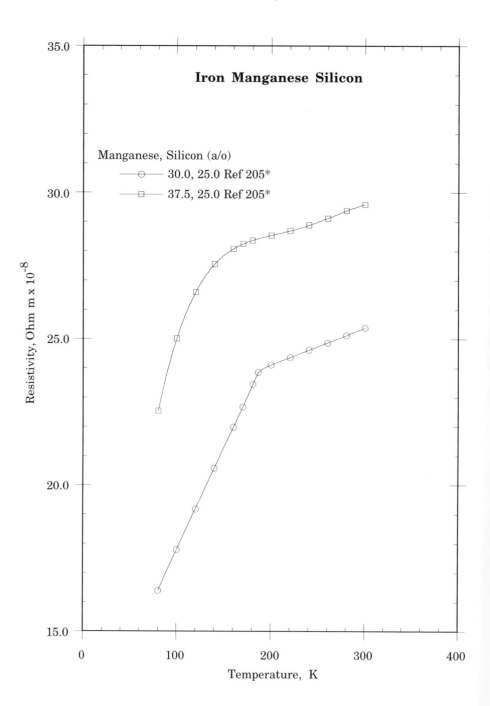

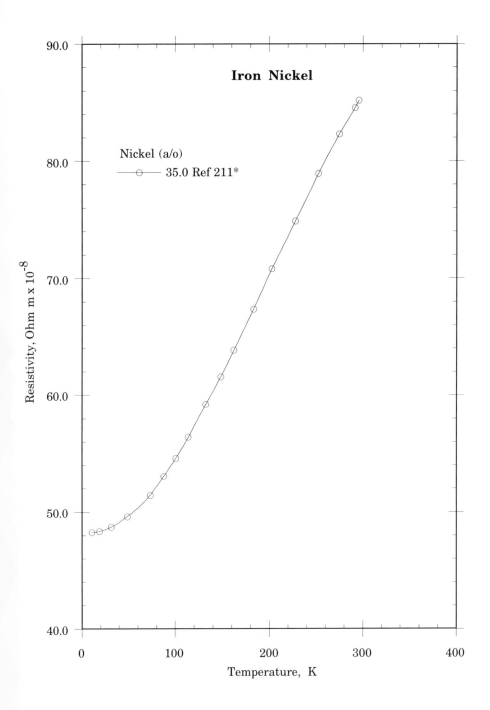

Iron Nickel

Nickel (a/o)
— ⊖ — 35.0 Ref 211*

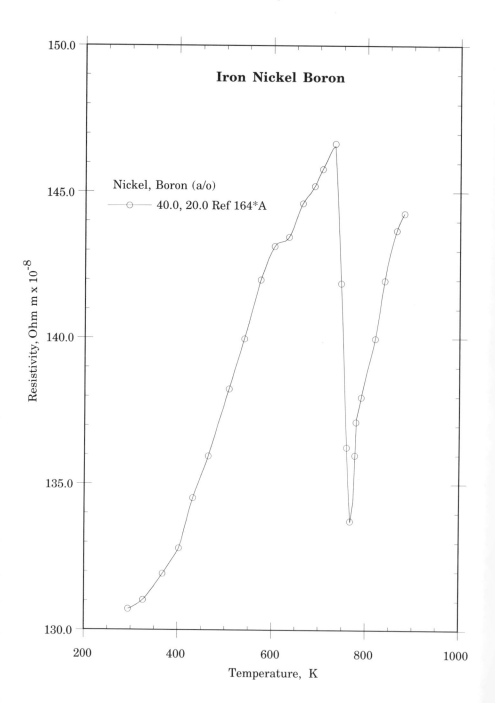

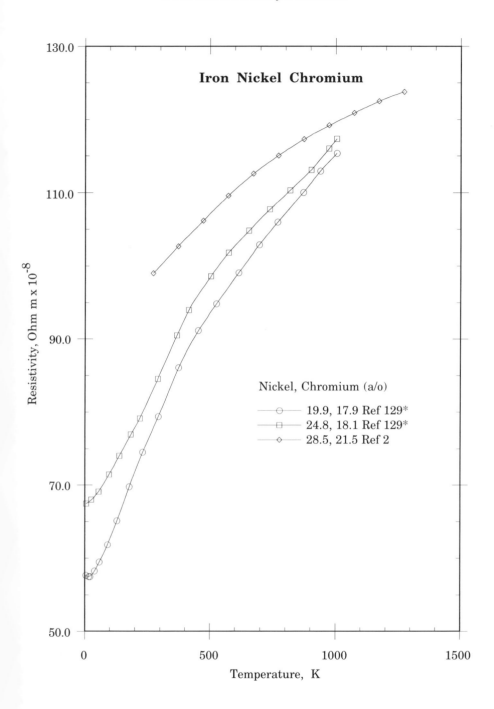

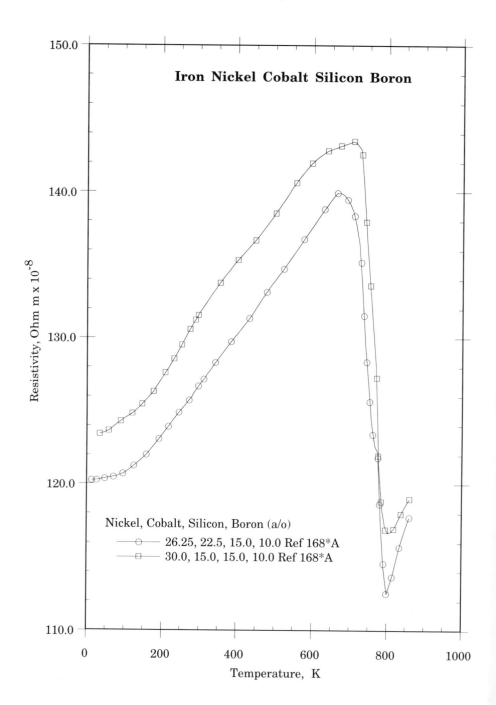

Iron Nickel Cobalt Silicon Boron

Nickel, Cobalt, Silicon, Boron (a/o)
- ⊖ 26.25, 22.5, 15.0, 10.0 Ref 168*A
- ⊟ 30.0, 15.0, 15.0, 10.0 Ref 168*A

Resistivity, Ohm m x 10^{-8}

Temperature, K

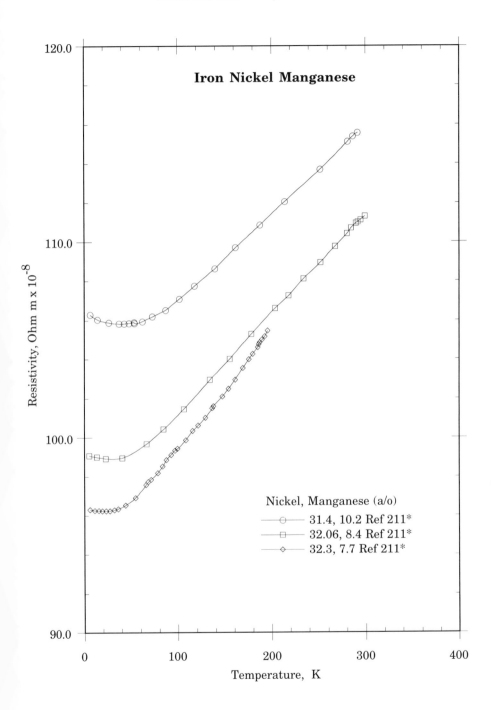

Iron Nickel Manganese

Resistivity, Ohm m x 10^{-8}

Temperature, K

Nickel, Manganese (a/o)
31.4, 10.2 Ref 211*
32.06, 8.4 Ref 211*
32.3, 7.7 Ref 211*

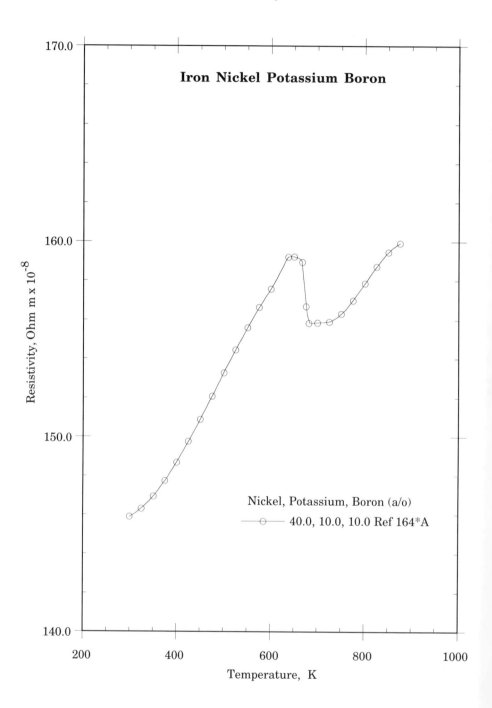

Iron Nickel Potassium Boron

Resistivity, Ohm m x 10^{-8}

Temperature, K

Nickel, Potassium, Boron (a/o)
—○— 40.0, 10.0, 10.0 Ref 164*A

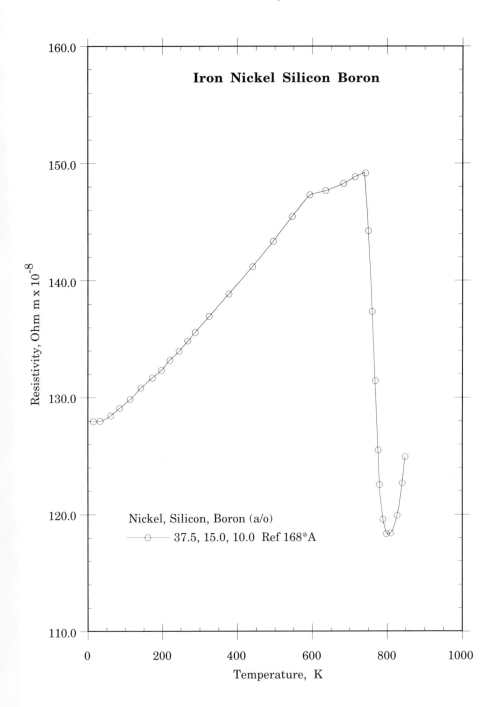

Iron Nickel Silicon Boron

Resistivity, Ohm m x 10^{-8}

Temperature, K

Nickel, Silicon, Boron (a/o)

37.5, 15.0, 10.0 Ref 168*A

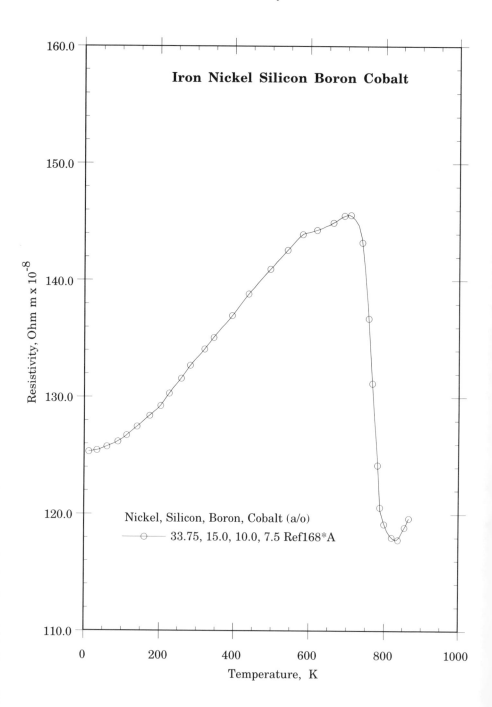

Iron Nickel Silicon Boron Cobalt

Nickel, Silicon, Boron, Cobalt (a/o)
33.75, 15.0, 10.0, 7.5 Ref168*A

Resistivity, Ohm m x 10^{-8}

Temperature, K

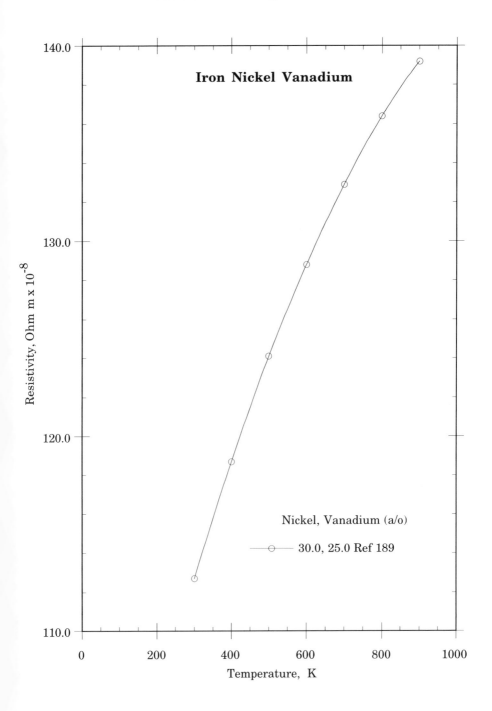

Iron Nickel Vanadium

Resistivity, Ohm m x 10^{-8}

Temperature, K

Nickel, Vanadium (a/o)

30.0, 25.0 Ref 189

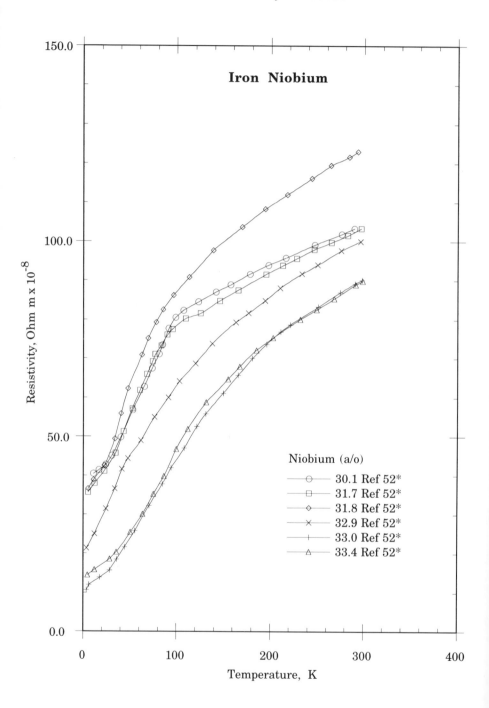

Iron Niobium

Resistivity, Ohm m x 10^{-8}

Temperature, K

Niobium (a/o)

- 30.1 Ref 52*
- 31.7 Ref 52*
- 31.8 Ref 52*
- 32.9 Ref 52*
- 33.0 Ref 52*
- 33.4 Ref 52*

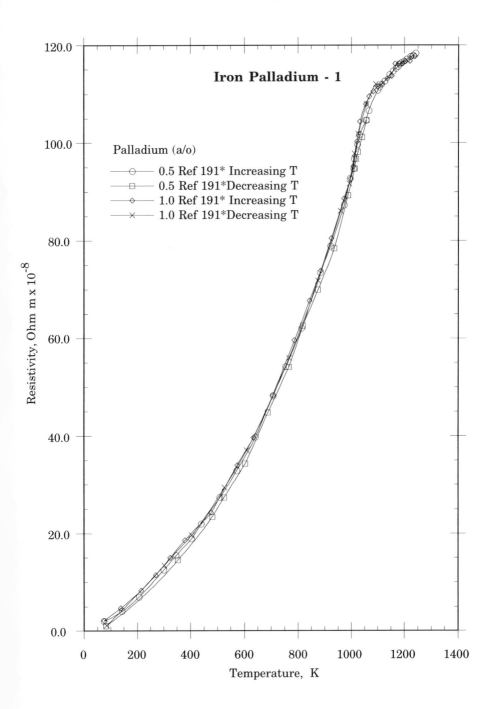

Iron Palladium - 1

Palladium (a/o)
- ⊖ 0.5 Ref 191* Increasing T
- ☐ 0.5 Ref 191*Decreasing T
- ◇ 1.0 Ref 191* Increasing T
- ✕ 1.0 Ref 191*Decreasing T

Resistivity, Ohm m x 10^{-8}

Temperature, K

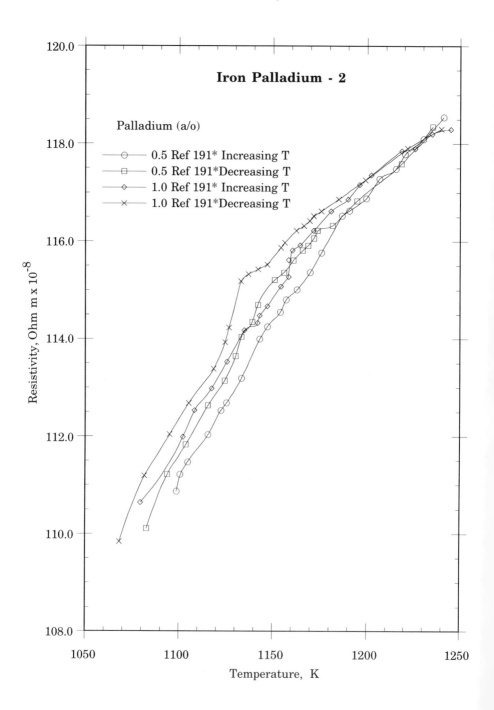

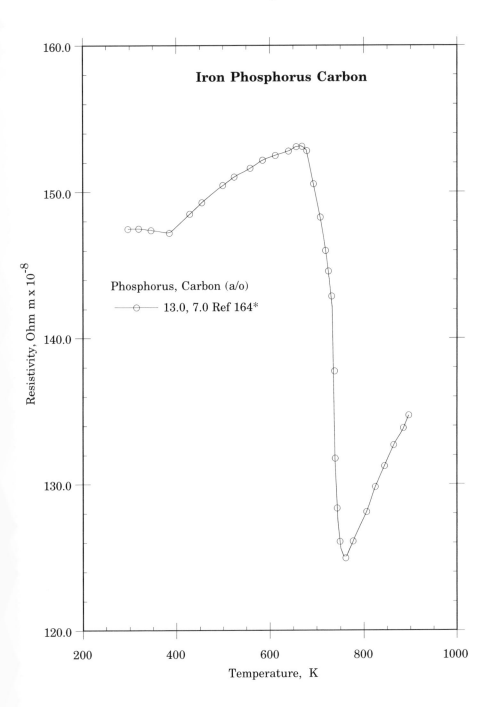

Iron Phosphorus Carbon

Phosphorus, Carbon (a/o)
——⊙—— 13.0, 7.0 Ref 164*

Resistivity, Ohm m x 10^{-8}

Temperature, K

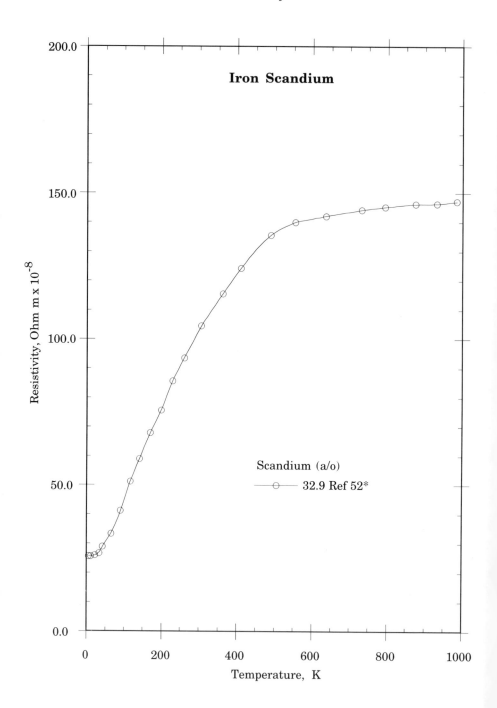

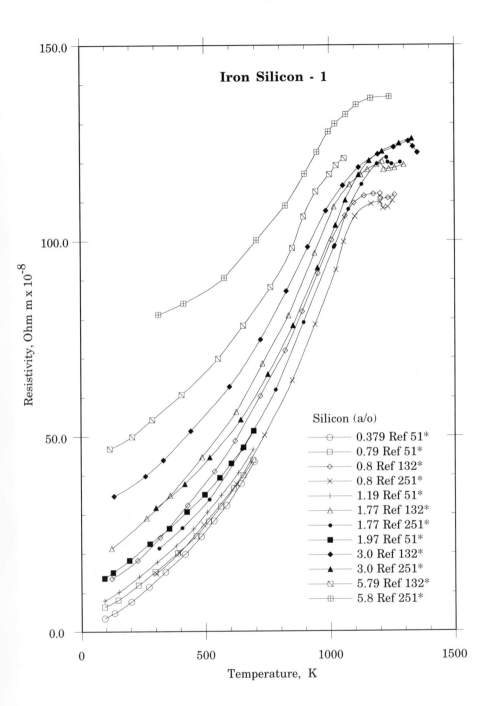

Iron Silicon - 1

Resistivity, Ohm m x 10⁻⁸

Temperature, K

Silicon (a/o)

- ⊙ 0.379 Ref 51*
- ⊟ 0.79 Ref 51*
- ◇ 0.8 Ref 132*
- × 0.8 Ref 251*
- + 1.19 Ref 51*
- △ 1.77 Ref 132*
- ● 1.77 Ref 251*
- ■ 1.97 Ref 51*
- ◆ 3.0 Ref 132*
- ▲ 3.0 Ref 251*
- ◁ 5.79 Ref 132*
- ⊞ 5.8 Ref 251*

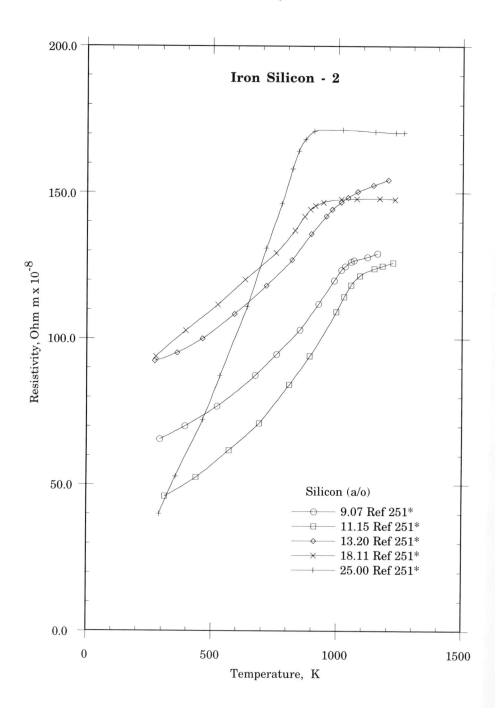

Iron Silicon - 2

Resistivity, Ohm m x 10^{-8}

Temperature, K

Silicon (a/o)

——◦—— 9.07 Ref 251*
——□—— 11.15 Ref 251*
——◇—— 13.20 Ref 251*
——×—— 18.11 Ref 251*
——+—— 25.00 Ref 251*

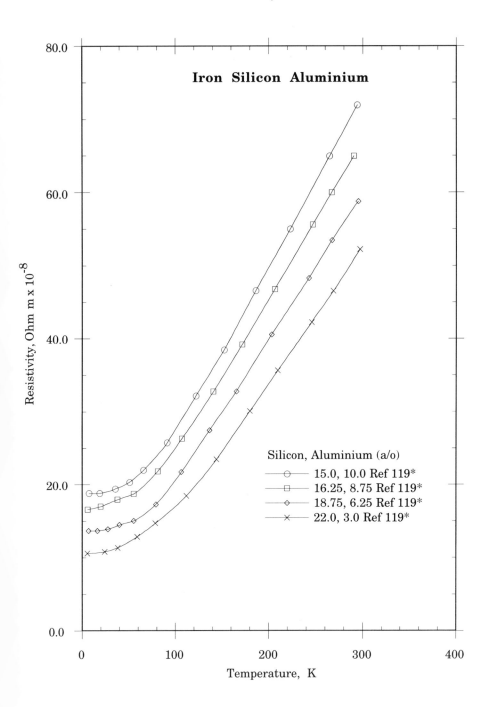

Iron Silicon Aluminium

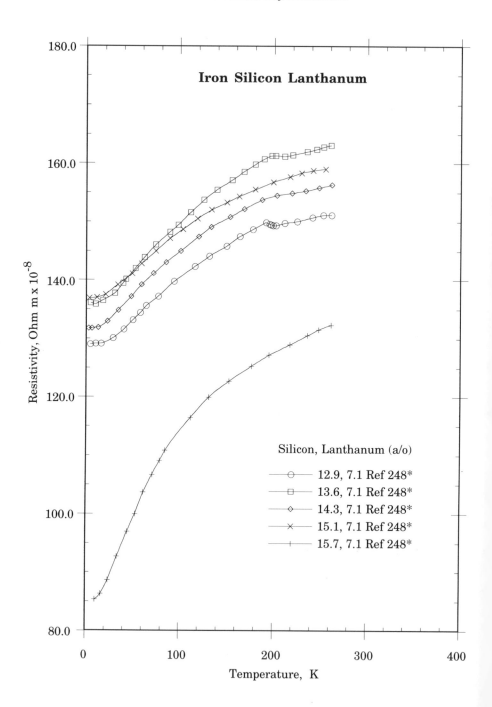

Iron Silicon Lanthanum

Silicon, Lanthanum (a/o)

⊖ 12.9, 7.1 Ref 248*
☐ 13.6, 7.1 Ref 248*
◇ 14.3, 7.1 Ref 248*
✕ 15.1, 7.1 Ref 248*
+ 15.7, 7.1 Ref 248*

Resistivity, Ohm m x 10⁻⁸

Temperature, K

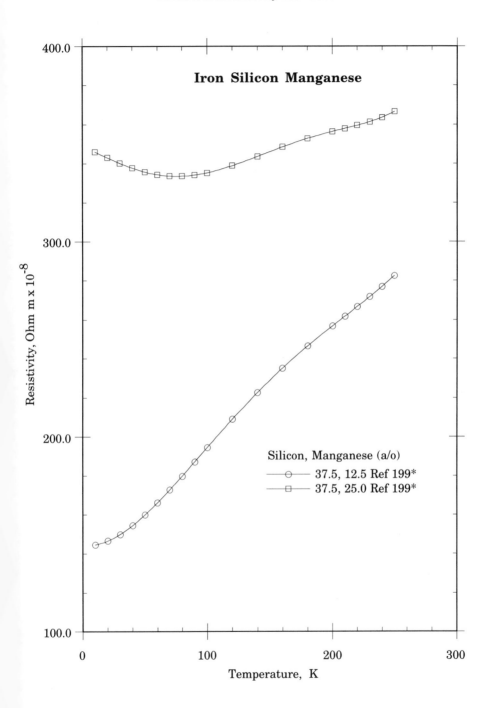

Iron Silicon Manganese

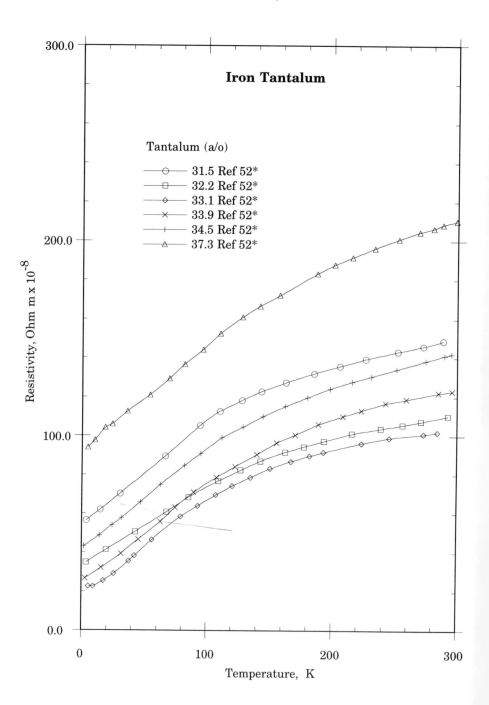

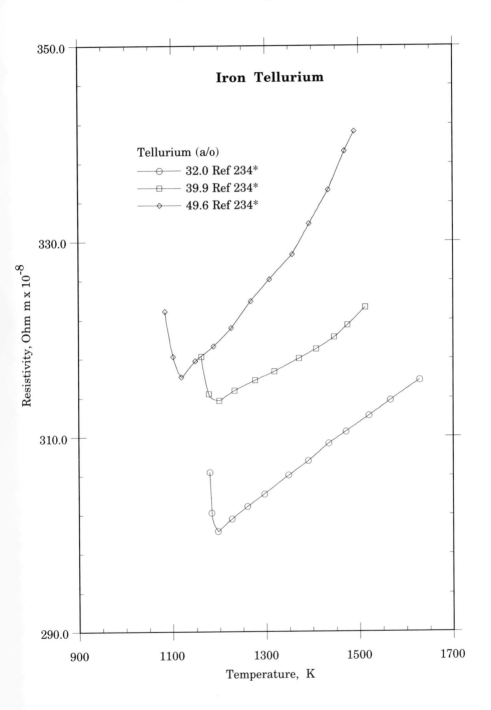

Iron Tellurium

Tellurium (a/o)
- 32.0 Ref 234*
- 39.9 Ref 234*
- 49.6 Ref 234*

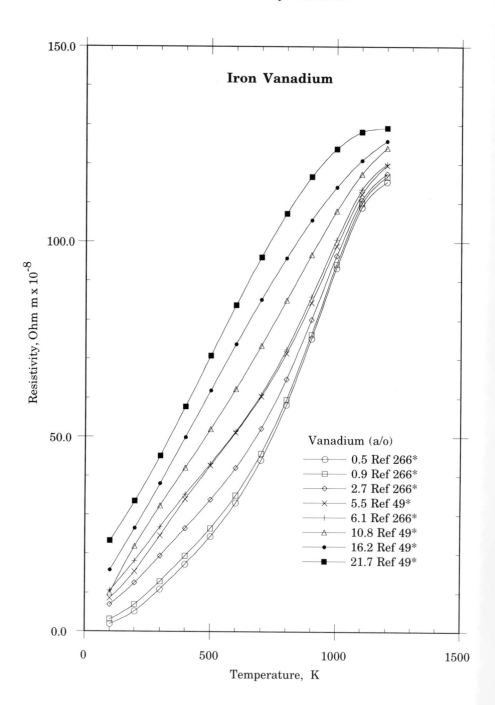

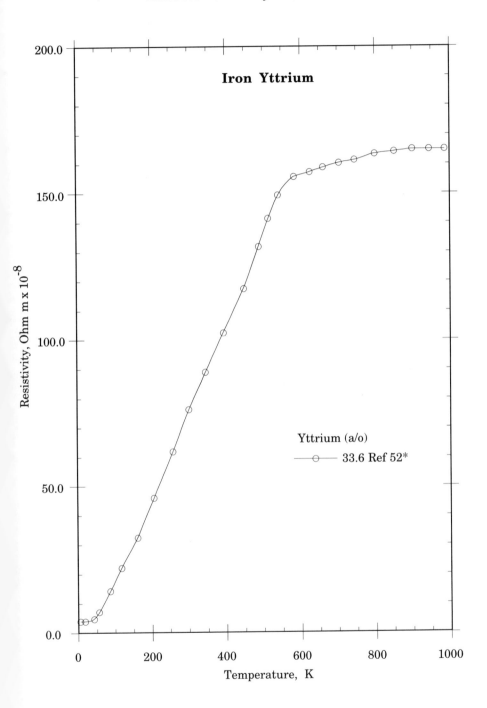

Iron Yttrium

Yttrium (a/o)
—o— 33.6 Ref 52*

Resistivity, Ohm m x 10^{-8}

Temperature, K

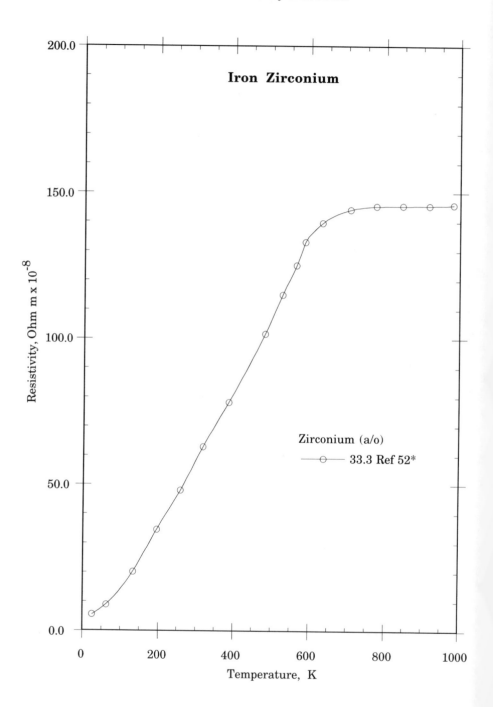

Iron Zirconium

Zirconium (a/o)
—⊙— 33.3 Ref 52*

Resistivity, Ohm m x 10⁻⁸

Temperature, K

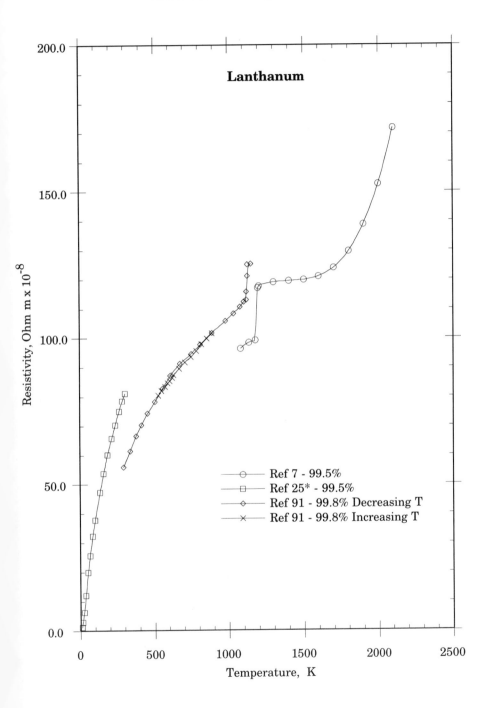

Lanthanum

Resistivity, Ohm m x 10^{-8}

Temperature, K

Ref 7 - 99.5%
Ref 25* - 99.5%
Ref 91 - 99.8% Decreasing T
Ref 91 - 99.8% Increasing T

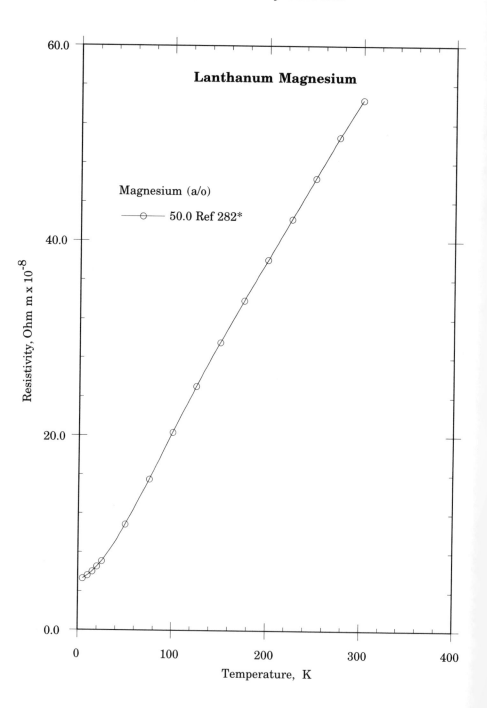

Lanthanum Magnesium

Magnesium (a/o)

—⊖— 50.0 Ref 282*

Resistivity, Ohm m x 10⁻⁸

Temperature, K

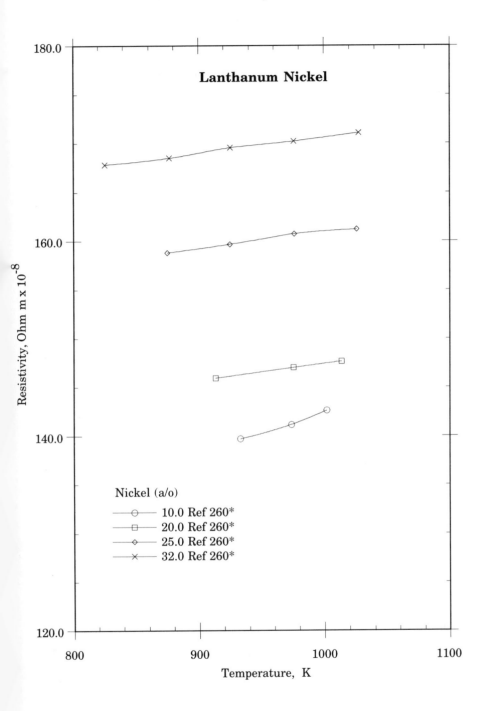

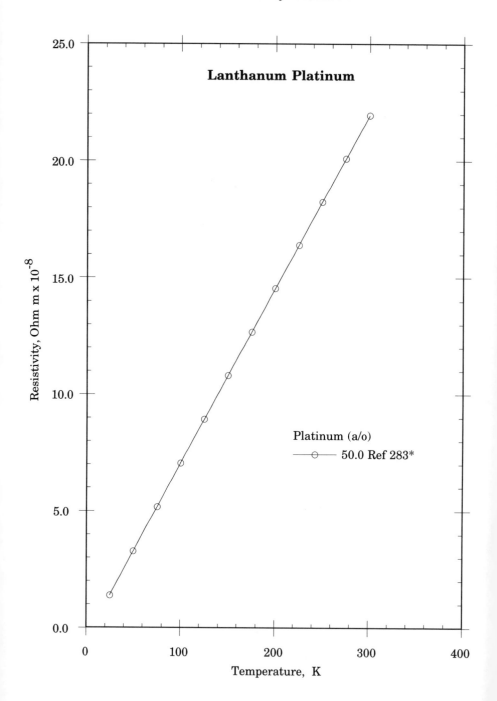

Lanthanum Platinum

Resistivity, Ohm m x 10^{-8}

Temperature, K

Platinum (a/o)
—⊙— 50.0 Ref 283*

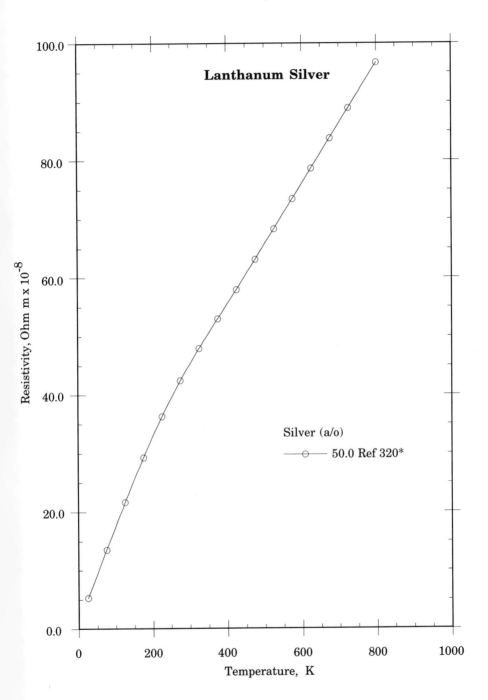

Lanthanum Silver

Resistivity, Ohm m x 10^{-8}

Temperature, K

Silver (a/o)
—⊖— 50.0 Ref 320*

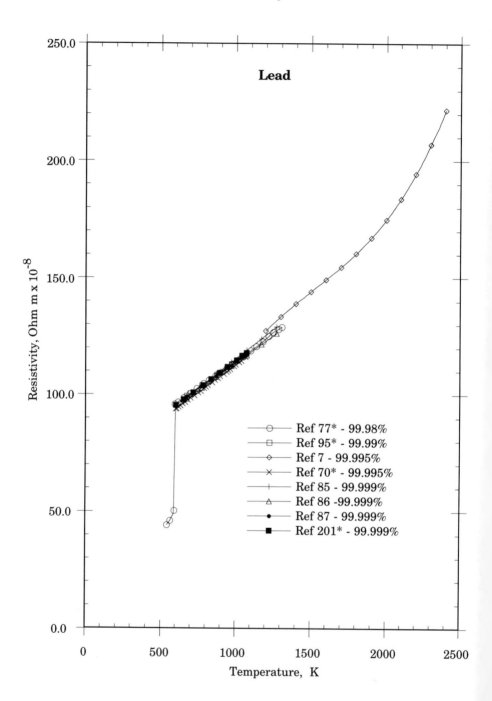

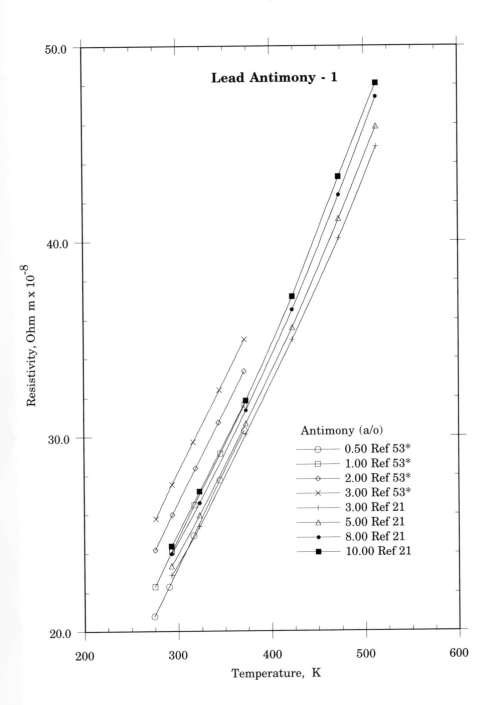

Lead Antimony - 1

Resistivity, Ohm m x 10^{-8}

Temperature, K

Antimony (a/o)
- —○— 0.50 Ref 53*
- —□— 1.00 Ref 53*
- —◇— 2.00 Ref 53*
- —×— 3.00 Ref 53*
- —+— 3.00 Ref 21
- —△— 5.00 Ref 21
- —●— 8.00 Ref 21
- —■— 10.00 Ref 21

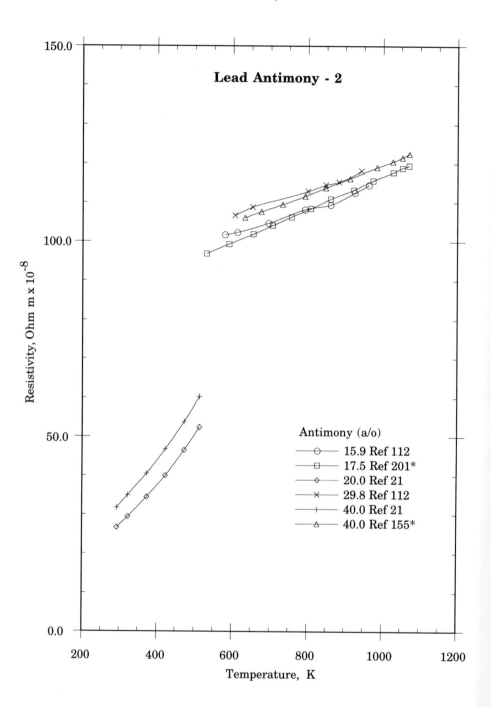

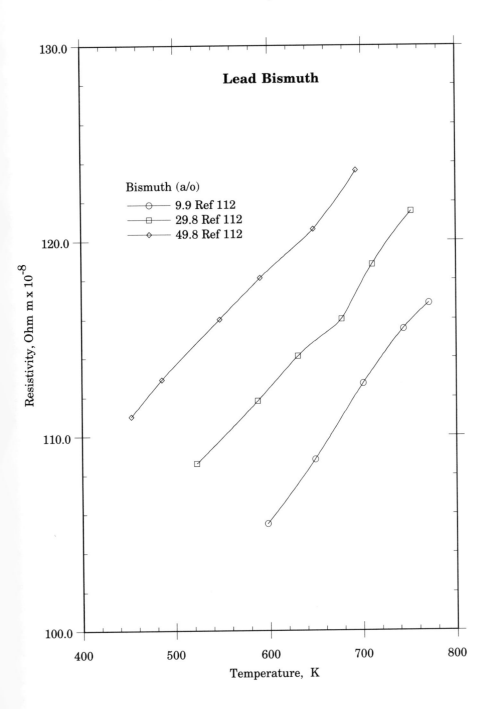

Lead Bismuth

Bismuth (a/o)
— ○ — 9.9 Ref 112
— □ — 29.8 Ref 112
— ◇ — 49.8 Ref 112

Resistivity, Ohm m x 10^{-8}

Temperature, K

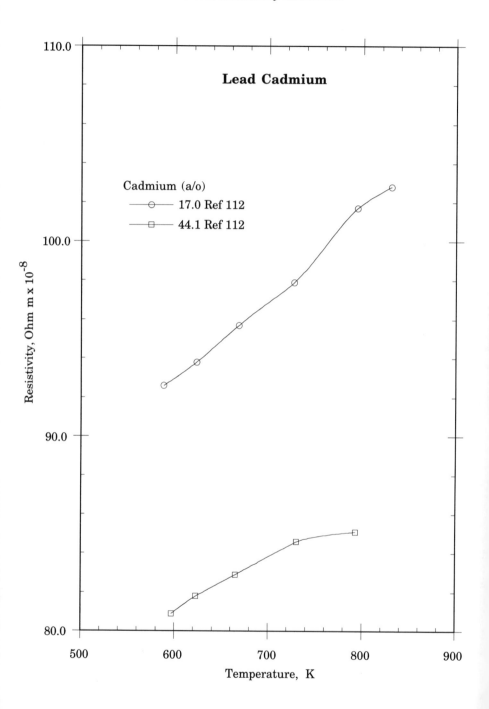

Lead Cadmium

Cadmium (a/o)
— ⊖ — 17.0 Ref 112
— ⊟ — 44.1 Ref 112

Resistivity, Ohm m x 10^{-8}

Temperature, K

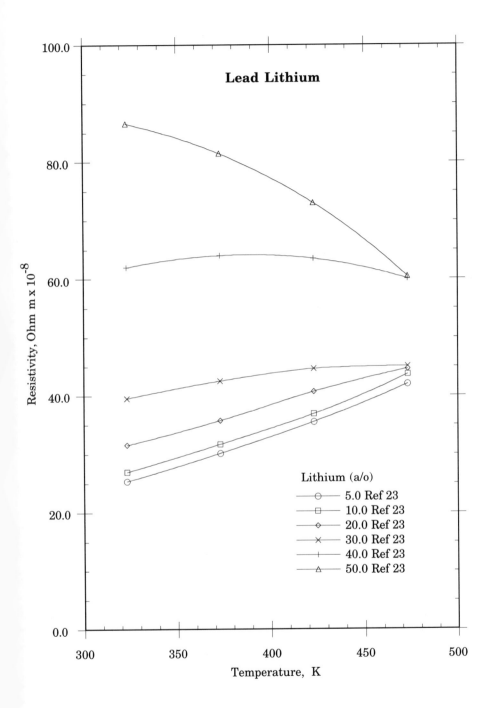

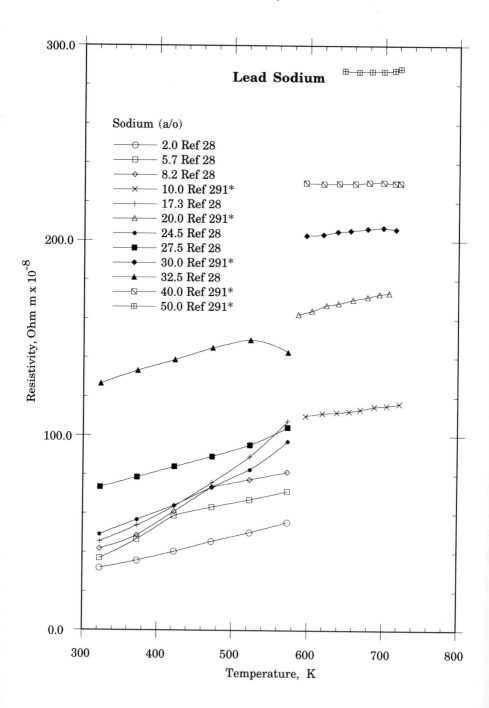

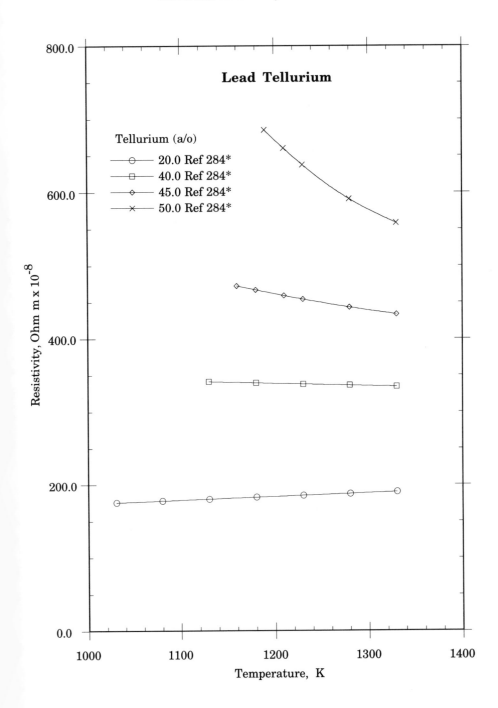

Lead Tellurium

Tellurium (a/o)
- ⊖ 20.0 Ref 284*
- ⊟ 40.0 Ref 284*
- ⬦ 45.0 Ref 284*
- ✕ 50.0 Ref 284*

Resistivity, Ohm m x 10^{-8}

Temperature, K

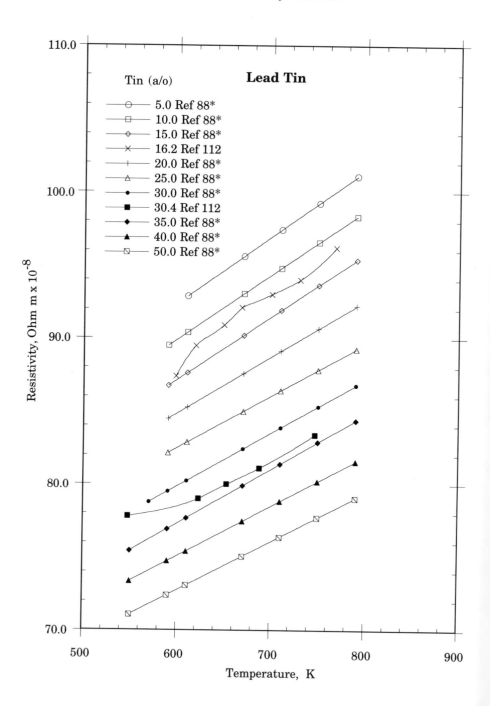

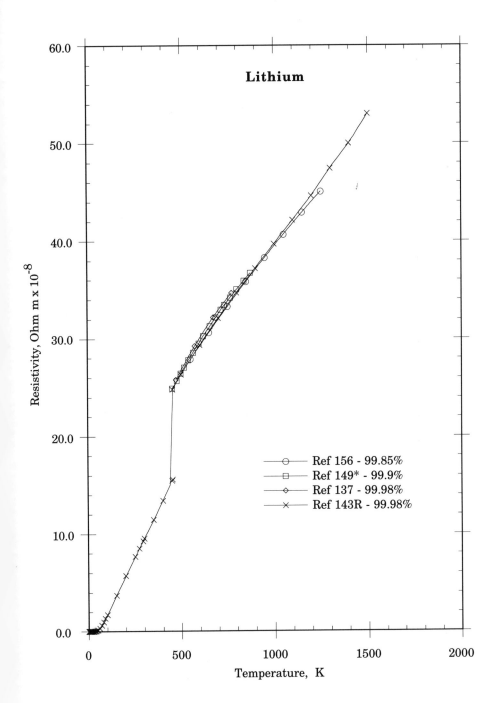

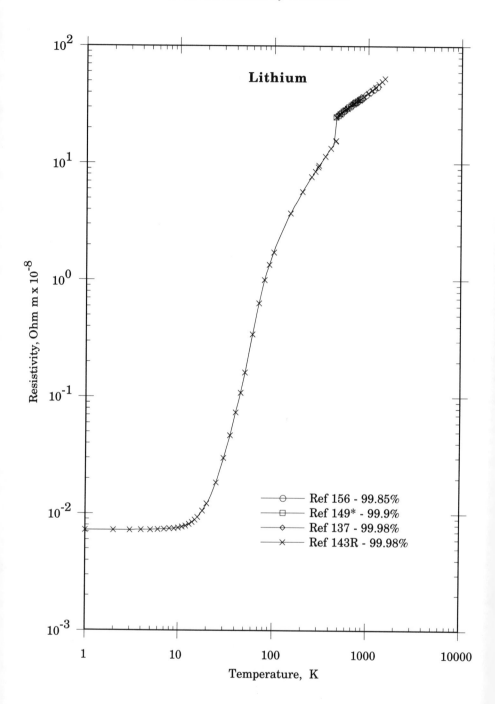

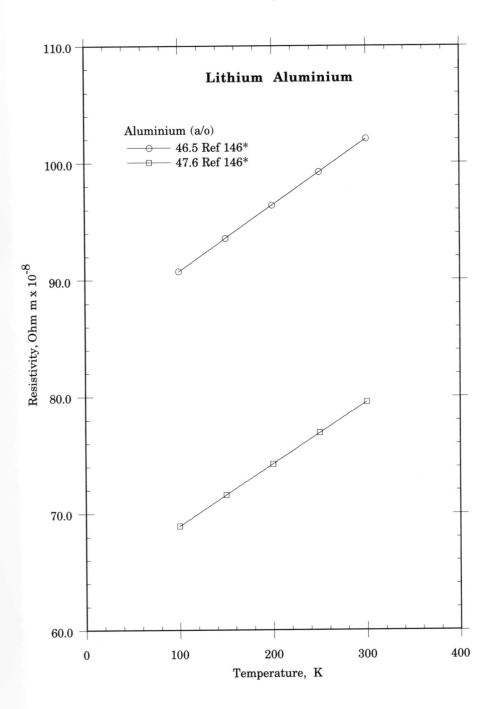

Lithium Aluminium

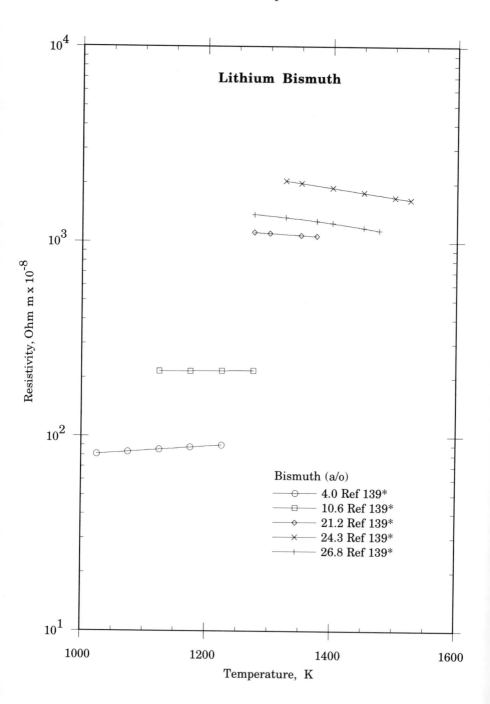

Lithium Bismuth

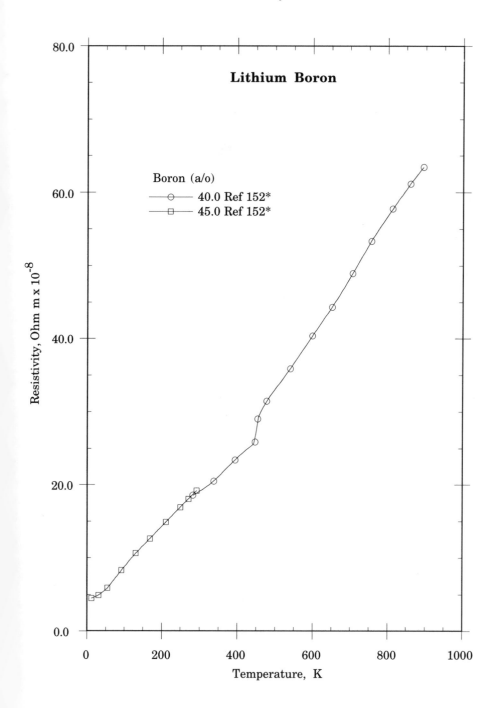

Lithium Boron

Boron (a/o)
- 40.0 Ref 152*
- 45.0 Ref 152*

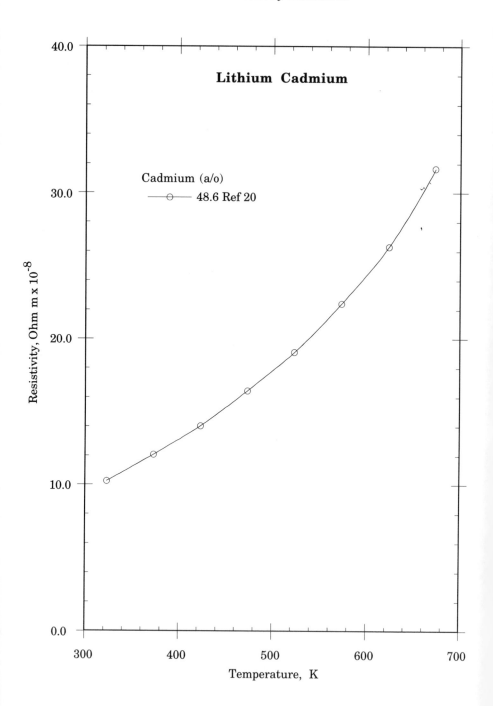

Lithium Cadmium

Cadmium (a/o)
—◦— 48.6 Ref 20

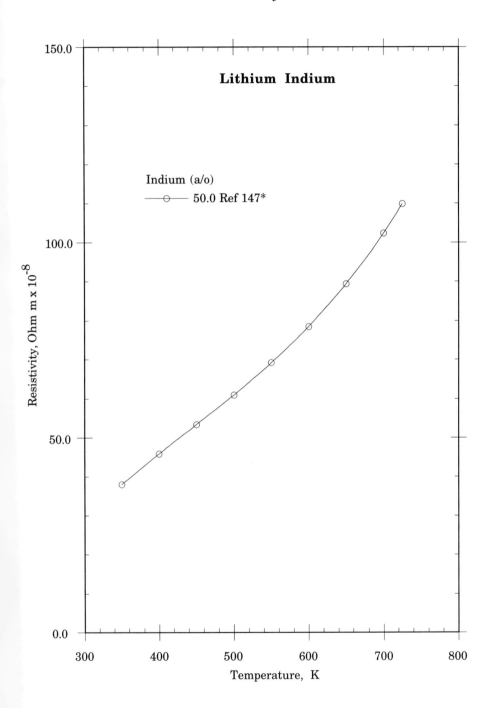

Lithium Indium

Indium (a/o)
—⊖— 50.0 Ref 147*

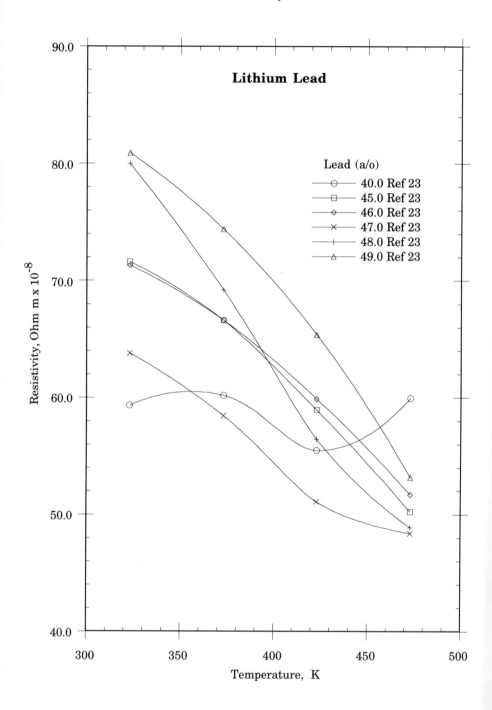

Lithium Lead

Lead (a/o)
—○— 40.0 Ref 23
—□— 45.0 Ref 23
—◇— 46.0 Ref 23
—×— 47.0 Ref 23
—+— 48.0 Ref 23
—△— 49.0 Ref 23

Resistivity, Ohm m x 10^{-8}

Temperature, K

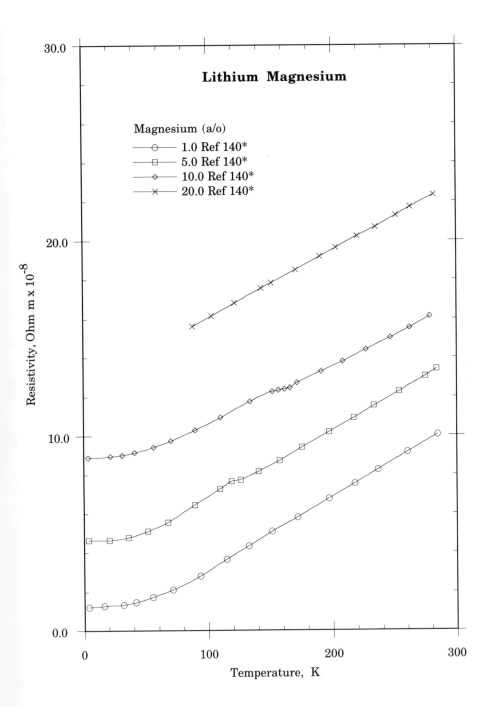

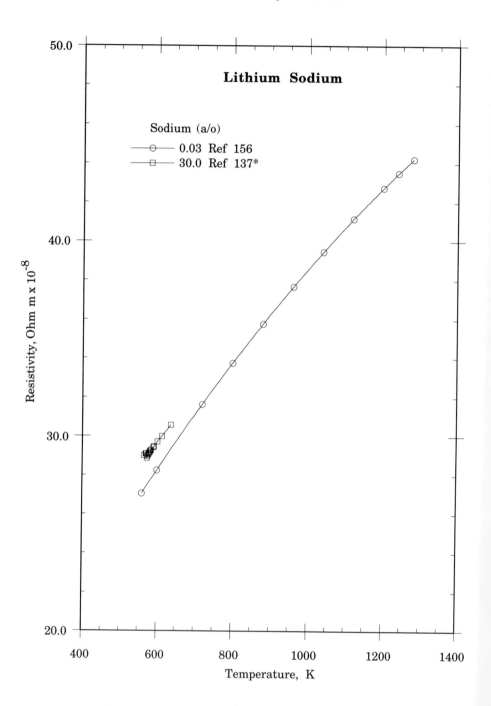

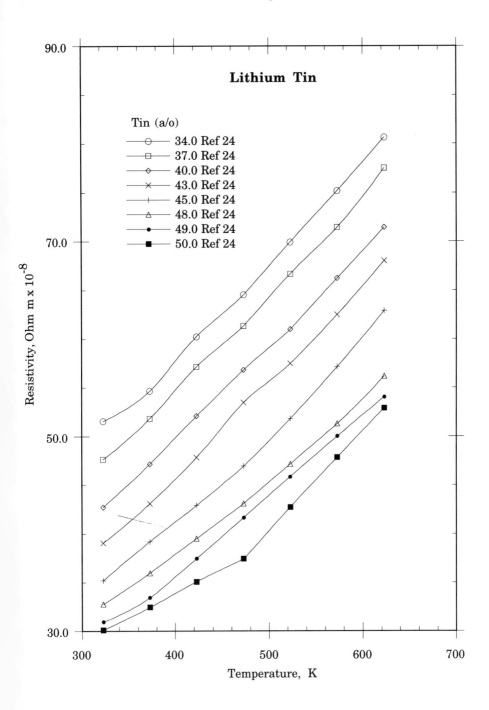

Lithium Tin

Tin (a/o)
- ○ 34.0 Ref 24
- □ 37.0 Ref 24
- ◇ 40.0 Ref 24
- × 43.0 Ref 24
- + 45.0 Ref 24
- △ 48.0 Ref 24
- ● 49.0 Ref 24
- ■ 50.0 Ref 24

Resistivity, Ohm m x 10^{-8}

Temperature, K

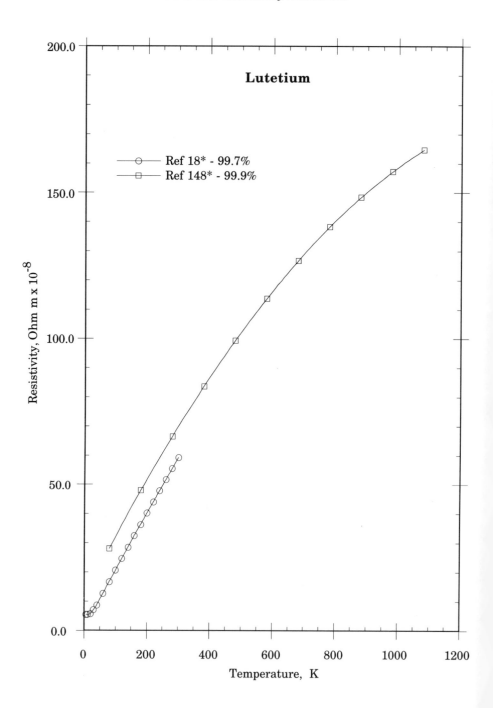

Lutetium

Ref 18* - 99.7%
Ref 148* - 99.9%

Resistivity, Ohm m x 10⁻⁸

Temperature, K

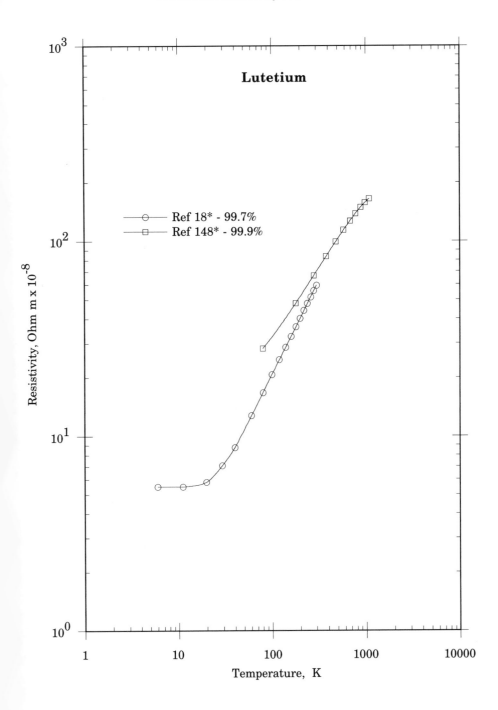

Lutetium

Ref 18* - 99.7%
Ref 148* - 99.9%

Resistivity, Ohm m x 10⁻⁸

Temperature, K

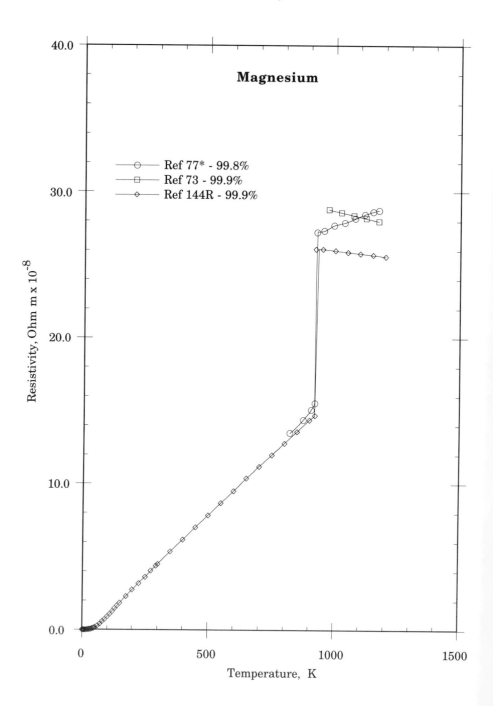

Magnesium

Ref 77* - 99.8%
Ref 73 - 99.9%
Ref 144R - 99.9%

Resistivity, Ohm m x 10^{-8}

Temperature, K

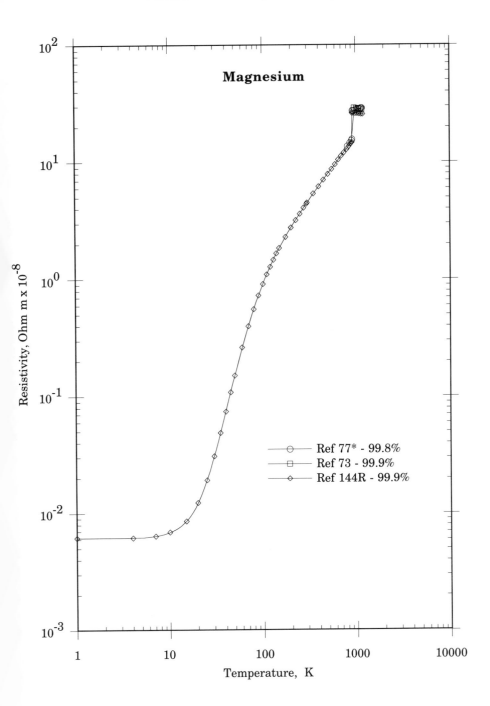

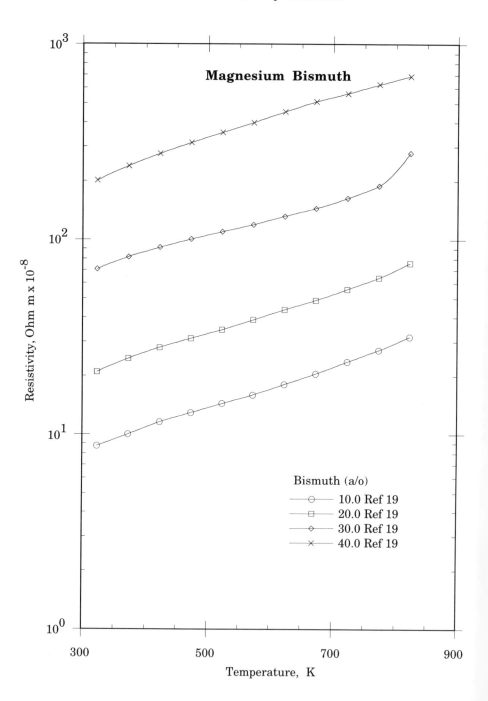

Magnesium Bismuth

Resistivity, Ohm m x 10⁻⁸

Temperature, K

Bismuth (a/o)
10.0 Ref 19
20.0 Ref 19
30.0 Ref 19
40.0 Ref 19

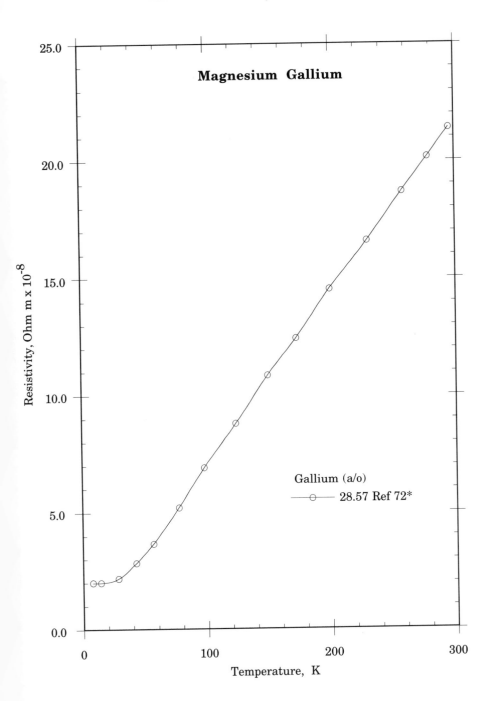

Magnesium Gallium

Resistivity, Ohm m x 10^{-8}

Temperature, K

Gallium (a/o)
28.57 Ref 72*

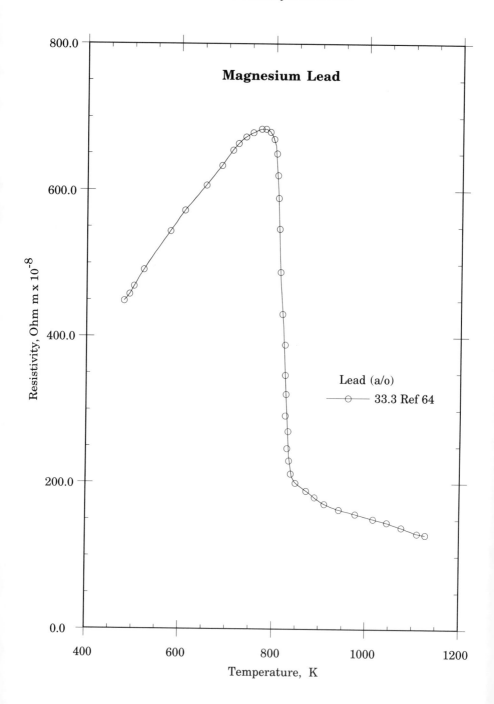

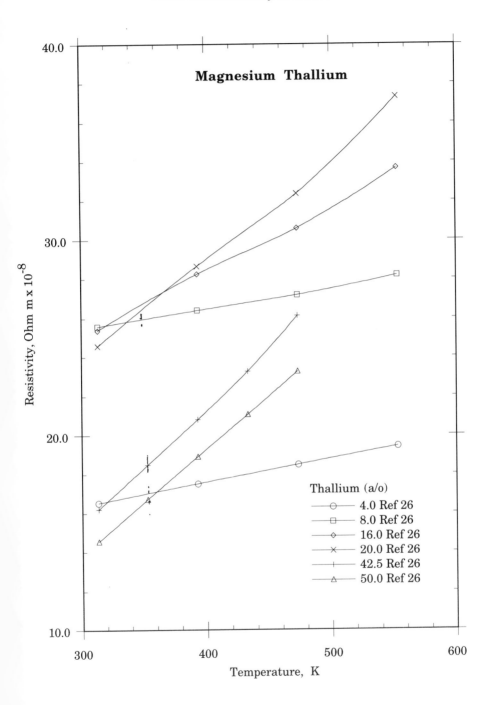

Magnesium Thallium

Resistivity, Ohm m x 10^{-8}

Temperature, K

Thallium (a/o)
- ⊖ 4.0 Ref 26
- ☐ 8.0 Ref 26
- ◇ 16.0 Ref 26
- ✕ 20.0 Ref 26
- + 42.5 Ref 26
- △ 50.0 Ref 26

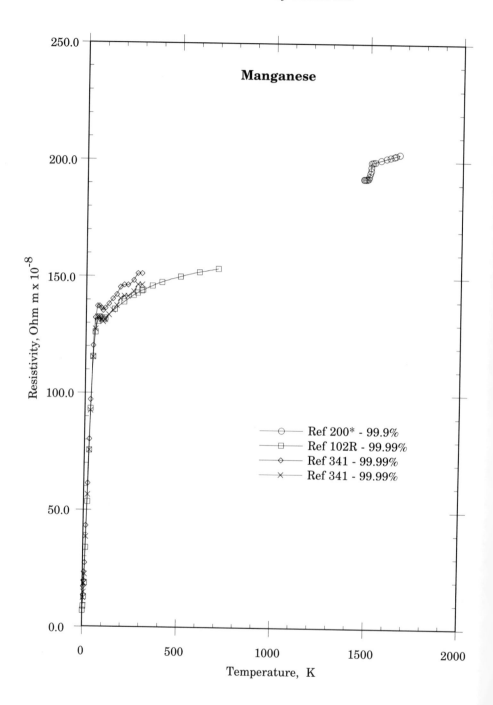

Manganese

Resistivity, Ohm m x 10^{-8}

Temperature, K

Ref 200* - 99.9%
Ref 102R - 99.99%
Ref 341 - 99.99%
Ref 341 - 99.99%

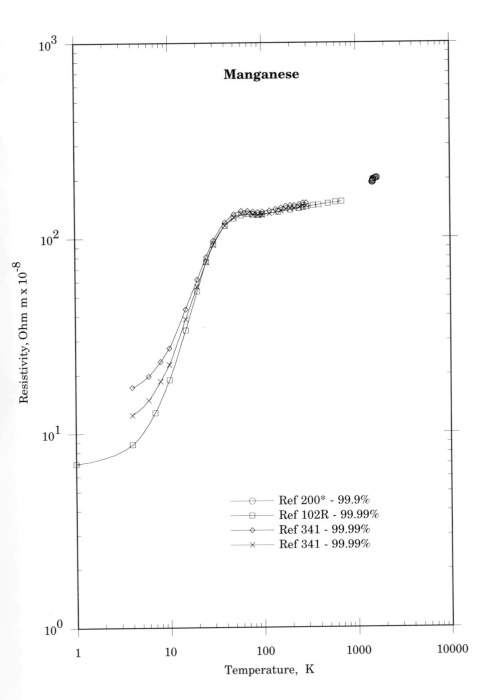

Manganese

Resistivity, Ohm m x 10^{-8}

Temperature, K

Ref 200* - 99.9%
Ref 102R - 99.99%
Ref 341 - 99.99%
Ref 341 - 99.99%

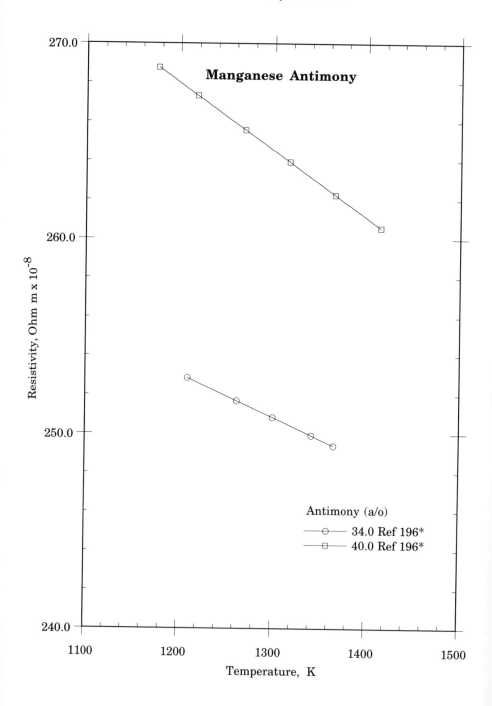

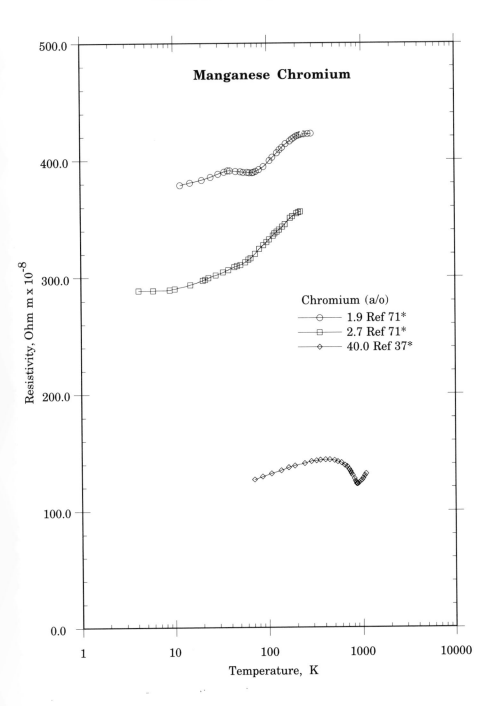

Manganese Chromium

Resistivity, Ohm m x 10⁻⁸

Temperature, K

Chromium (a/o)
- —⊙— 1.9 Ref 71*
- —☐— 2.7 Ref 71*
- —◇— 40.0 Ref 37*

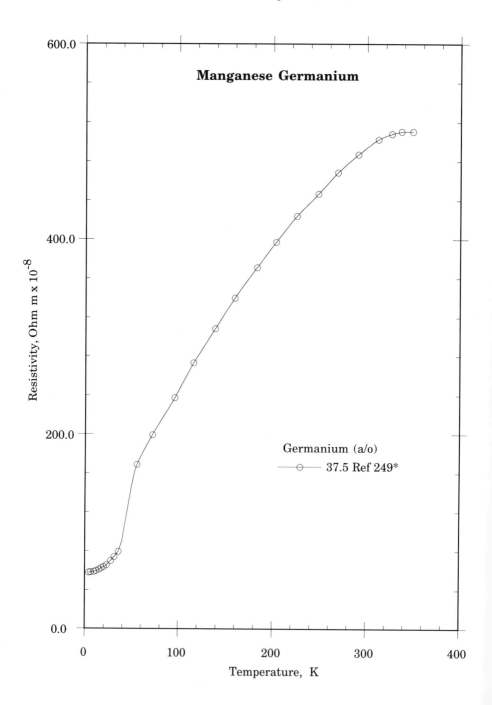

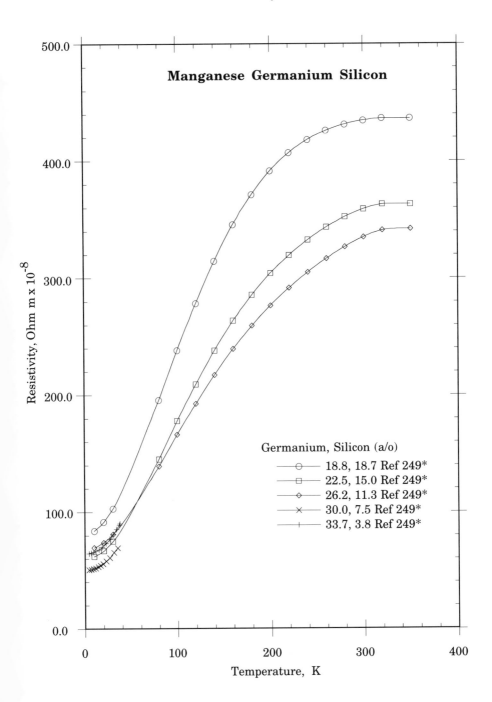

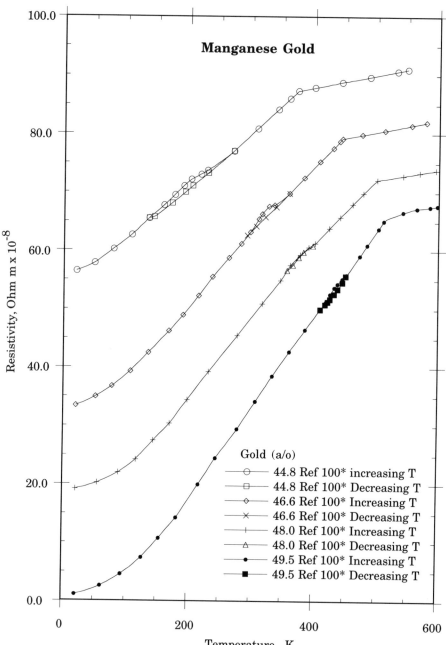

Manganese Gold

Resistivity, Ohm m x 10^{-8}

Temperature, K

Gold (a/o)
—⊙— 44.8 Ref 100* increasing T
—□— 44.8 Ref 100* Decreasing T
—◇— 46.6 Ref 100* Increasing T
—✕— 46.6 Ref 100* Decreasing T
—+— 48.0 Ref 100* Increasing T
—△— 48.0 Ref 100* Decreasing T
—•— 49.5 Ref 100* Increasing T
—■— 49.5 Ref 100* Decreasing T

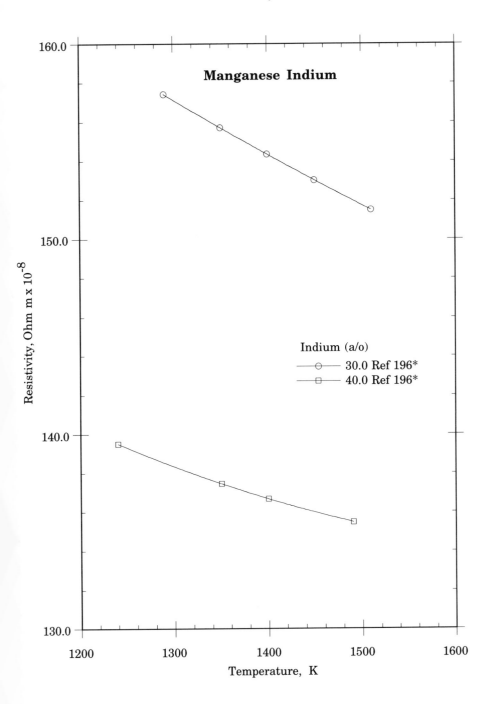

Manganese Indium

Resistivity, Ohm m x 10⁻⁸ (y-axis)

Temperature, K (x-axis)

Indium (a/o)
─○─ 30.0 Ref 196*
─□─ 40.0 Ref 196*

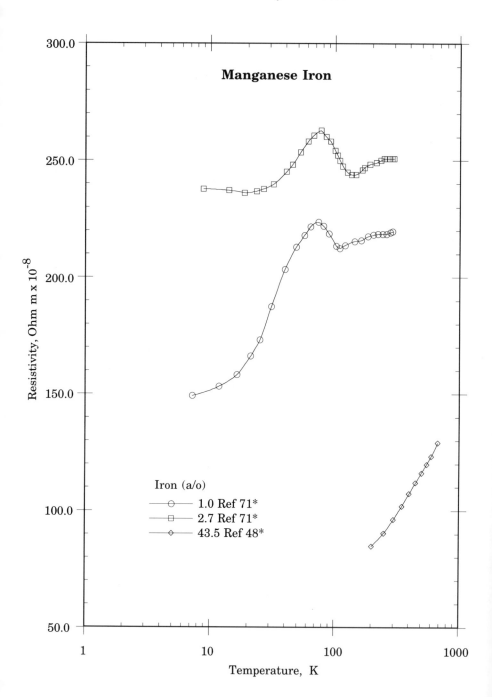

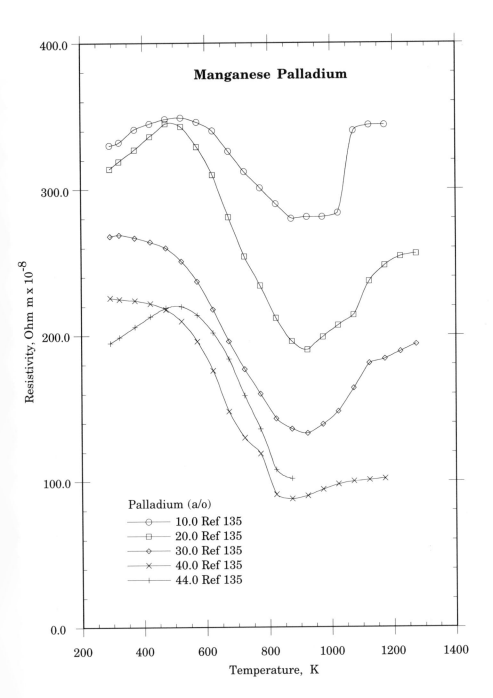

Manganese Palladium

Resistivity, Ohm m x 10^{-8}

Temperature, K

Palladium (a/o)
- 10.0 Ref 135
- 20.0 Ref 135
- 30.0 Ref 135
- 40.0 Ref 135
- 44.0 Ref 135

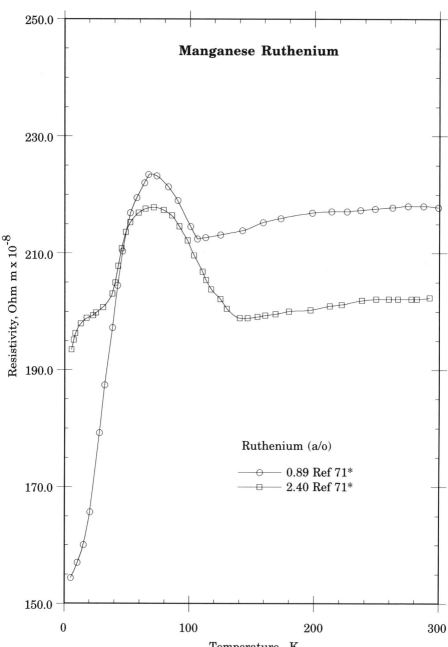

Manganese Ruthenium

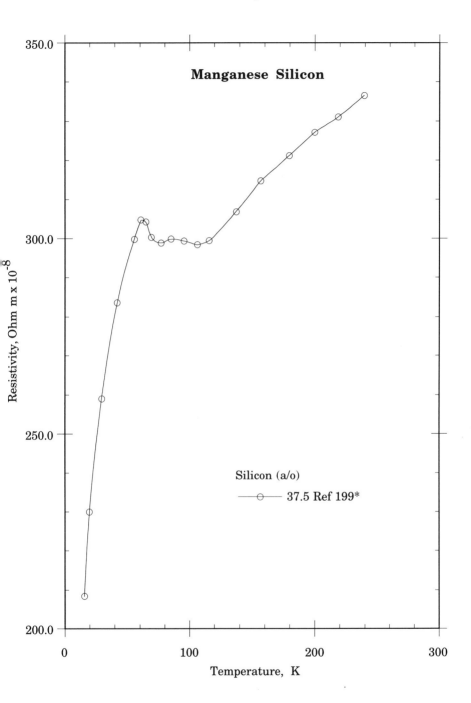

Manganese Silicon

Silicon (a/o)

———⊙——— 37.5 Ref 199*

Resistivity, Ohm m x 10^{-8}

Temperature, K

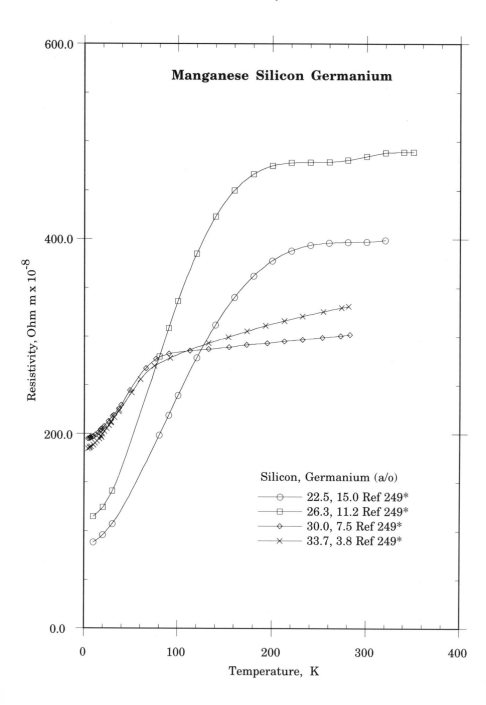

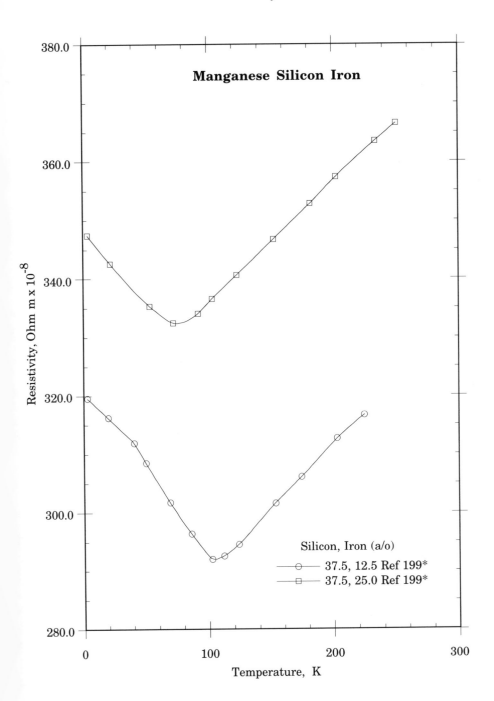

Manganese Silicon Iron

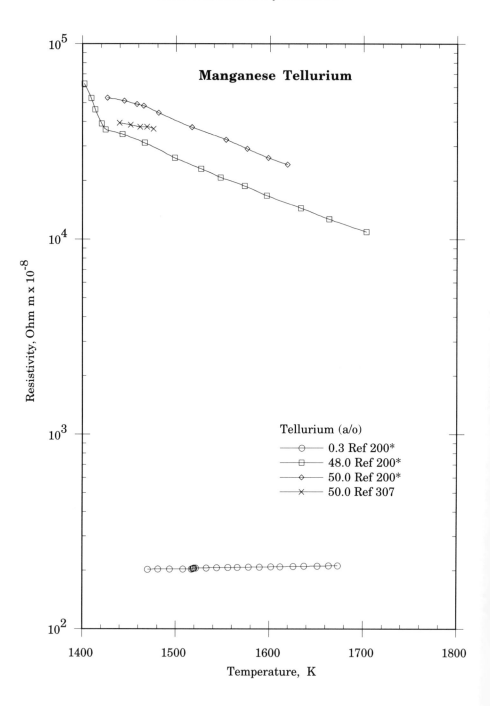

Manganese Tellurium

Resistivity, Ohm m x 10^{-8}

Temperature, K

Tellurium (a/o)
—○— 0.3 Ref 200*
—□— 48.0 Ref 200*
—◇— 50.0 Ref 200*
—×— 50.0 Ref 307

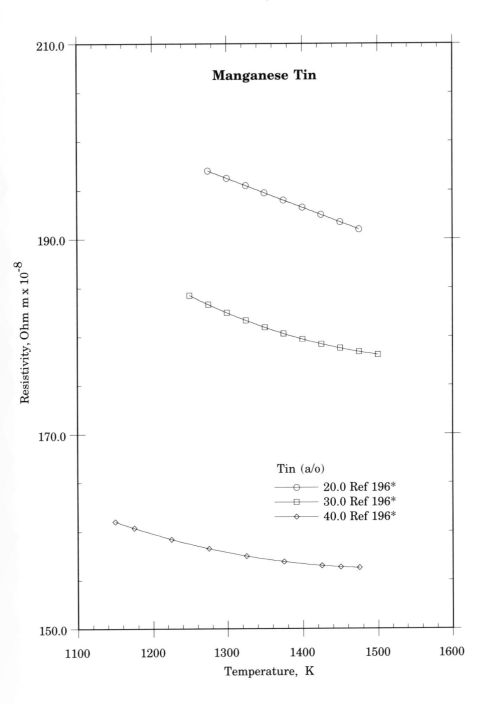

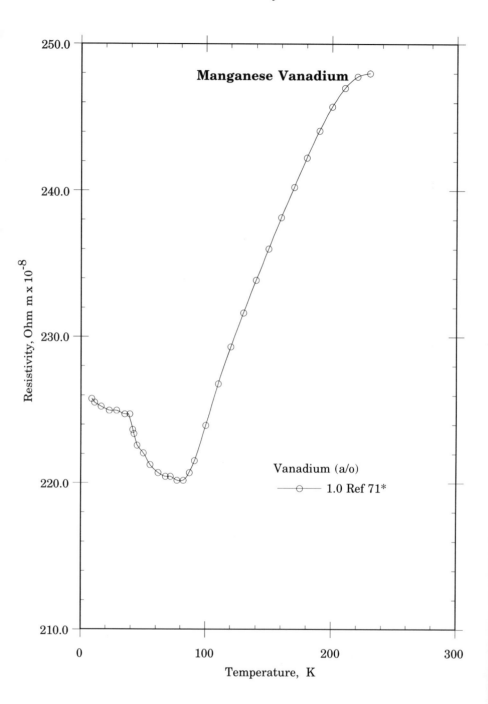

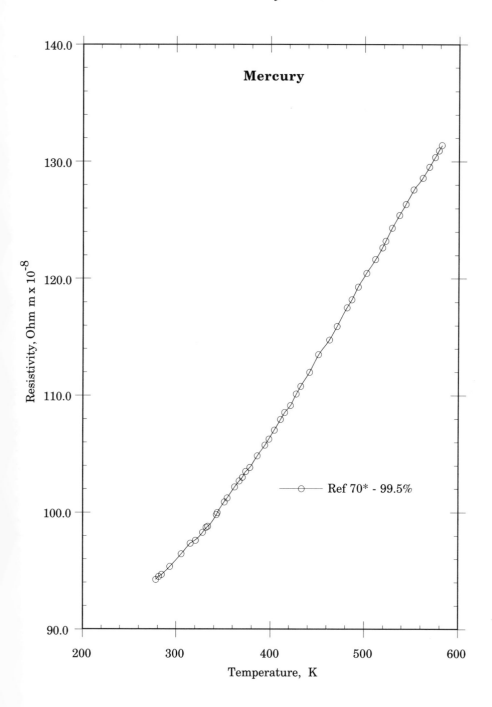

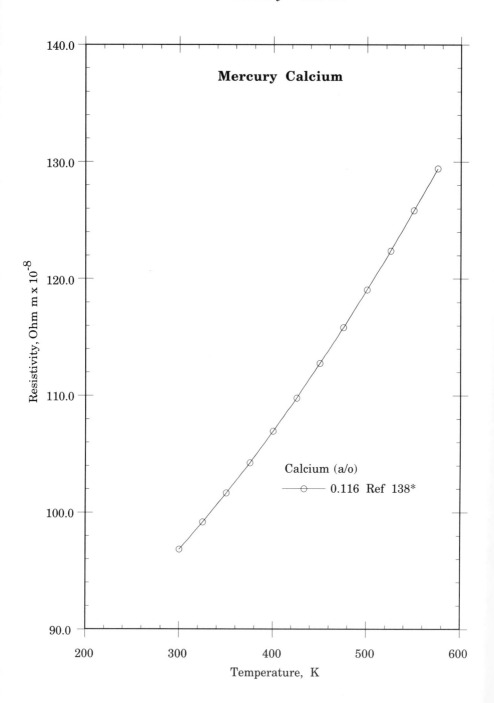

Mercury Calcium

Resistivity, Ohm m x 10⁻⁸

Temperature, K

Calcium (a/o)
—o— 0.116 Ref 138*

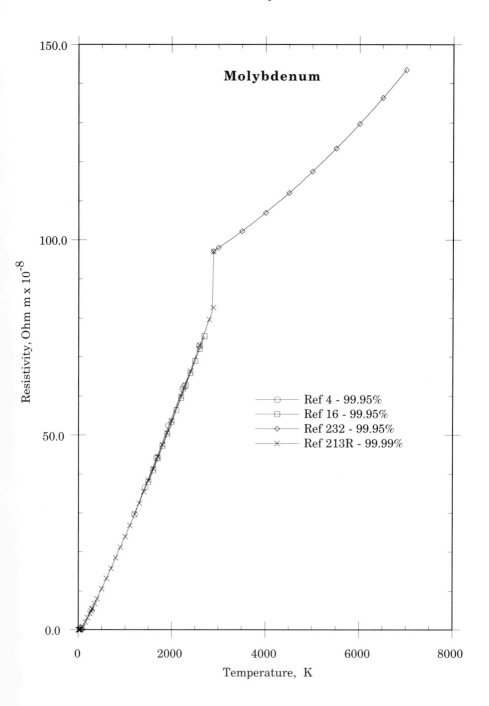

Molybdenum

Resistivity, Ohm m x 10^{-8}

Temperature, K

Ref 4 - 99.95%
Ref 16 - 99.95%
Ref 232 - 99.95%
Ref 213R - 99.99%

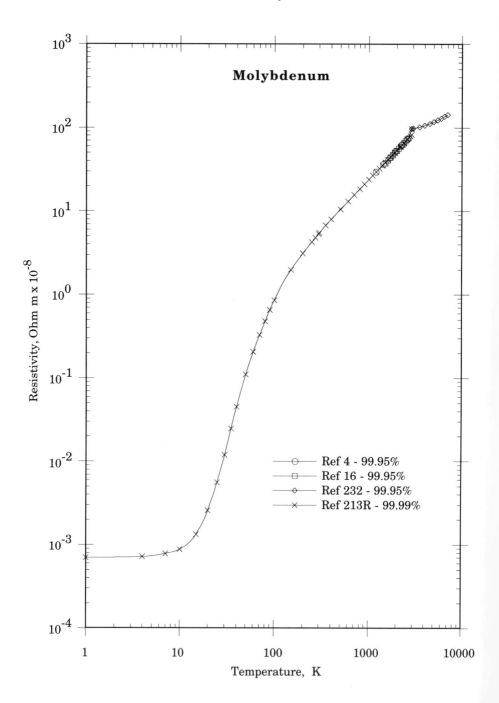

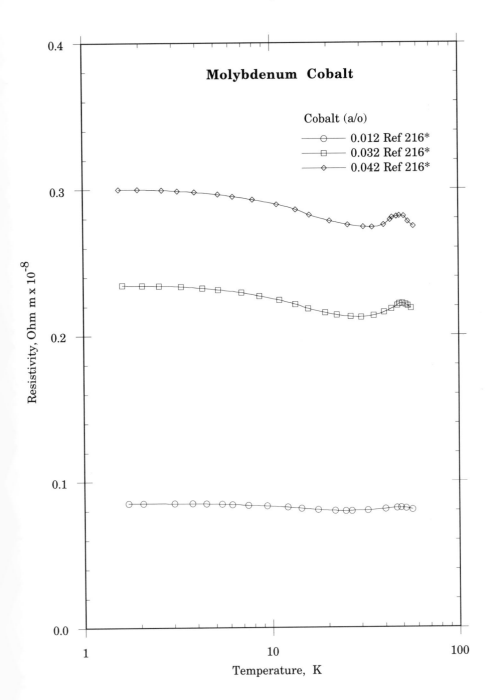

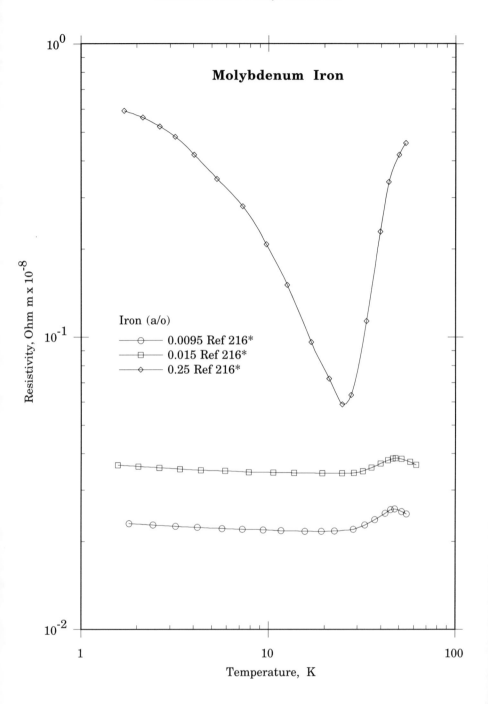

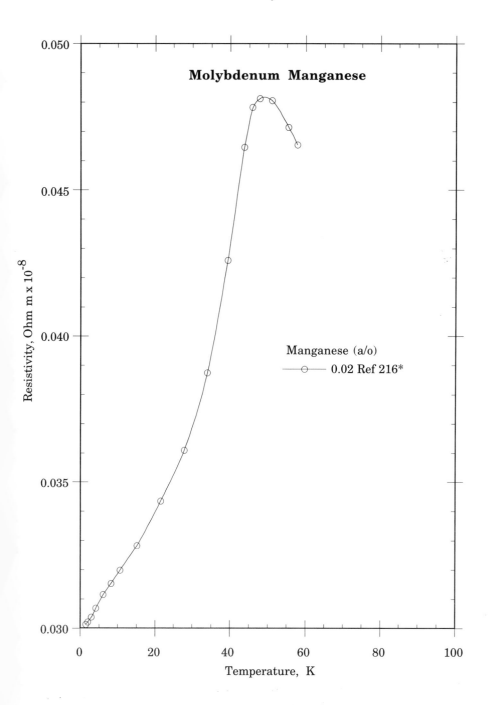

Molybdenum Manganese

Manganese (a/o)
—⊙— 0.02 Ref 216*

Resistivity, Ohm m x 10^{-8}

Temperature, K

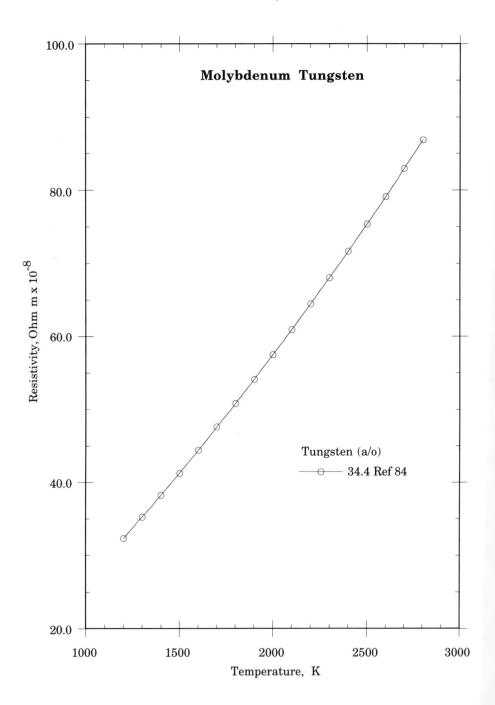

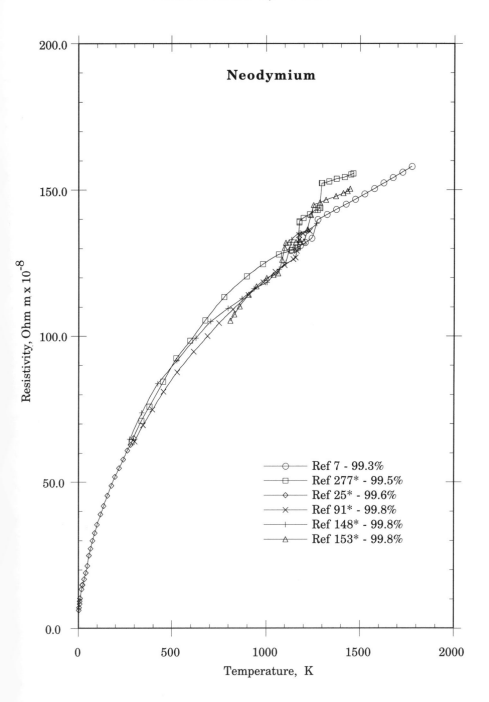

Neodymium

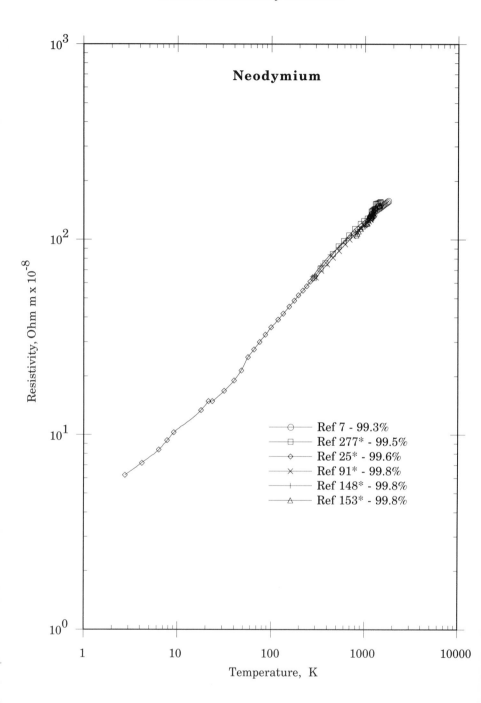

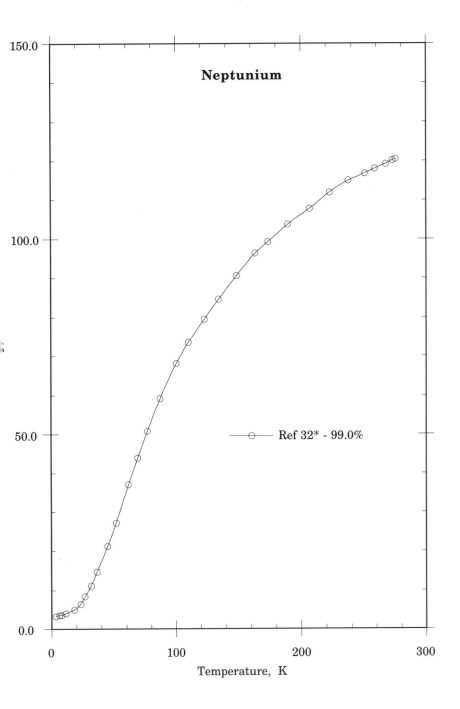

Neptunium

Ref 32* - 99.0%

Temperature, K

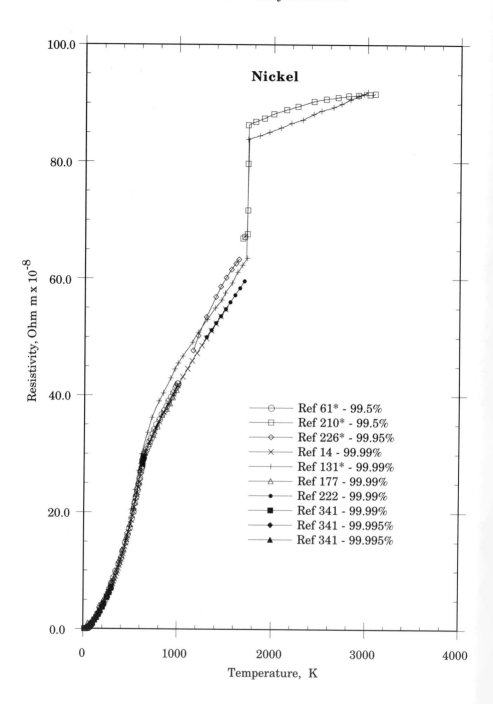

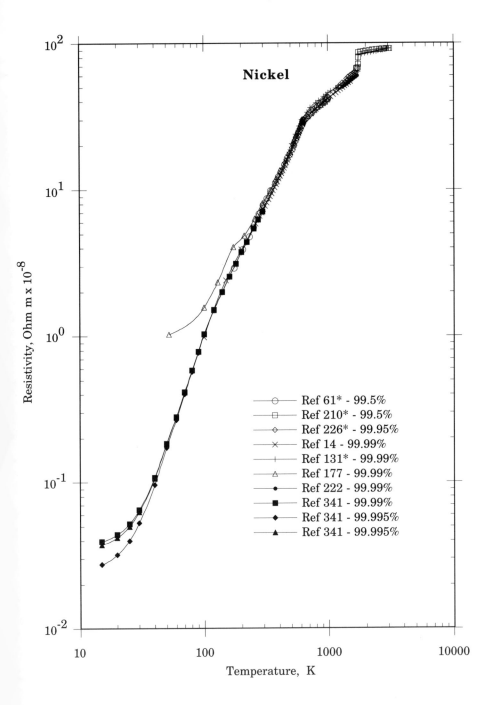

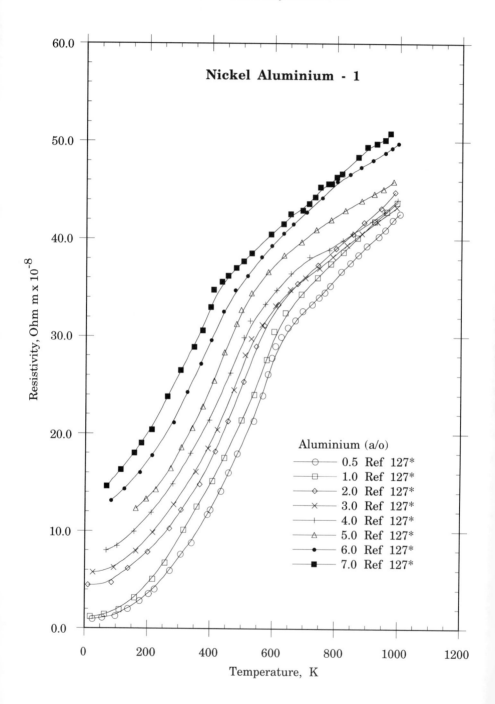

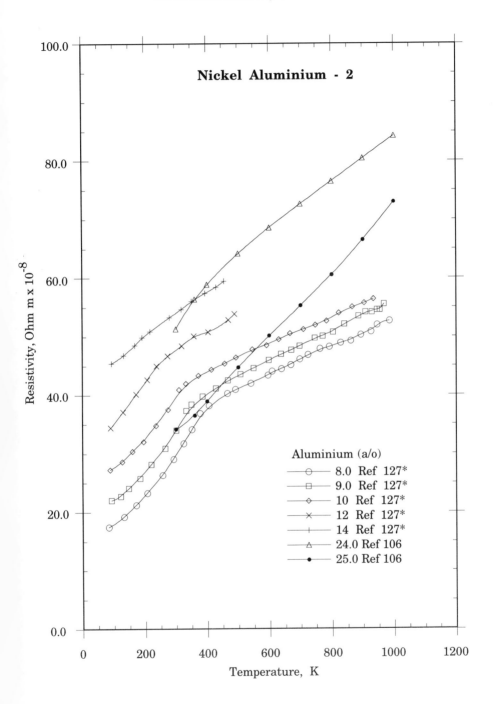

Nickel Aluminium - 2

Resistivity, Ohm m x 10^{-8}

Temperature, K

Aluminium (a/o)

—◦— 8.0 Ref 127*
—□— 9.0 Ref 127*
—◇— 10 Ref 127*
—×— 12 Ref 127*
—+— 14 Ref 127*
—△— 24.0 Ref 106
—•— 25.0 Ref 106

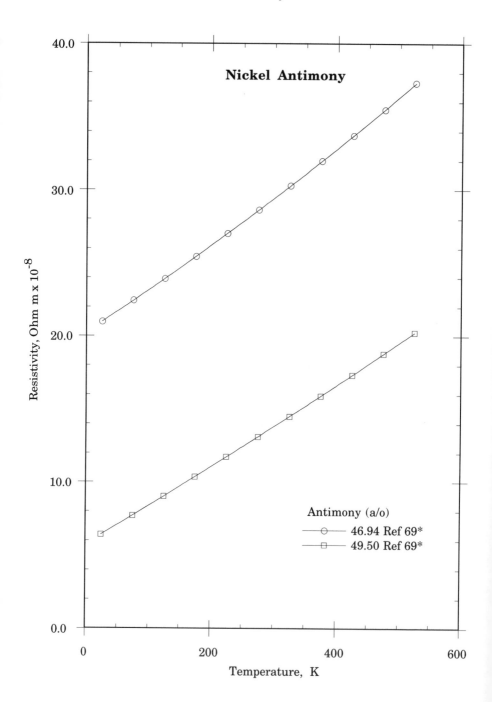

Nickel Antimony

Resistivity, Ohm m x 10^{-8}

Temperature, K

Antimony (a/o)
—○— 46.94 Ref 69*
—□— 49.50 Ref 69*

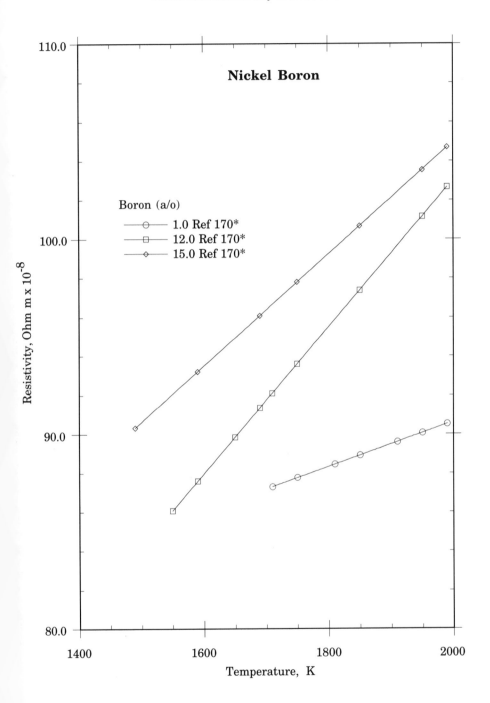

Nickel Boron

Boron (a/o)

———○——— 1.0 Ref 170*
———□——— 12.0 Ref 170*
———◇——— 15.0 Ref 170*

Resistivity, Ohm m x 10^{-8}

Temperature, K

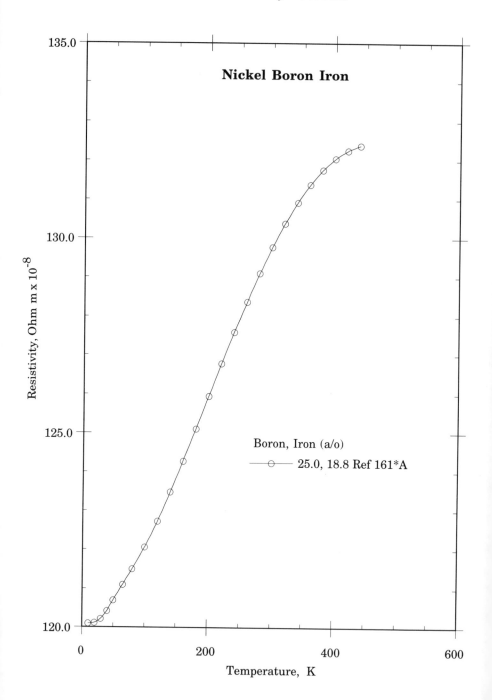

Nickel Boron Iron

Resistivity, Ohm m x 10⁻⁸

Temperature, K

Boron, Iron (a/o)

25.0, 18.8 Ref 161*A

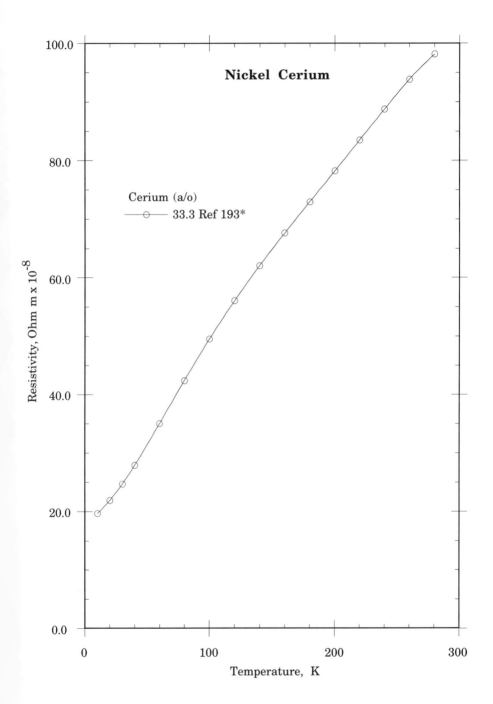

Nickel Cerium

Cerium (a/o)
—⊖— 33.3 Ref 193*

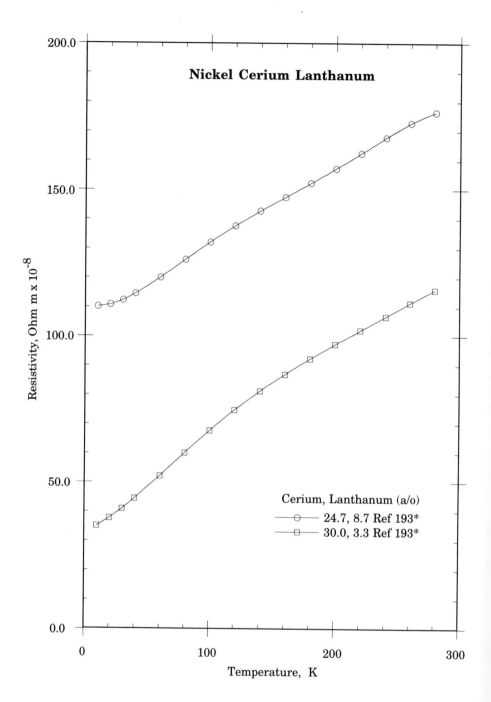

Nickel Cerium Lanthanum

Cerium, Lanthanum (a/o)
—○— 24.7, 8.7 Ref 193*
—□— 30.0, 3.3 Ref 193*

Resistivity, Ohm m x 10^{-8}

Temperature, K

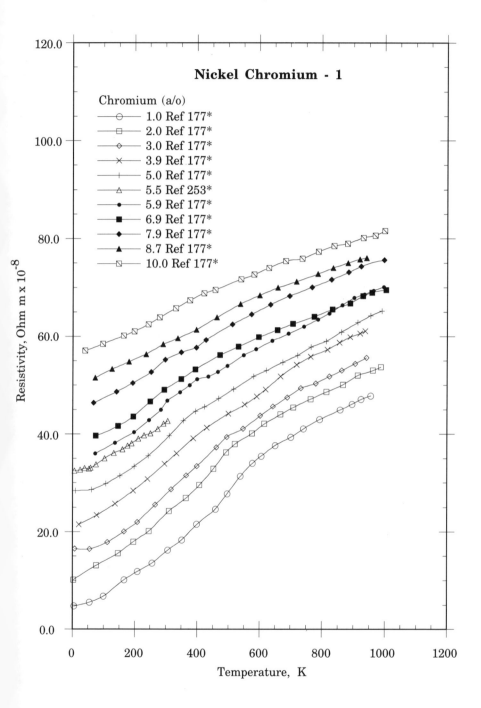

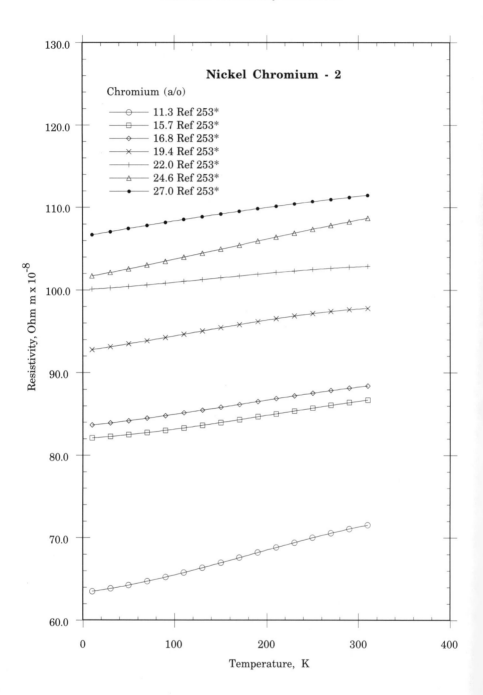

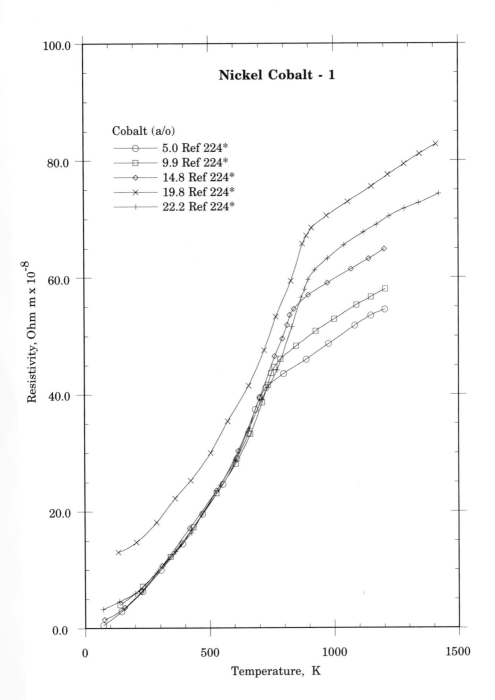

Nickel Cobalt - 1

Cobalt (a/o)
5.0 Ref 224*
9.9 Ref 224*
14.8 Ref 224*
19.8 Ref 224*
22.2 Ref 224*

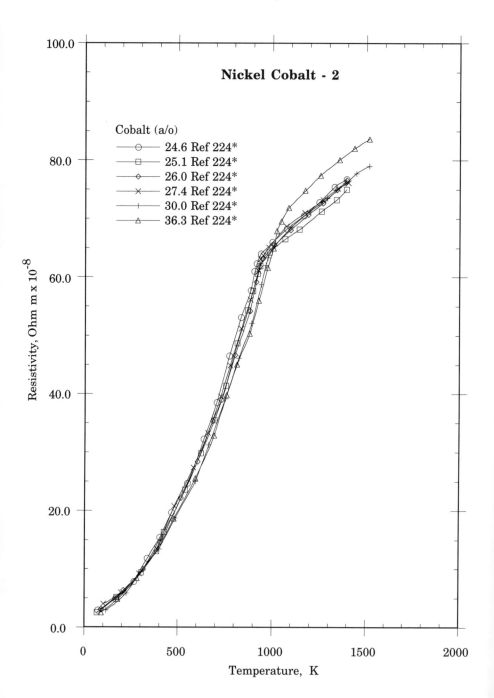

Nickel Cobalt - 2

Cobalt (a/o)
- ○ 24.6 Ref 224*
- □ 25.1 Ref 224*
- ◇ 26.0 Ref 224*
- × 27.4 Ref 224*
- + 30.0 Ref 224*
- △ 36.3 Ref 224*

Resistivity, Ohm m x 10^{-8}

Temperature, K

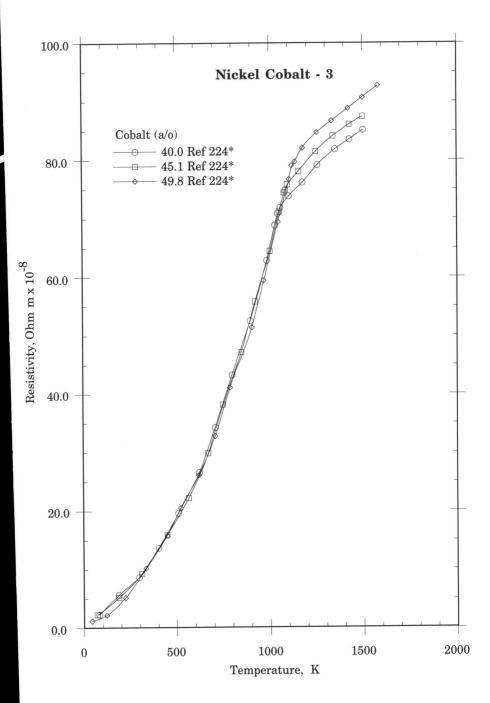

Nickel Cobalt - 3

Cobalt (a/o)
—○— 40.0 Ref 224*
—□— 45.1 Ref 224*
—◇— 49.8 Ref 224*

Resistivity, Ohm m x 10⁻⁸

Temperature, K

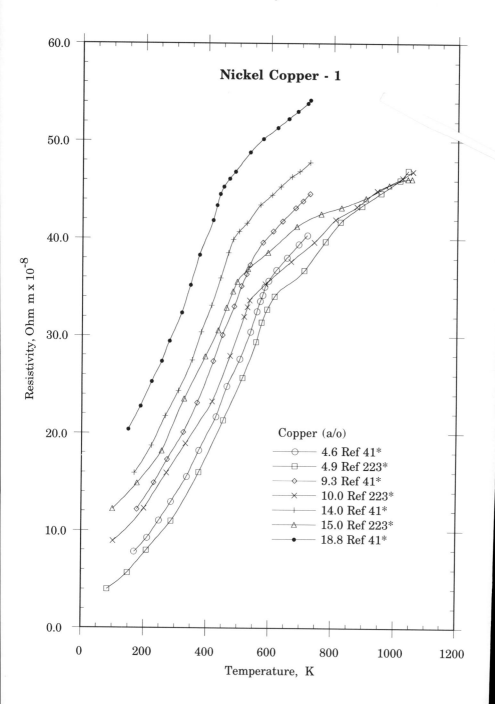

Nickel Copper - 1

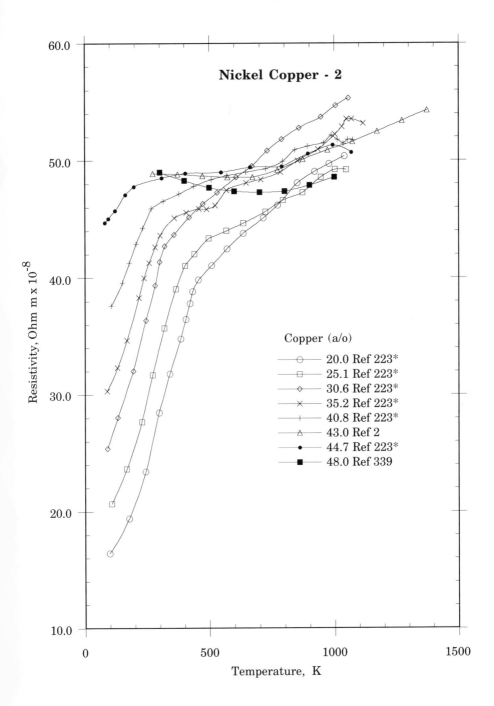

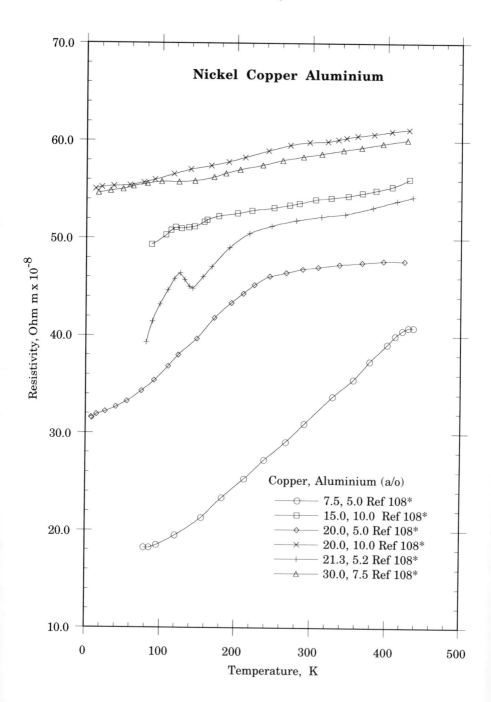

Nickel Copper Aluminium

Copper, Aluminium (a/o)

- ——⊖—— 7.5, 5.0 Ref 108*
- ——☐—— 15.0, 10.0 Ref 108*
- ——◇—— 20.0, 5.0 Ref 108*
- ——×—— 20.0, 10.0 Ref 108*
- ——+—— 21.3, 5.2 Ref 108*
- ——△—— 30.0, 7.5 Ref 108*

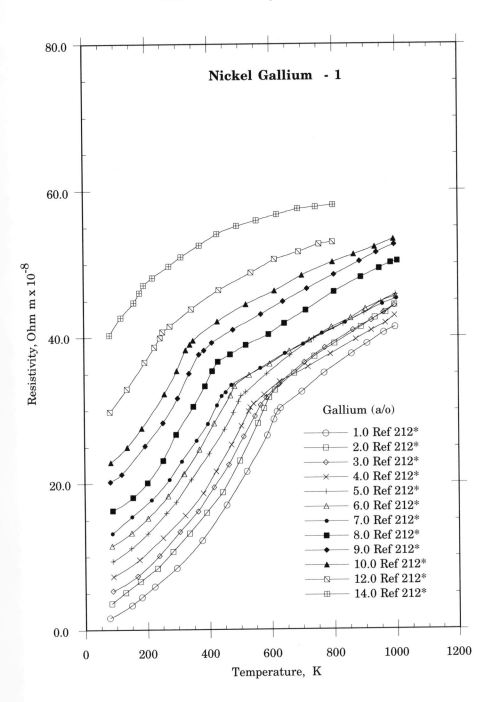

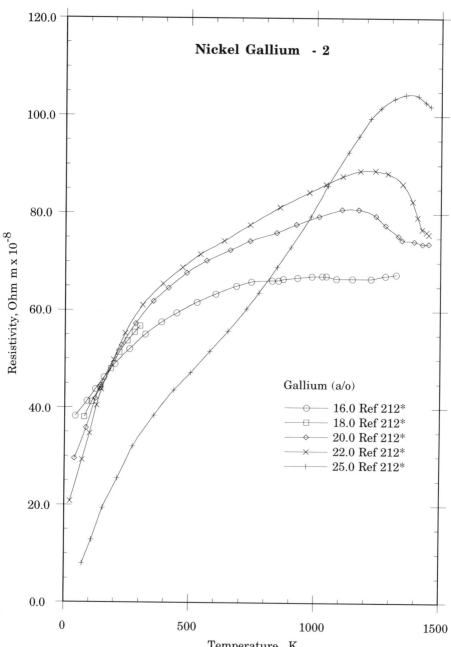

Nickel Gallium - 2

Gallium (a/o)

⊖ 16.0 Ref 212*
□ 18.0 Ref 212*
◇ 20.0 Ref 212*
× 22.0 Ref 212*
+ 25.0 Ref 212*

Resistivity, Ohm m x 10⁻⁸

Temperature, K

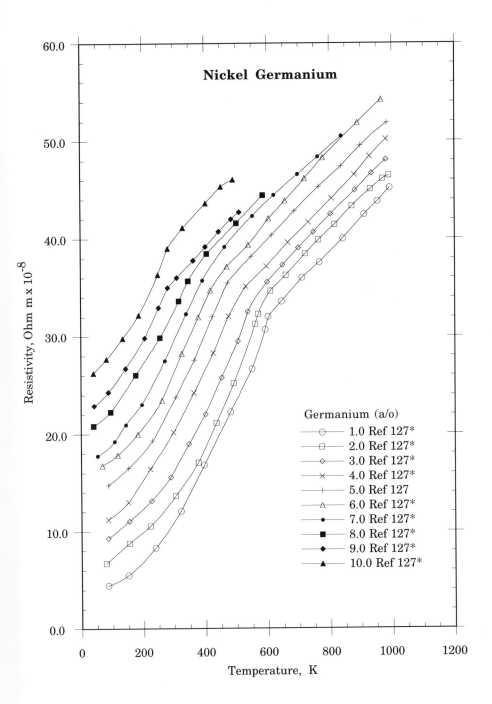

Nickel Germanium

Germanium (a/o)
- ○ 1.0 Ref 127*
- □ 2.0 Ref 127*
- ◇ 3.0 Ref 127*
- × 4.0 Ref 127*
- + 5.0 Ref 127
- △ 6.0 Ref 127*
- ● 7.0 Ref 127*
- ■ 8.0 Ref 127*
- ◆ 9.0 Ref 127*
- ▲ 10.0 Ref 127*

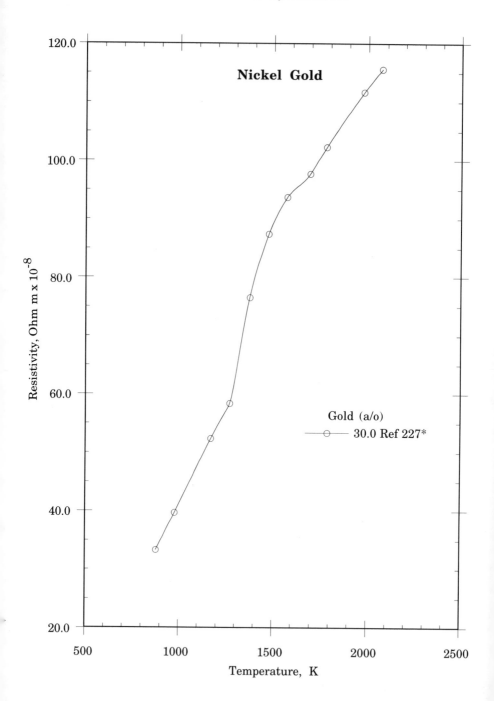

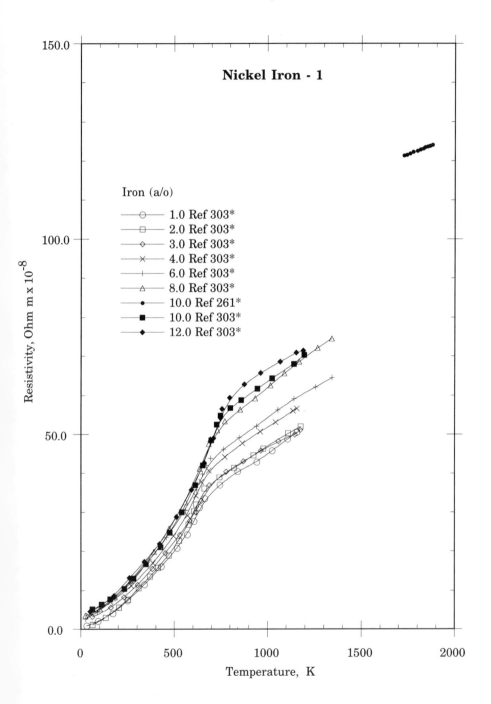

Nickel Iron - 1

Iron (a/o)

- ○ 1.0 Ref 303*
- □ 2.0 Ref 303*
- ◇ 3.0 Ref 303*
- × 4.0 Ref 303*
- + 6.0 Ref 303*
- △ 8.0 Ref 303*
- ● 10.0 Ref 261*
- ■ 10.0 Ref 303*
- ◆ 12.0 Ref 303*

Resistivity, Ohm m x 10^{-8}

Temperature, K

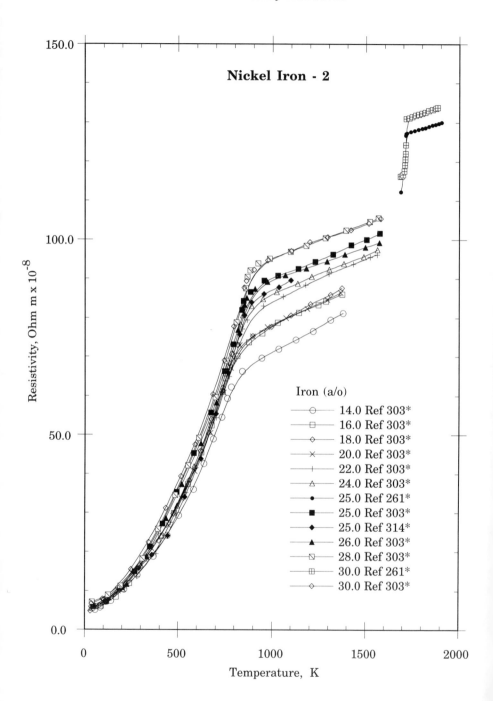

Nickel Iron - 2

Iron (a/o)

⊖ 14.0 Ref 303*
⊟ 16.0 Ref 303*
◇ 18.0 Ref 303*
✕ 20.0 Ref 303*
+ 22.0 Ref 303*
△ 24.0 Ref 303*
● 25.0 Ref 261*
■ 25.0 Ref 303*
◆ 25.0 Ref 314*
▲ 26.0 Ref 303*
◁ 28.0 Ref 303*
⊞ 30.0 Ref 261*
◇ 30.0 Ref 303*

Resistivity, Ohm m x 10^{-8}

Temperature, K

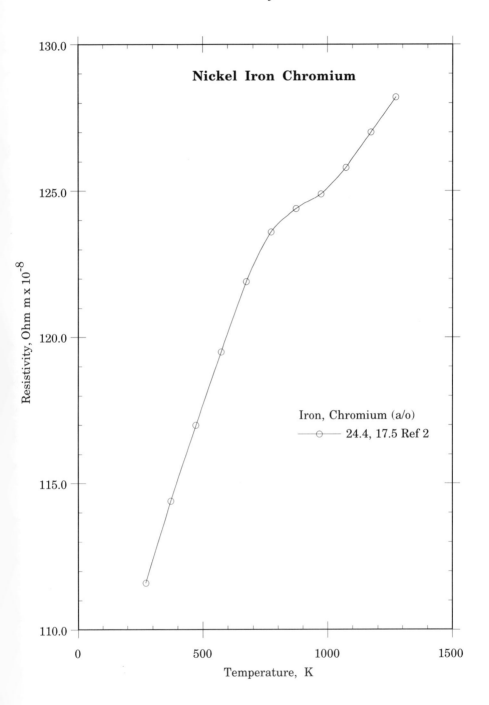

Nickel Iron Chromium

Iron, Chromium (a/o)

─○─ 24.4, 17.5 Ref 2

Resistivity, Ohm m x 10^{-8}

Temperature, K

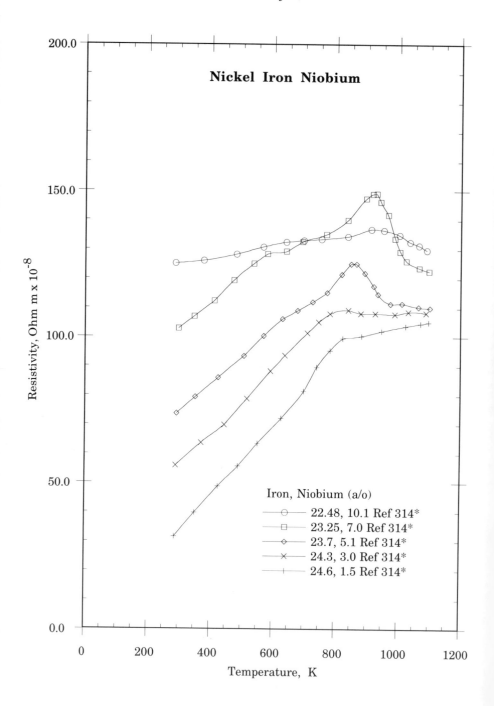

Nickel Iron Niobium

Resistivity, Ohm m x 10⁻⁸

Temperature, K

Iron, Niobium (a/o)
22.48, 10.1 Ref 314*
23.25, 7.0 Ref 314*
23.7, 5.1 Ref 314*
24.3, 3.0 Ref 314*
24.6, 1.5 Ref 314*

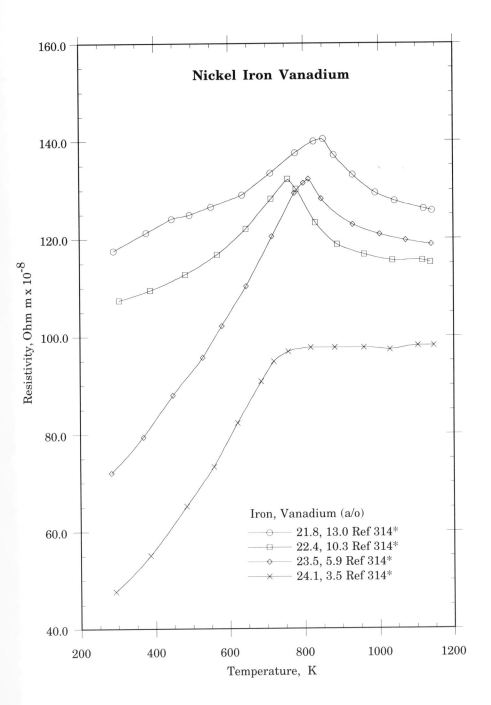

Nickel Iron Vanadium

Resistivity, Ohm m x 10^{-8}

Temperature, K

Iron, Vanadium (a/o)
- ⊙— 21.8, 13.0 Ref 314*
- ☐— 22.4, 10.3 Ref 314*
- ◇— 23.5, 5.9 Ref 314*
- ✕— 24.1, 3.5 Ref 314*

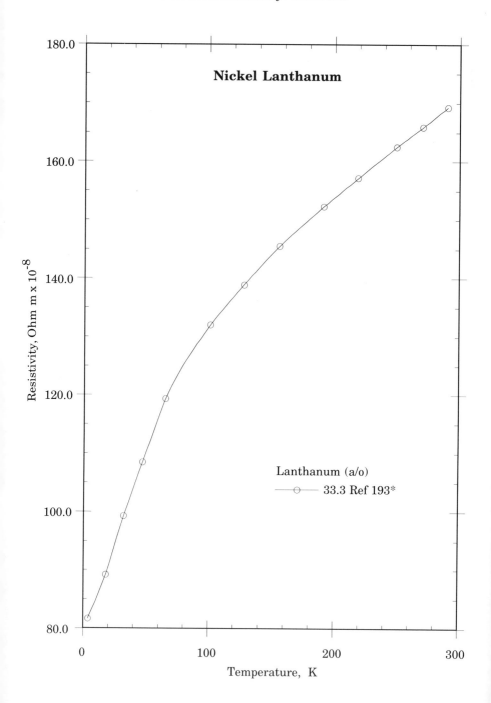

Nickel Lanthanum

Resistivity, Ohm m x 10^{-8}

Temperature, K

Lanthanum (a/o)
33.3 Ref 193*

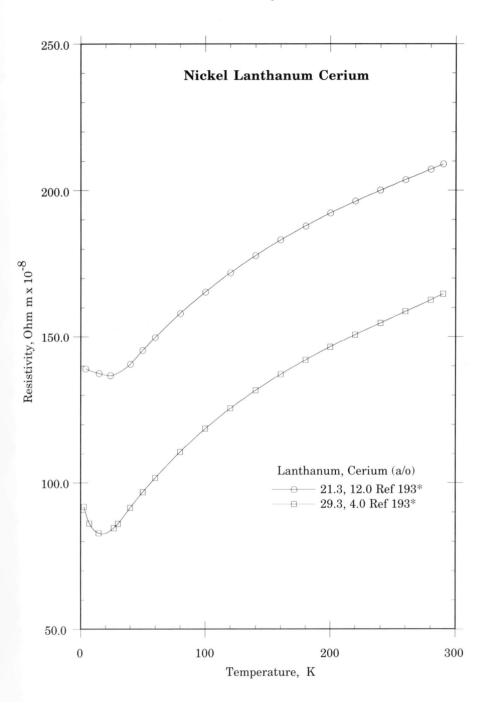

Nickel Lanthanum Cerium

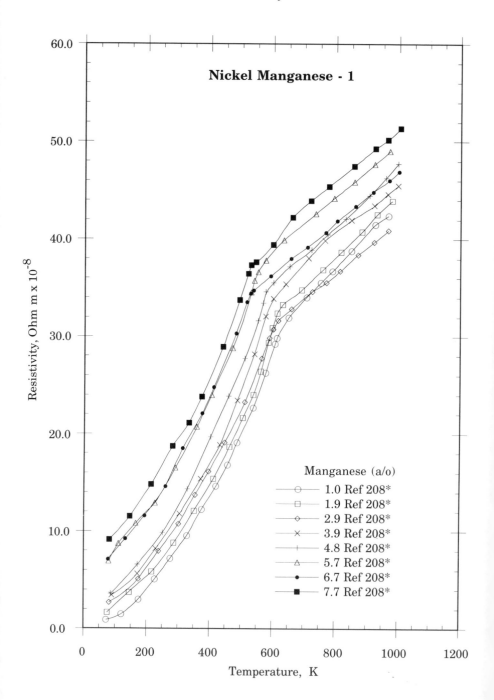

Nickel Manganese - 1

Resistivity, Ohm m x 10^{-8}

Temperature, K

Manganese (a/o)
- 1.0 Ref 208*
- 1.9 Ref 208*
- 2.9 Ref 208*
- 3.9 Ref 208*
- 4.8 Ref 208*
- 5.7 Ref 208*
- 6.7 Ref 208*
- 7.7 Ref 208*

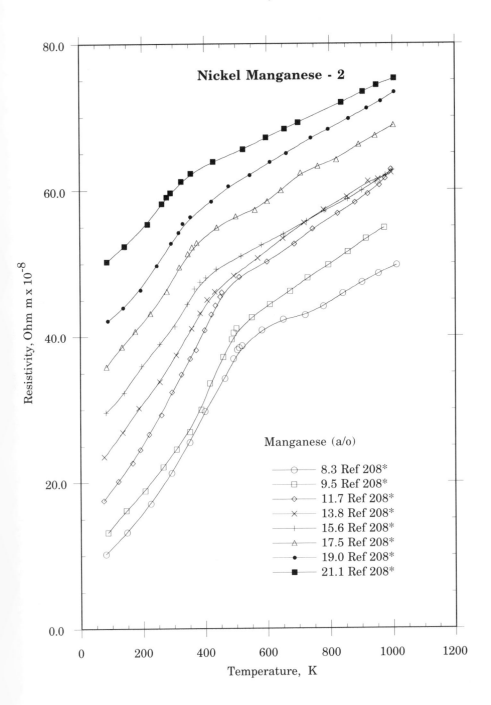

Nickel Manganese - 2

Manganese (a/o)

○	8.3 Ref 208*
□	9.5 Ref 208*
◇	11.7 Ref 208*
×	13.8 Ref 208*
+	15.6 Ref 208*
△	17.5 Ref 208*
●	19.0 Ref 208*
■	21.1 Ref 208*

Resistivity, Ohm m x 10^{-8}

Temperature, K

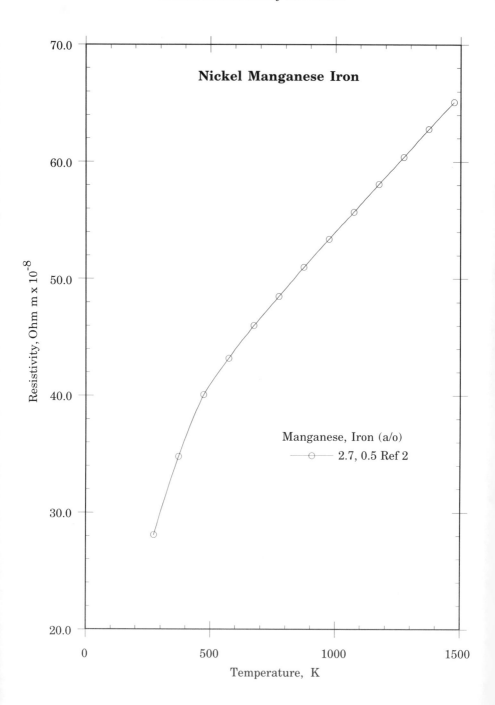

Nickel Manganese Iron

Resistivity, Ohm m x 10⁻⁸

Temperature, K

Manganese, Iron (a/o)
2.7, 0.5 Ref 2

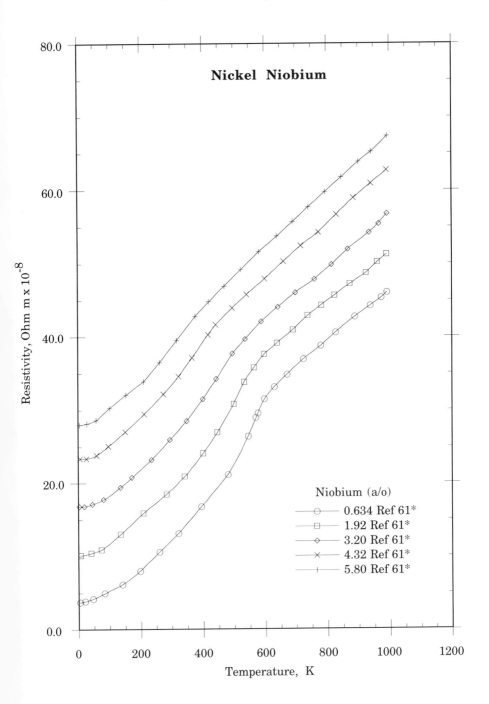

Nickel Niobium

Resistivity, Ohm m x 10^{-8}

Temperature, K

Niobium (a/o)

— ⊙ — 0.634 Ref 61*
— □ — 1.92 Ref 61*
— ◇ — 3.20 Ref 61*
— × — 4.32 Ref 61*
— + — 5.80 Ref 61*

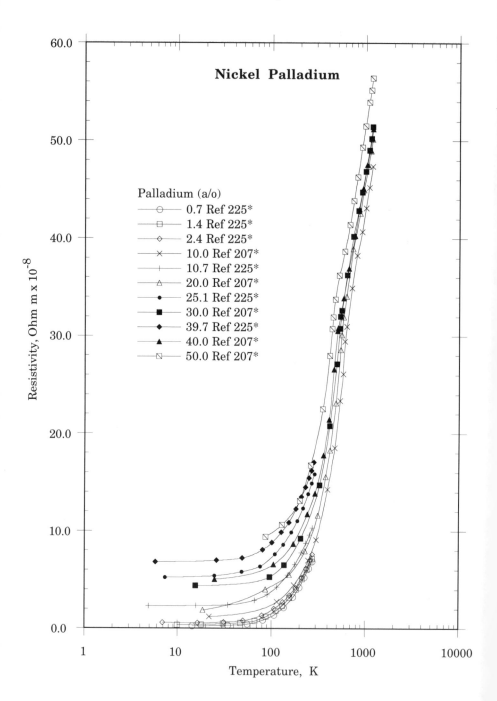

Nickel Palladium

Palladium (a/o)
- ───○─── 0.7 Ref 225*
- ───□─── 1.4 Ref 225*
- ───◇─── 2.4 Ref 225*
- ───×─── 10.0 Ref 207*
- ───+─── 10.7 Ref 225*
- ───△─── 20.0 Ref 207*
- ───•─── 25.1 Ref 225*
- ───■─── 30.0 Ref 207*
- ───◆─── 39.7 Ref 225*
- ───▲─── 40.0 Ref 207*
- ───◁─── 50.0 Ref 207*

Resistivity, Ohm m x 10^{-8}

Temperature, K

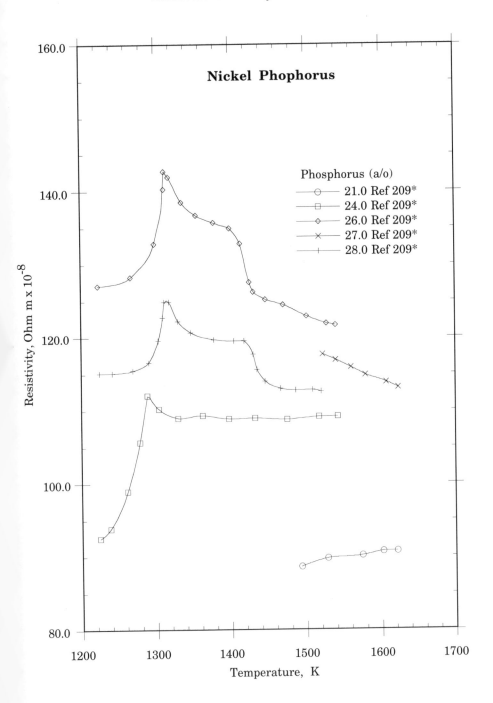

Nickel Phophorus

Phosphorus (a/o)
- ──⊖── 21.0 Ref 209*
- ──□── 24.0 Ref 209*
- ──◇── 26.0 Ref 209*
- ──✕── 27.0 Ref 209*
- ──+── 28.0 Ref 209*

Resistivity, Ohm m x 10⁻⁸

Temperature, K

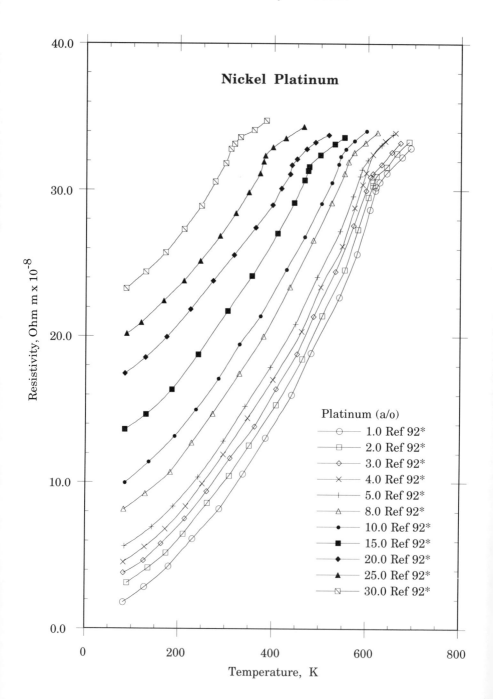

Nickel Platinum

Platinum (a/o)
- —⊙— 1.0 Ref 92*
- —□— 2.0 Ref 92*
- —◇— 3.0 Ref 92*
- —×— 4.0 Ref 92*
- —+— 5.0 Ref 92*
- —△— 8.0 Ref 92*
- —•— 10.0 Ref 92*
- —■— 15.0 Ref 92*
- —◆— 20.0 Ref 92*
- —▲— 25.0 Ref 92*
- —◹— 30.0 Ref 92*

Resistivity, Ohm m x 10^{-8}

Temperature, K

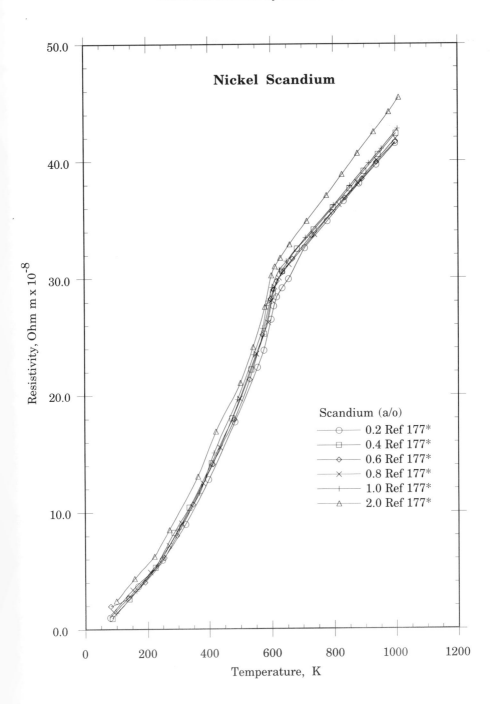

Nickel Scandium

Scandium (a/o)
- ─○─ 0.2 Ref 177*
- ─□─ 0.4 Ref 177*
- ─◇─ 0.6 Ref 177*
- ─×─ 0.8 Ref 177*
- ─+─ 1.0 Ref 177*
- ─△─ 2.0 Ref 177*

Resistivity, Ohm m x 10^{-8}

Temperature, K

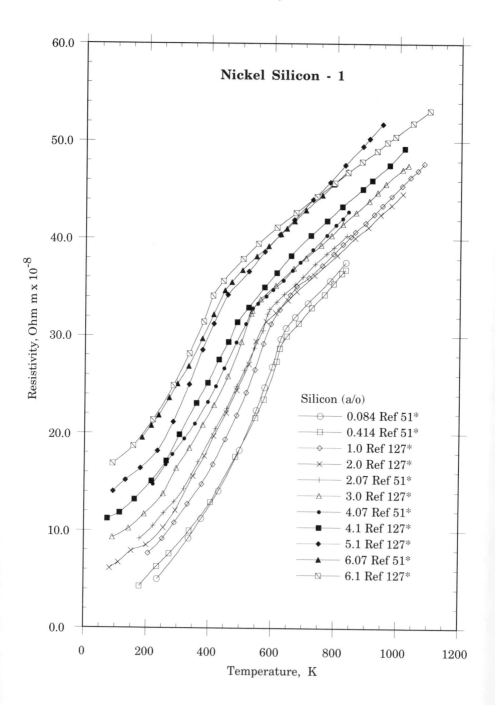

Nickel Silicon - 1

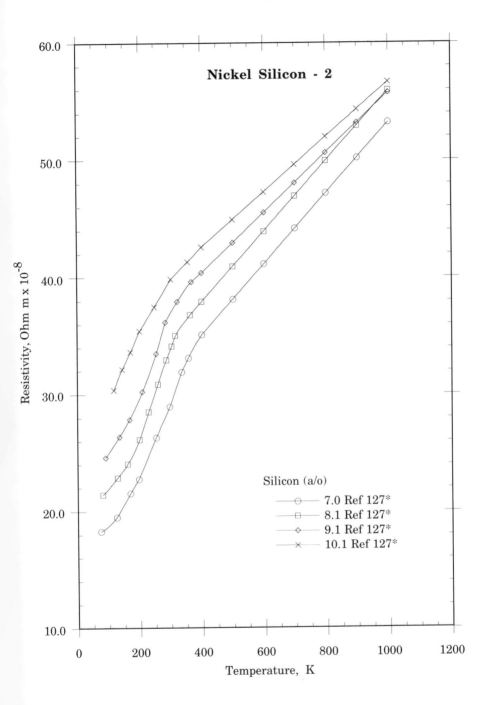

Nickel Silicon - 2

Silicon (a/o)

⎯⎯◯⎯⎯ 7.0 Ref 127*
⎯⎯□⎯⎯ 8.1 Ref 127*
⎯⎯◇⎯⎯ 9.1 Ref 127*
⎯⎯✕⎯⎯ 10.1 Ref 127*

Resistivity, Ohm m x 10⁻⁸

Temperature, K

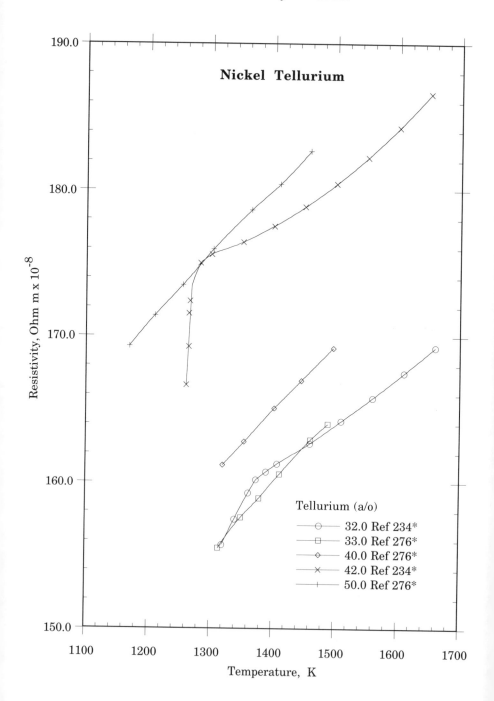

Nickel Tellurium

Tellurium (a/o)

- ─○─ 32.0 Ref 234*
- ─□─ 33.0 Ref 276*
- ─◇─ 40.0 Ref 276*
- ─×─ 42.0 Ref 234*
- ─+─ 50.0 Ref 276*

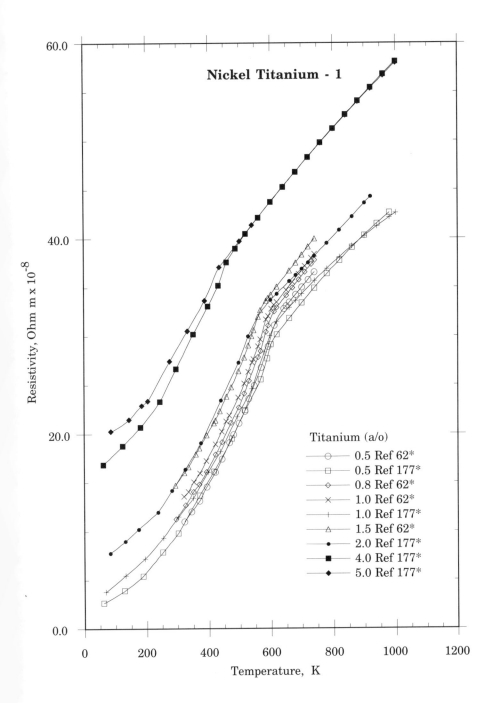

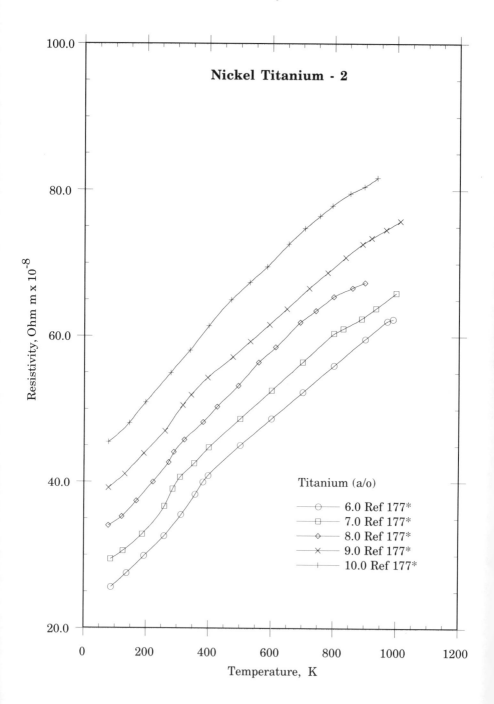

Nickel Titanium - 2

Resistivity, Ohm m x 10^{-8}

Temperature, K

Titanium (a/o)

- —○— 6.0 Ref 177*
- —□— 7.0 Ref 177*
- —◇— 8.0 Ref 177*
- —×— 9.0 Ref 177*
- —+— 10.0 Ref 177*

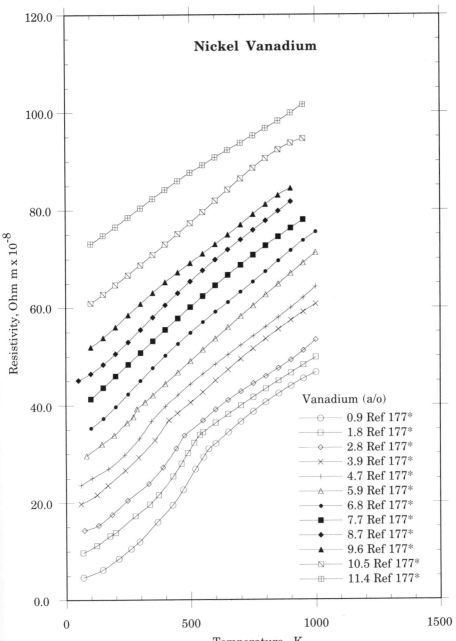

Nickel Vanadium

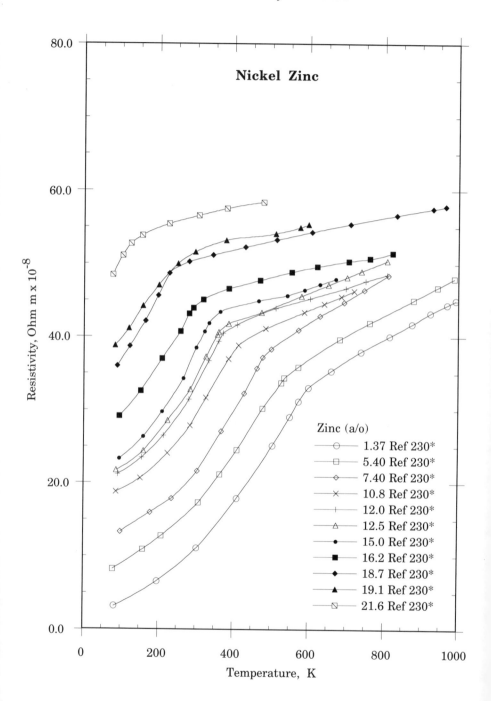

Nickel Zinc

Resistivity, Ohm m x 10^{-8}

Temperature, K

Zinc (a/o)
- ⊖ 1.37 Ref 230*
- ⊟ 5.40 Ref 230*
- ⬦ 7.40 Ref 230*
- ✕ 10.8 Ref 230*
- + 12.0 Ref 230*
- △ 12.5 Ref 230*
- ● 15.0 Ref 230*
- ■ 16.2 Ref 230*
- ◆ 18.7 Ref 230*
- ▲ 19.1 Ref 230*
- ⬠ 21.6 Ref 230*

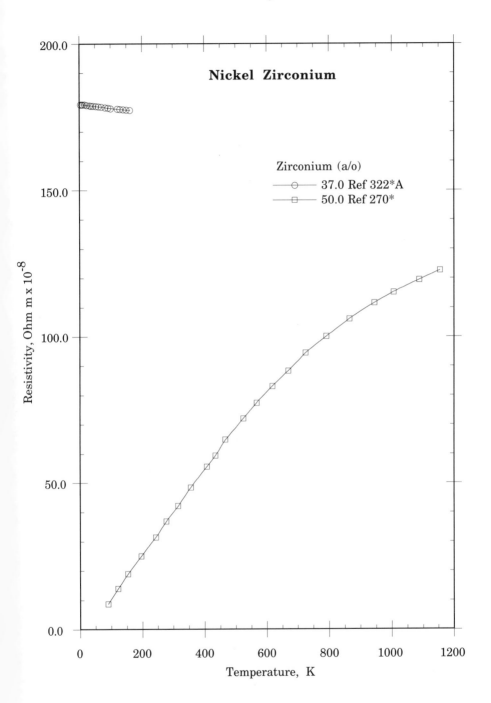

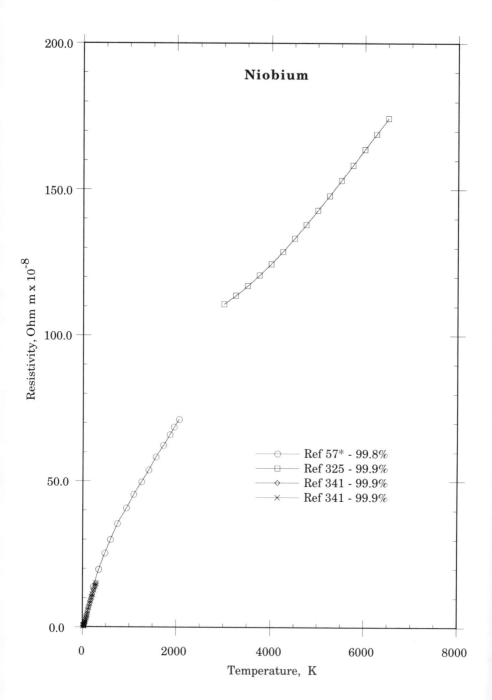

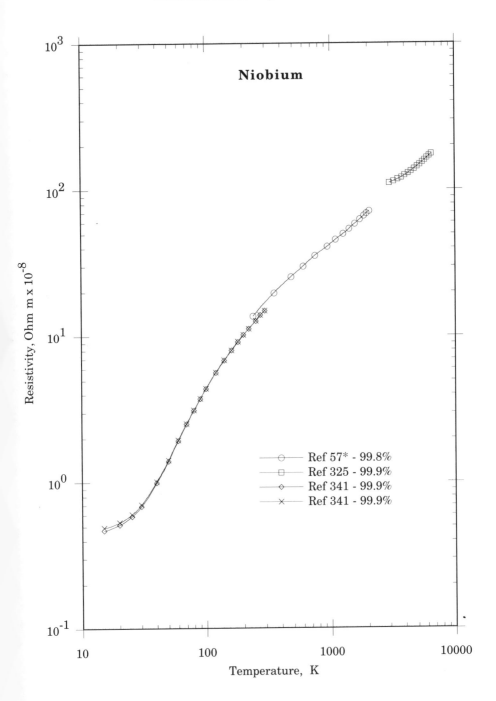

Niobium

Ref 57* - 99.8%
Ref 325 - 99.9%
Ref 341 - 99.9%
Ref 341 - 99.9%

Resistivity, Ohm m x 10^{-8}

Temperature, K

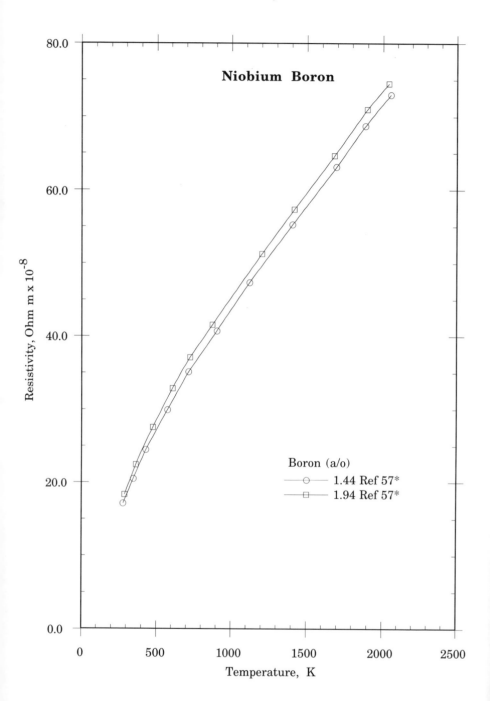

Niobium Boron

Boron (a/o)
○ 1.44 Ref 57*
□ 1.94 Ref 57*

Resistivity, Ohm m x 10^{-8}

Temperature, K

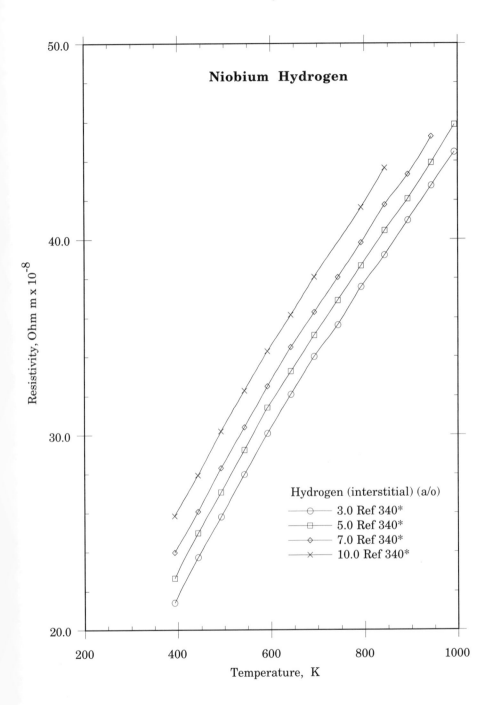

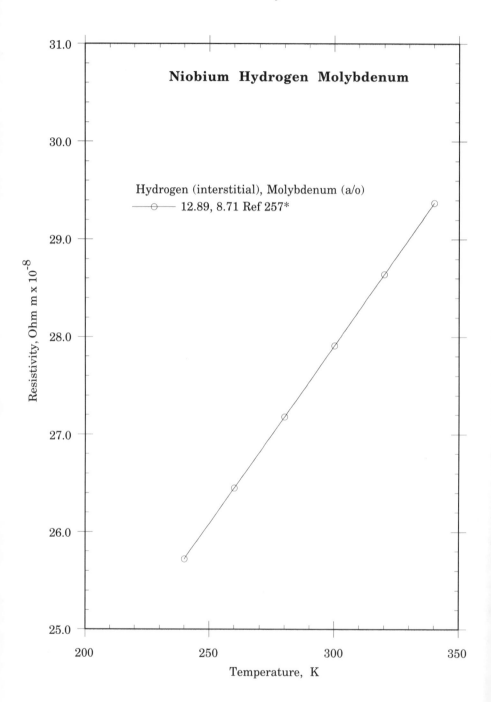

Niobium Hydrogen Molybdenum

Hydrogen (interstitial), Molybdenum (a/o)
—o— 12.89, 8.71 Ref 257*

Resistivity, Ohm m x 10^{-8}

Temperature, K

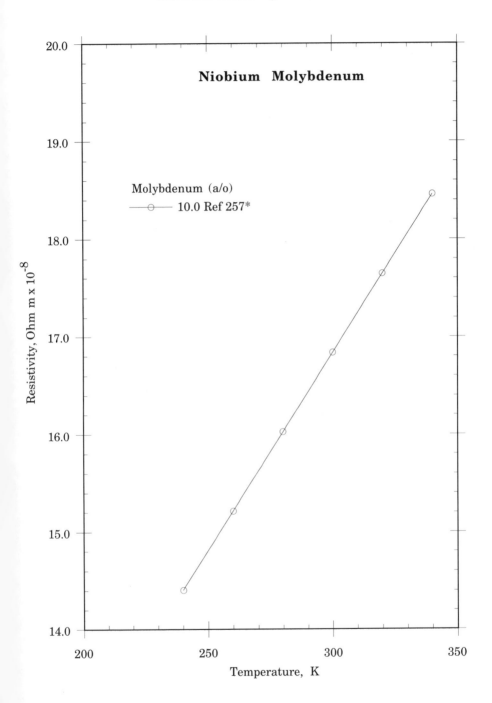

Niobium Molybdenum

Molybdenum (a/o)
———⊖——— 10.0 Ref 257*

Resistivity, Ohm m x 10⁻⁸

Temperature, K

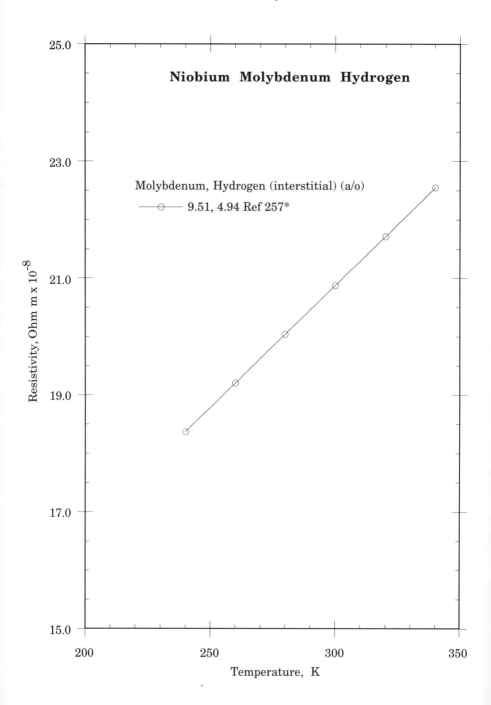

Niobium Molybdenum Hydrogen

Molybdenum, Hydrogen (interstitial) (a/o)

———⊖——— 9.51, 4.94 Ref 257*

Resistivity, Ohm m x 10⁻⁸

Temperature, K

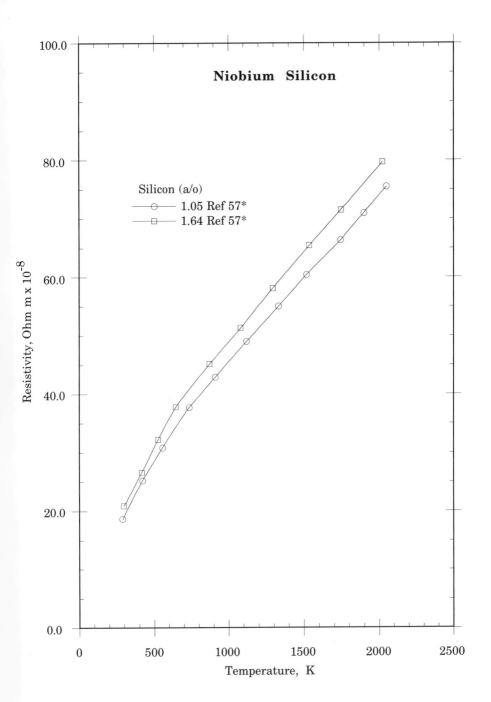

Niobium   Silicon

Silicon (a/o)
- ⊖— 1.05 Ref 57*
- ⊟— 1.64 Ref 57*

Resistivity, Ohm m x 10^{-8}

Temperature, K

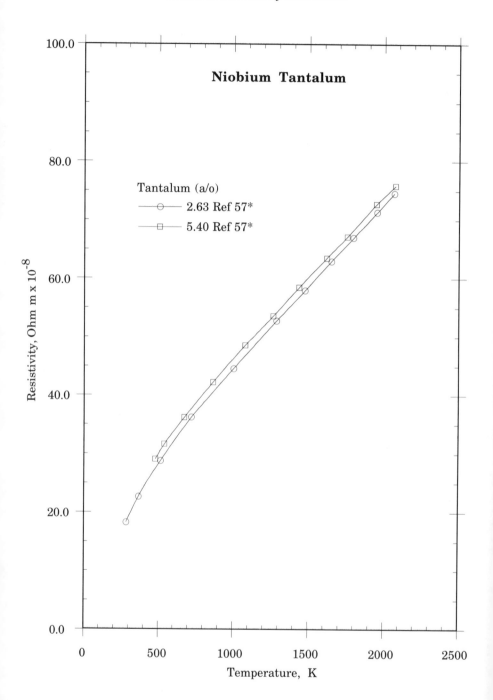

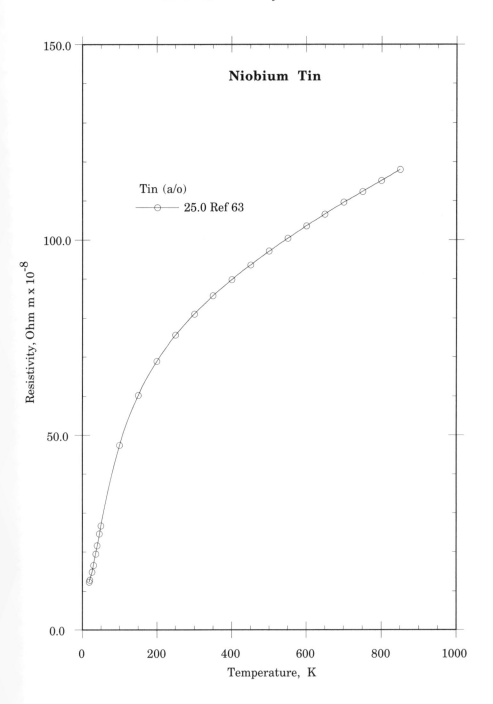

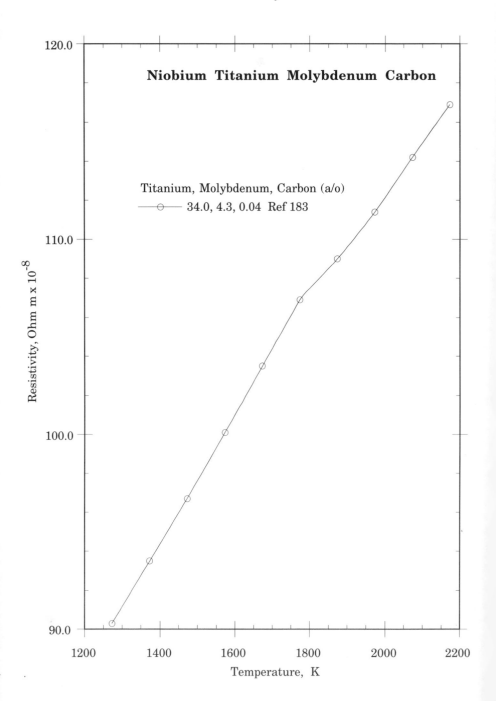

Niobium Titanium Molybdenum Carbon

Titanium, Molybdenum, Carbon (a/o)
—⊖— 34.0, 4.3, 0.04 Ref 183

Resistivity, Ohm m x 10^{-8}

Temperature, K

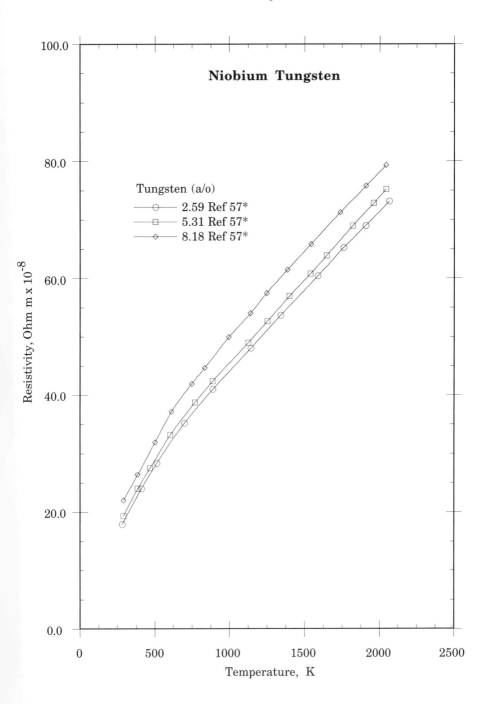

Niobium Tungsten

Tungsten (a/o)
- ──○── 2.59 Ref 57*
- ──□── 5.31 Ref 57*
- ──◇── 8.18 Ref 57*

Resistivity, Ohm m x 10^{-8}

Temperature, K

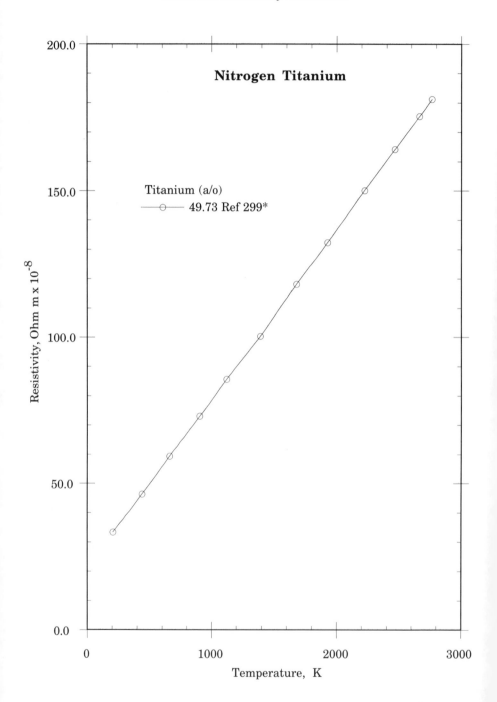

Nitrogen Titanium

Titanium (a/o)
—⊖— 49.73 Ref 299*

Resistivity, Ohm m x 10^{-8}

Temperature, K

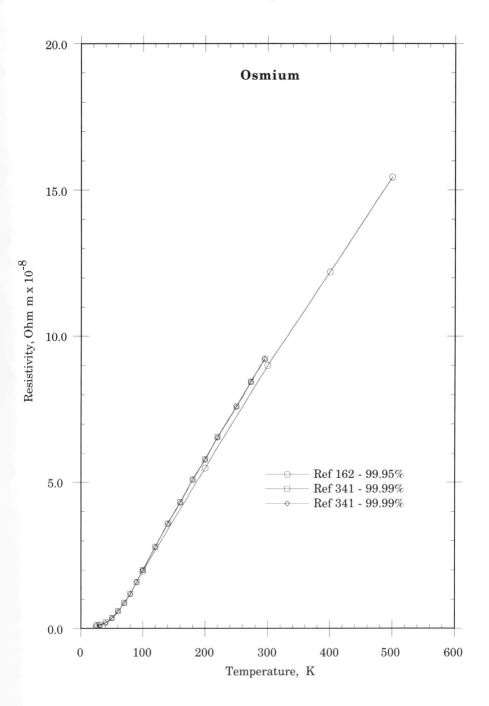

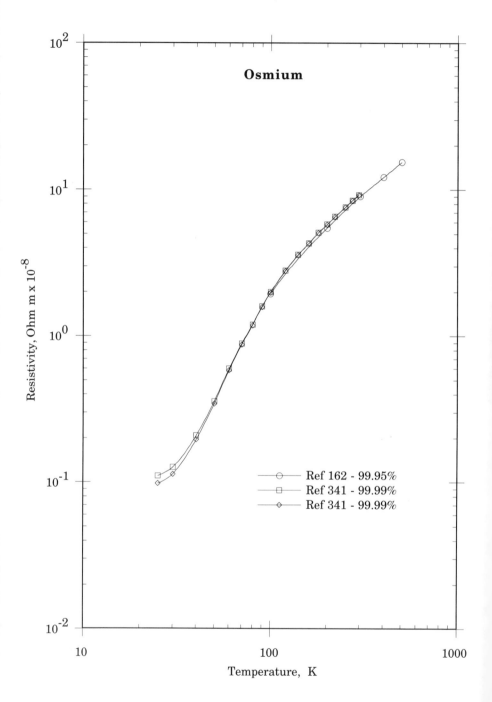

Osmium

Resistivity, Ohm m x 10⁻⁸

Temperature, K

Ref 162 - 99.95%
Ref 341 - 99.99%
Ref 341 - 99.99%

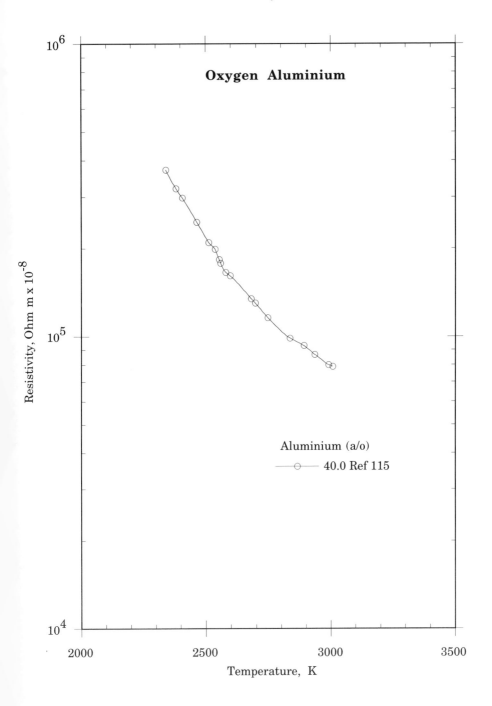

Oxygen Aluminium

Resistivity, Ohm m x 10^{-8}

Temperature, K

Aluminium (a/o)
—⊙— 40.0 Ref 115

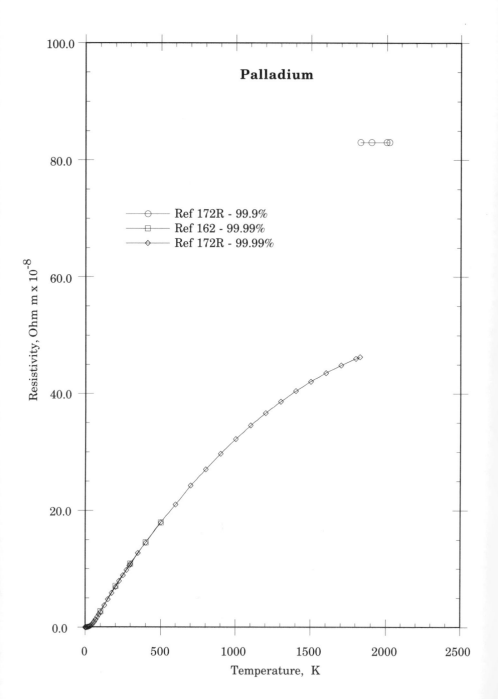

Palladium

Ref 172R - 99.9%
Ref 162 - 99.99%
Ref 172R - 99.99%

Resistivity, Ohm m x 10^{-8}

Temperature, K

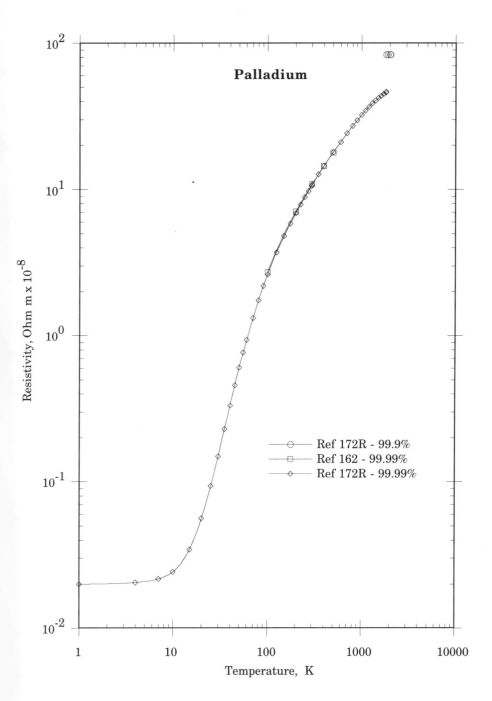

Palladium

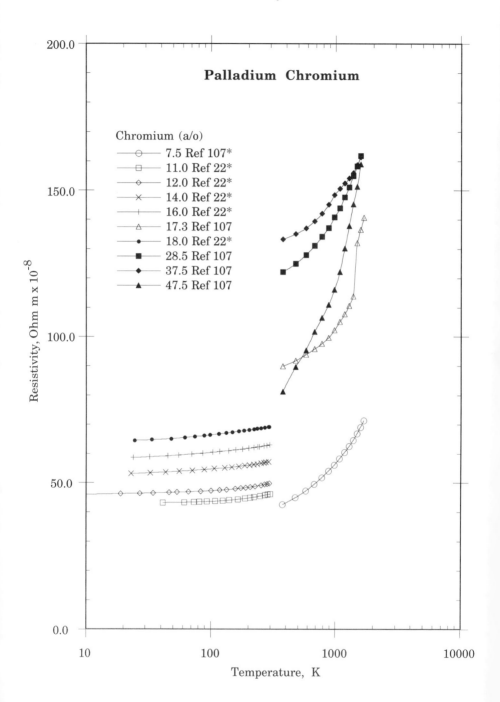

Palladium Chromium

Chromium (a/o)
- ⊖ 7.5 Ref 107*
- ☐ 11.0 Ref 22*
- ◇ 12.0 Ref 22*
- × 14.0 Ref 22*
- + 16.0 Ref 22*
- △ 17.3 Ref 107
- ● 18.0 Ref 22*
- ■ 28.5 Ref 107
- ◆ 37.5 Ref 107
- ▲ 47.5 Ref 107

Resistivity, Ohm m x 10^{-8}

Temperature, K

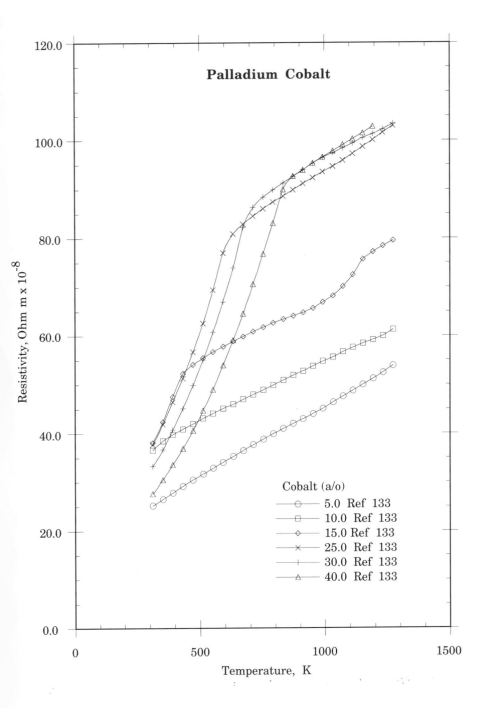

Palladium Cobalt

Resistivity, Ohm m x 10⁻⁸

Temperature, K

Cobalt (a/o)
- ⊙ 5.0 Ref 133
- ☐ 10.0 Ref 133
- ◇ 15.0 Ref 133
- ✕ 25.0 Ref 133
- + 30.0 Ref 133
- △ 40.0 Ref 133

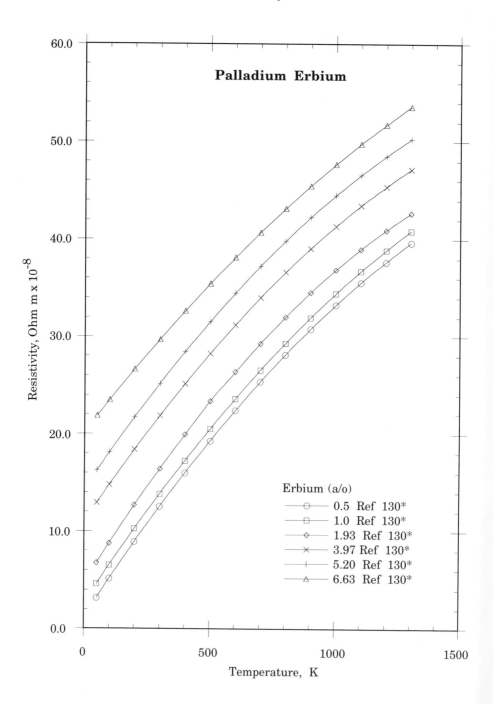

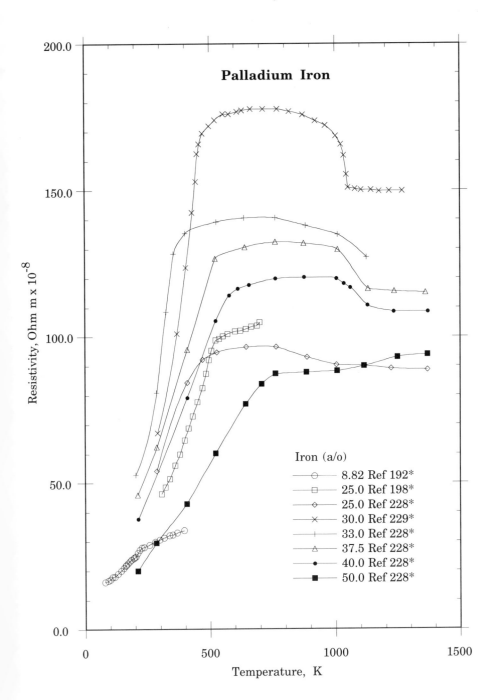

Palladium Iron

Resistivity, Ohm m x 10^{-8}

Temperature, K

Iron (a/o)
- 8.82 Ref 192*
- 25.0 Ref 198*
- 25.0 Ref 228*
- 30.0 Ref 229*
- 33.0 Ref 228*
- 37.5 Ref 228*
- 40.0 Ref 228*
- 50.0 Ref 228*

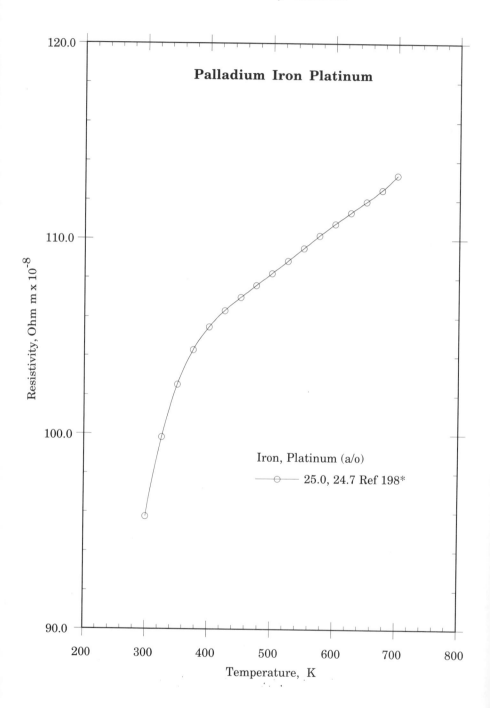

Palladium Iron Platinum

Iron, Platinum (a/o)

25.0, 24.7 Ref 198*

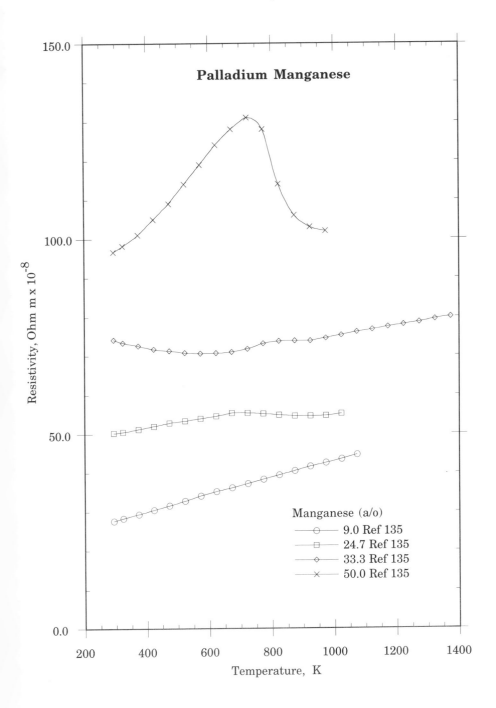

Palladium Manganese

Resistivity, Ohm m x 10^{-8}

Temperature, K

Manganese (a/o)
- —○— 9.0 Ref 135
- —□— 24.7 Ref 135
- —◇— 33.3 Ref 135
- —×— 50.0 Ref 135

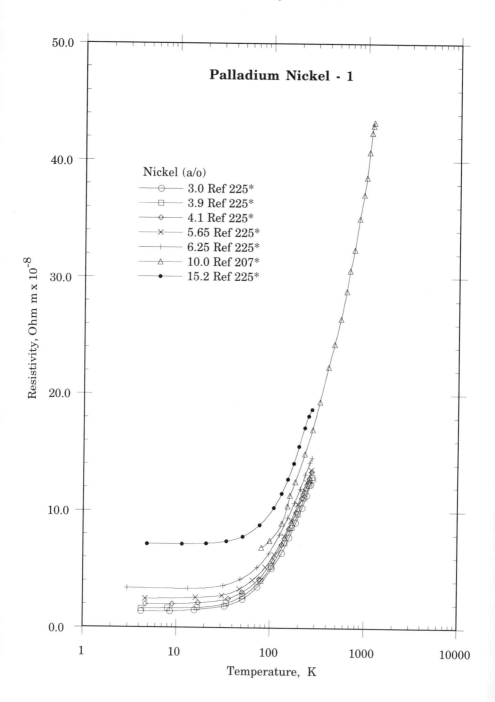

Palladium Nickel - 1

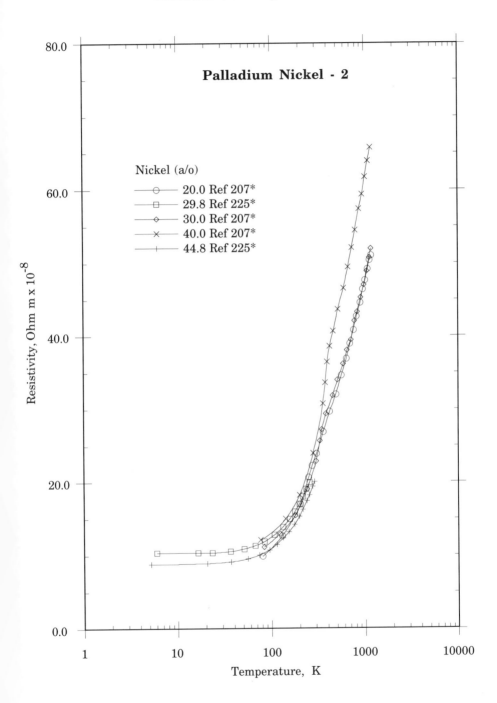

Palladium Nickel - 2

Nickel (a/o)

⊖ 20.0 Ref 207*
□ 29.8 Ref 225*
◇ 30.0 Ref 207*
✕ 40.0 Ref 207*
+ 44.8 Ref 225*

Resistivity, Ohm m x 10⁻⁸

Temperature, K

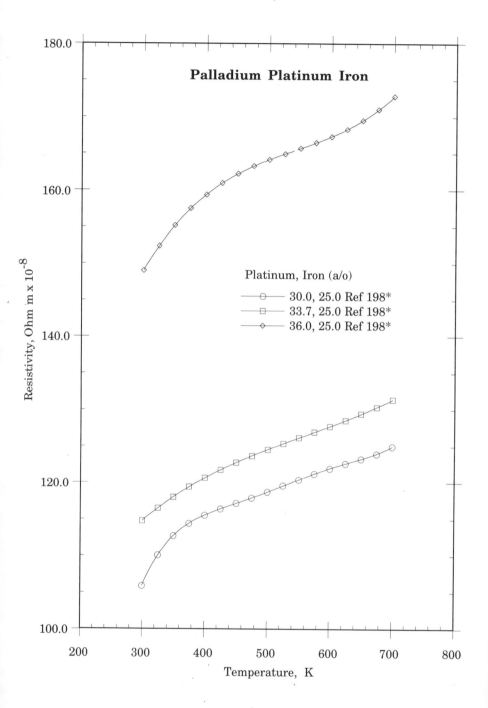

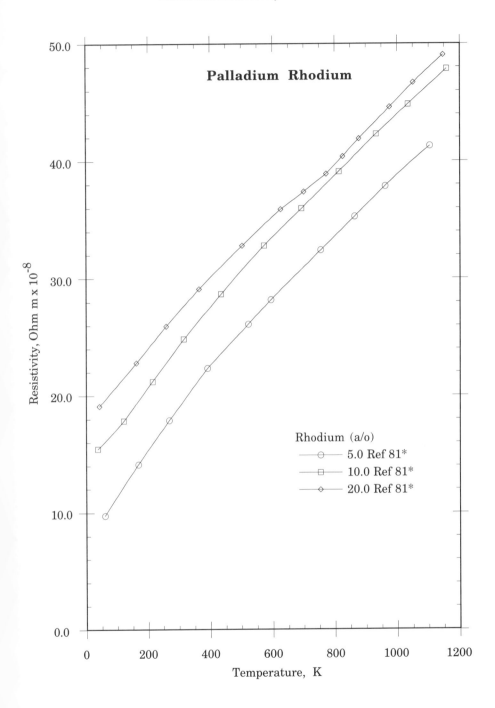

Palladium Rhodium

Rhodium (a/o)
— ○ — 5.0 Ref 81*
— □ — 10.0 Ref 81*
— ◇ — 20.0 Ref 81*

Resistivity, Ohm m x 10⁻⁸

Temperature, K

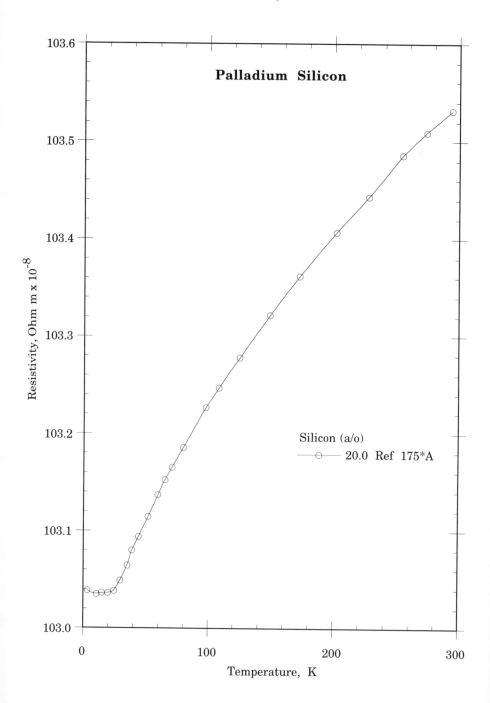

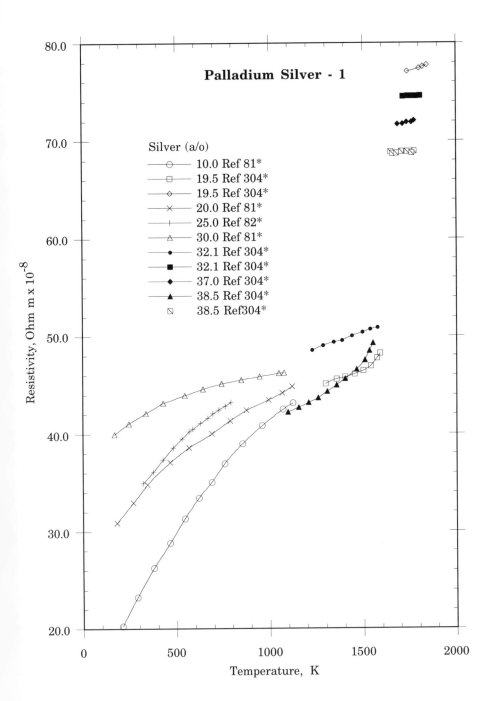

Palladium Silver - 1

Silver (a/o)
— ○ — 10.0 Ref 81*
— □ — 19.5 Ref 304*
— ◇ — 19.5 Ref 304*
— × — 20.0 Ref 81*
— + — 25.0 Ref 82*
— △ — 30.0 Ref 81*
— • — 32.1 Ref 304*
— ■ — 32.1 Ref 304*
— ◆ — 37.0 Ref 304*
— ▲ — 38.5 Ref 304*
    ⊡   38.5 Ref304*

Resistivity, Ohm m x 10⁻⁸

Temperature, K

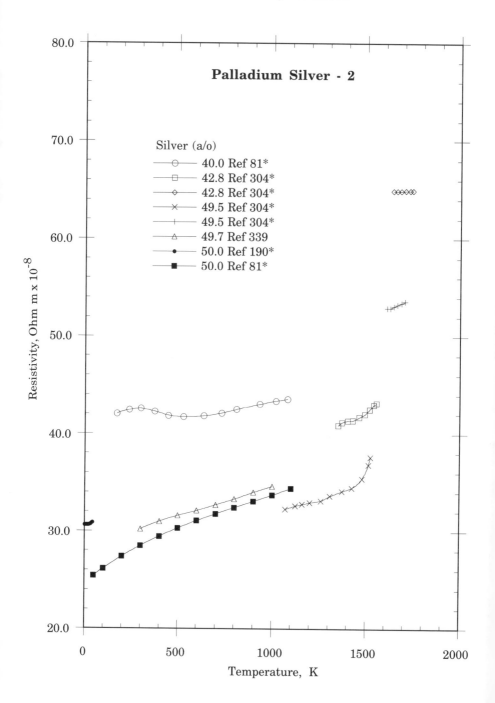

Palladium Silver - 2

Silver (a/o)
- ——⊖—— 40.0 Ref 81*
- ——□—— 42.8 Ref 304*
- ——◇—— 42.8 Ref 304*
- ——×—— 49.5 Ref 304*
- ——+—— 49.5 Ref 304*
- ——△—— 49.7 Ref 339
- ——•—— 50.0 Ref 190*
- ——■—— 50.0 Ref 81*

Resistivity, Ohm m x 10^{-8}

Temperature, K

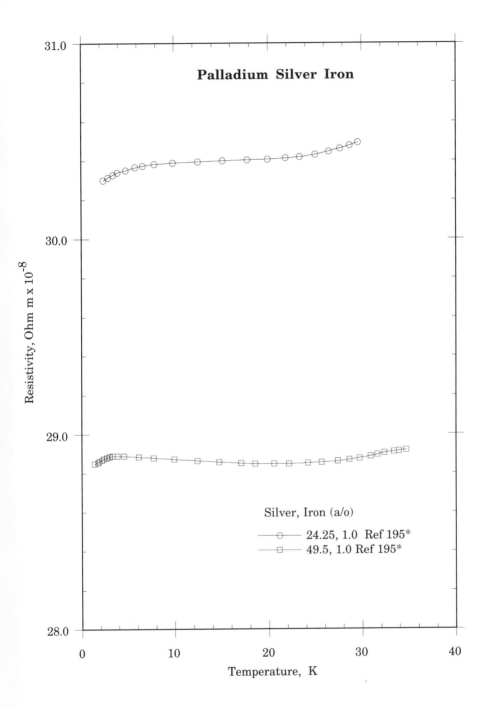

Palladium Silver Iron

Silver, Iron (a/o)

———⊙——— 24.25, 1.0 Ref 195*
———□——— 49.5, 1.0 Ref 195*

Resistivity, Ohm m x 10⁻⁸

Temperature, K

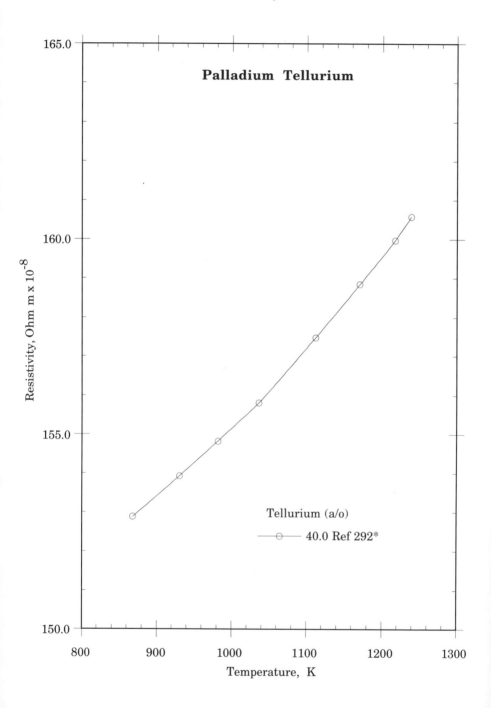

Palladium Tellurium

Resistivity, Ohm m x 10⁻⁸

Temperature, K

Tellurium (a/o)
40.0 Ref 292*

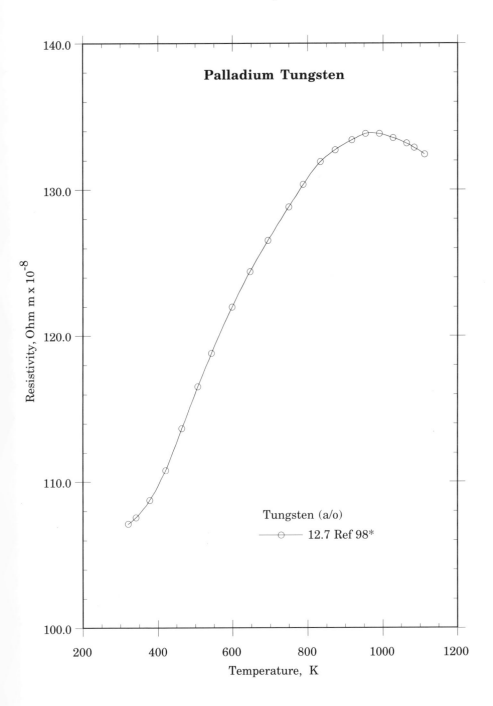

Palladium Tungsten

Resistivity, Ohm m x 10⁻⁸ — Temperature, K

Tungsten (a/o)
—⊖— 12.7 Ref 98*

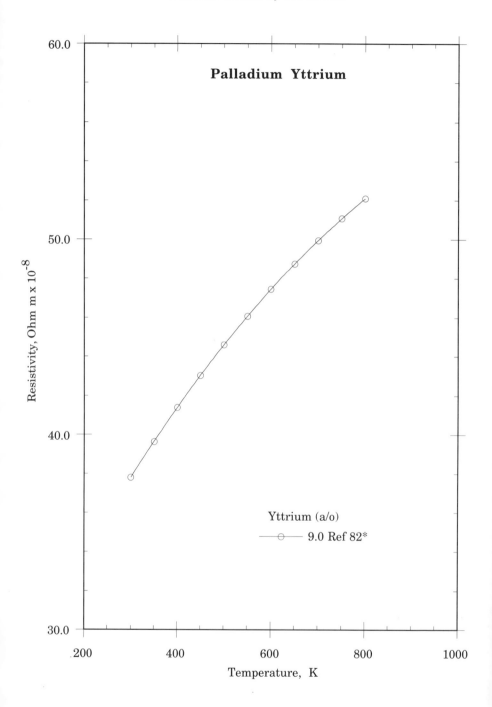

Palladium Yttrium

Resistivity, Ohm m x 10^{-8}

Temperature, K

Yttrium (a/o)
—o— 9.0 Ref 82*

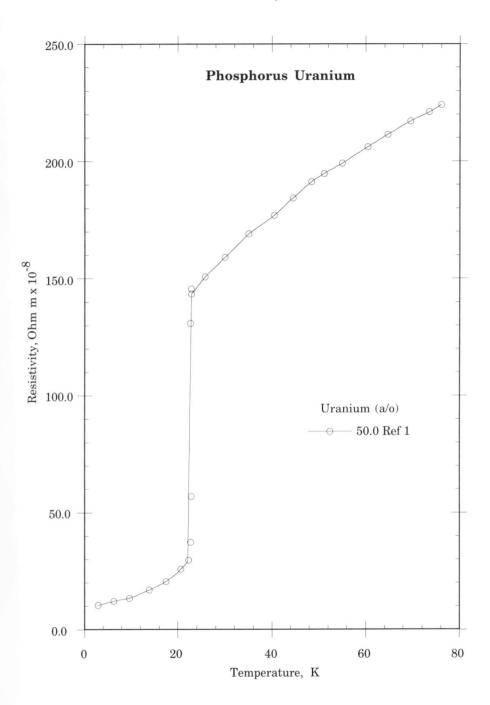

Phosphorus Uranium

Resistivity, Ohm m x 10⁻⁸

Temperature, K

Uranium (a/o)

50.0 Ref 1

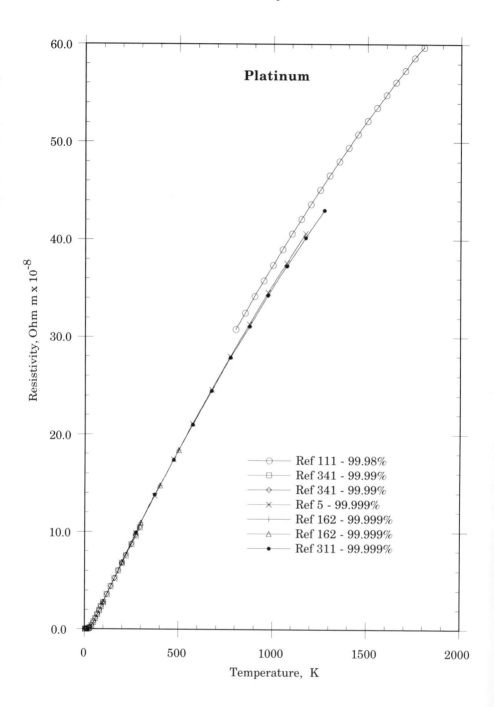

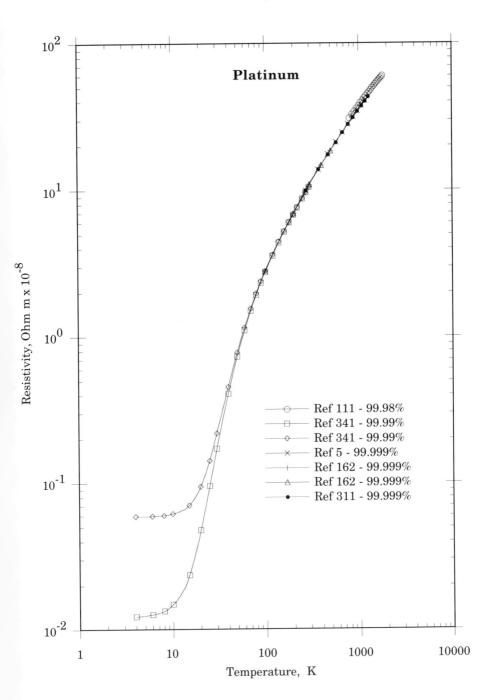

Platinum

Resistivity, Ohm m x 10⁻⁸

Temperature, K

Ref 111 - 99.98%
Ref 341 - 99.99%
Ref 341 - 99.99%
Ref 5 - 99.999%
Ref 162 - 99.999%
Ref 162 - 99.999%
Ref 311 - 99.999%

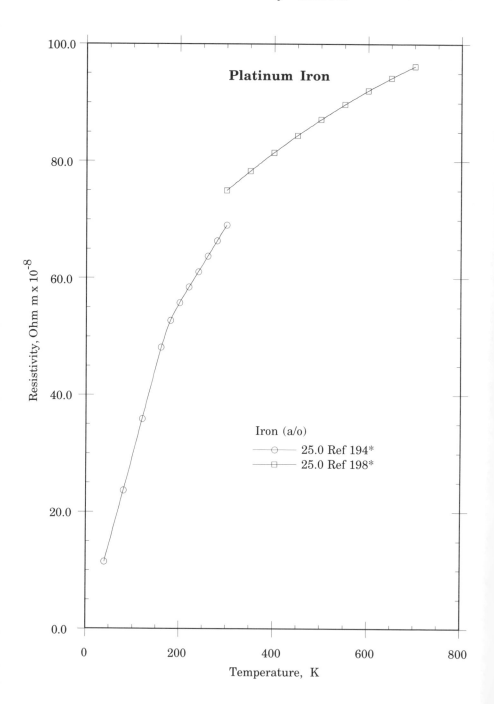

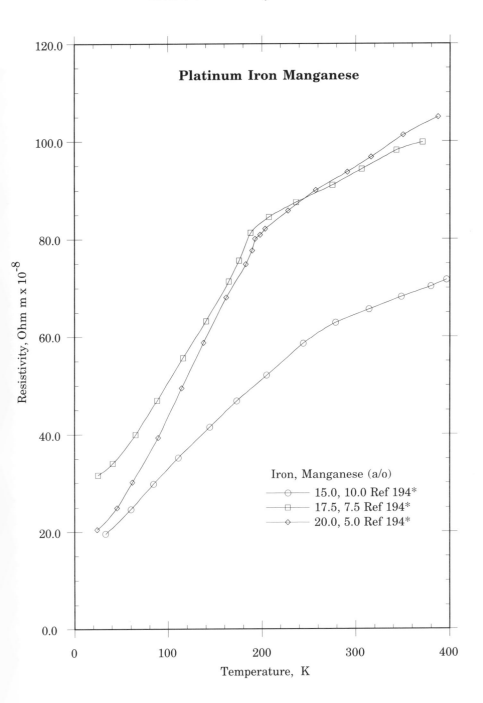

Platinum Iron Manganese

Iron, Manganese (a/o)
—○— 15.0, 10.0 Ref 194*
—□— 17.5, 7.5 Ref 194*
—◇— 20.0, 5.0 Ref 194*

Resistivity, Ohm m x 10^{-8}

Temperature, K

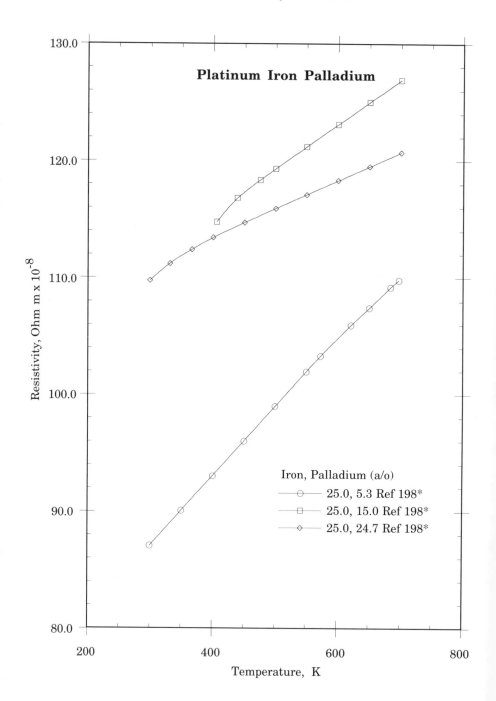

Platinum Iron Palladium

Iron, Palladium (a/o)
— ⊖ — 25.0, 5.3 Ref 198*
— ☐ — 25.0, 15.0 Ref 198*
— ◇ — 25.0, 24.7 Ref 198*

Resistivity, Ohm m x 10^{-8}

Temperature, K

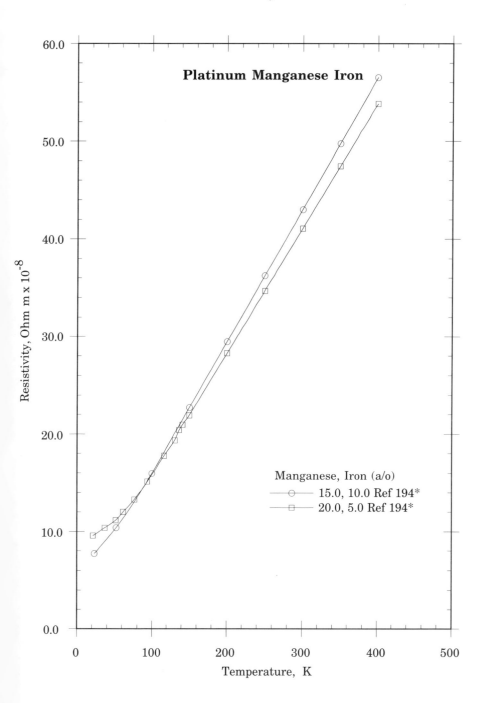

Platinum Manganese Iron

Manganese, Iron (a/o)
—o— 15.0, 10.0 Ref 194*
—□— 20.0, 5.0 Ref 194*

Resistivity, Ohm m x 10⁻⁸

Temperature, K

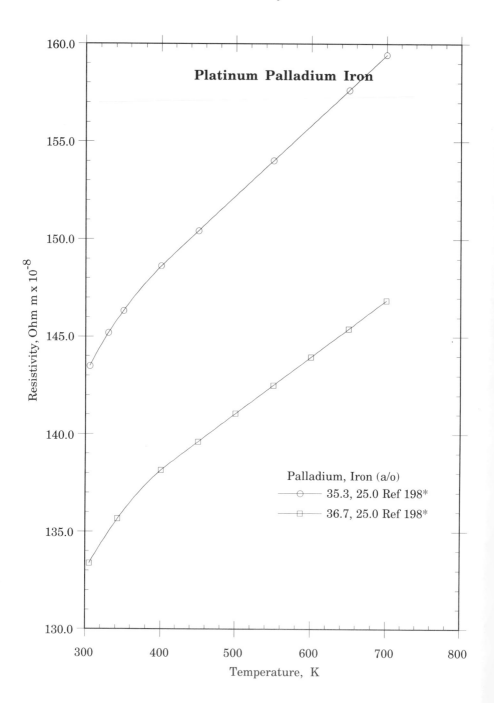

Platinum Palladium Iron

Palladium, Iron (a/o)
- ○ 35.3, 25.0 Ref 198*
- □ 36.7, 25.0 Ref 198*

Resistivity, Ohm m x 10^{-8}

Temperature, K

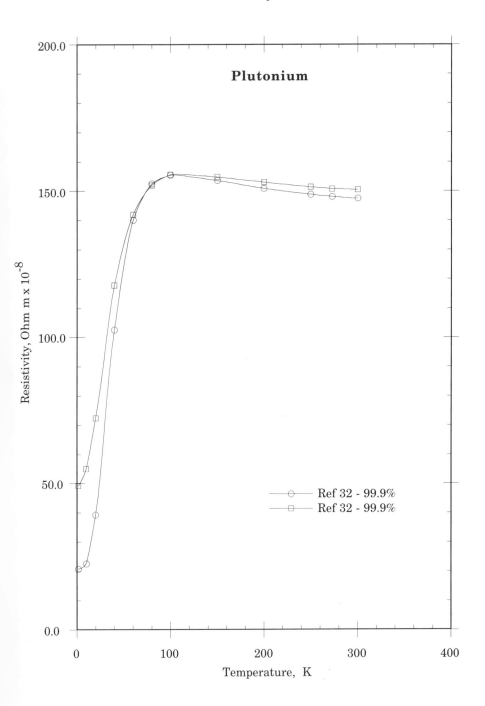

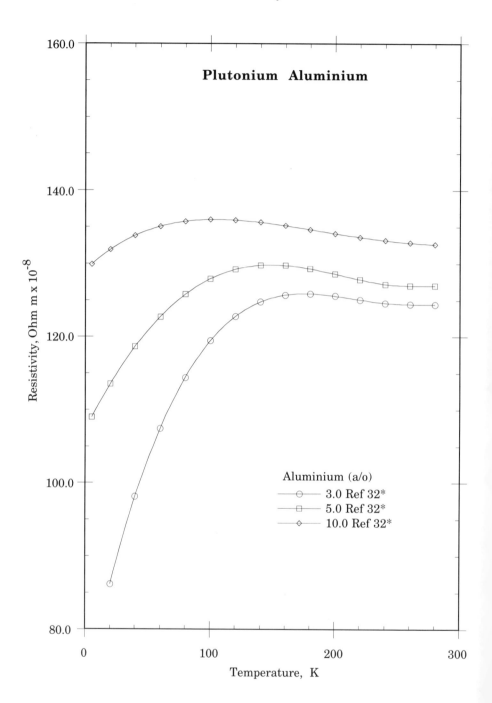

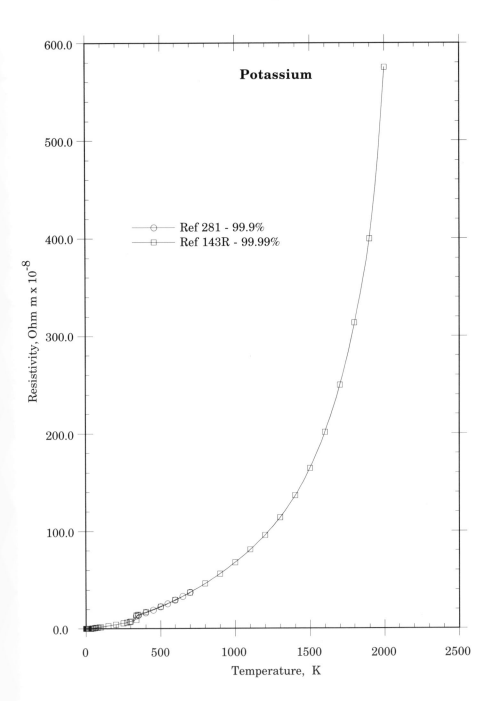

Potassium

Ref 281 - 99.9%
Ref 143R - 99.99%

Resistivity, Ohm m x 10^{-8}

Temperature, K

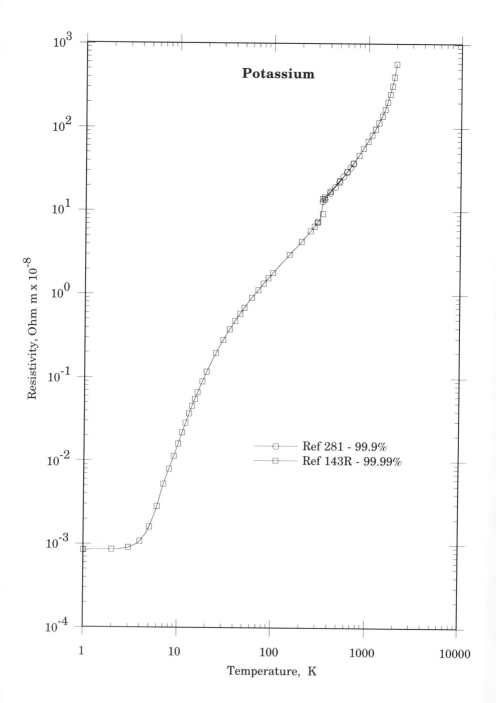

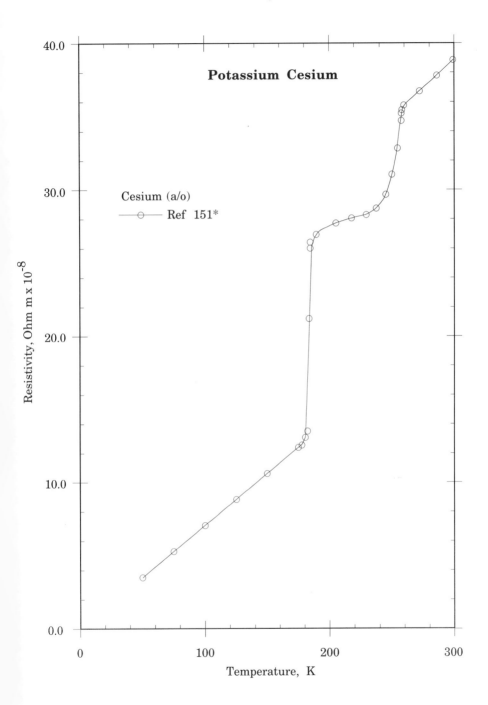

Potassium Cesium

Cesium (a/o)
———⊙——— Ref 151*

Resistivity, Ohm m x 10^{-8}

Temperature, K

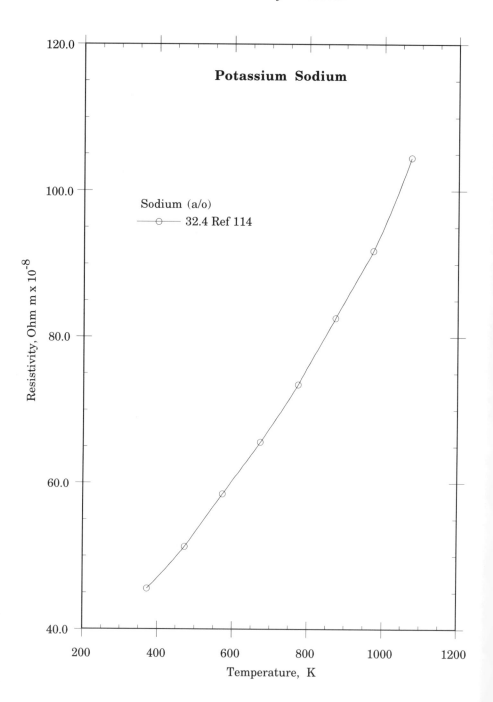

Potassium Sodium

Sodium (a/o)
———o——— 32.4 Ref 114

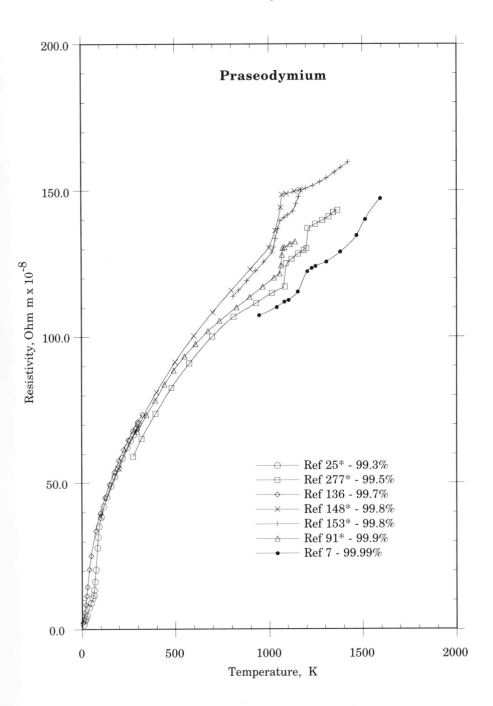

Praseodymium

Resistivity, Ohm m x 10^{-8}

Temperature, K

Ref 25* - 99.3%
Ref 277* - 99.5%
Ref 136 - 99.7%
Ref 148* - 99.8%
Ref 153* - 99.8%
Ref 91* - 99.9%
Ref 7 - 99.99%

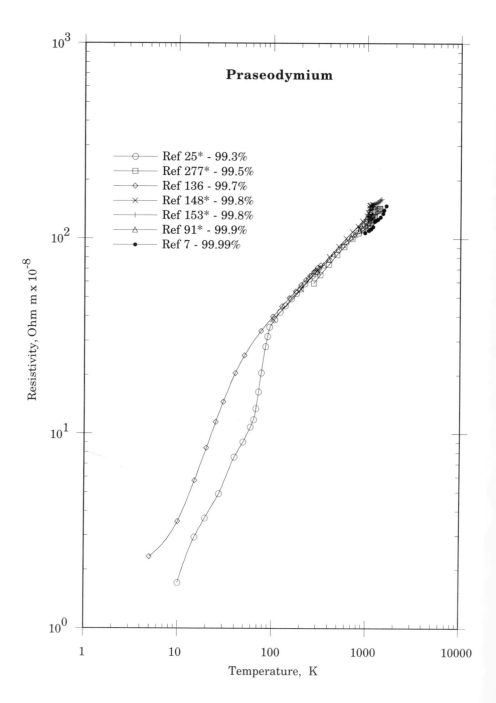

Praseodymium

Ref 25* - 99.3%
Ref 277* - 99.5%
Ref 136 - 99.7%
Ref 148* - 99.8%
Ref 153* - 99.8%
Ref 91* - 99.9%
Ref 7 - 99.99%

Resistivity, Ohm m x 10^{-8}

Temperature, K

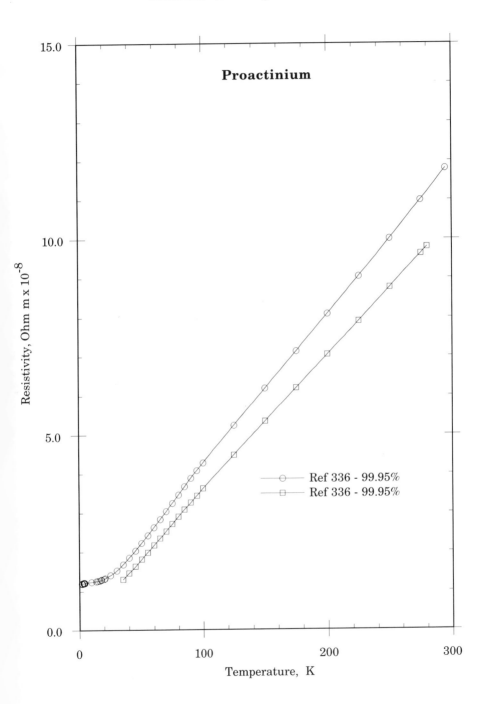

Proactinium

Resistivity, Ohm m x 10^{-8}

Temperature, K

Ref 336 - 99.95%
Ref 336 - 99.95%

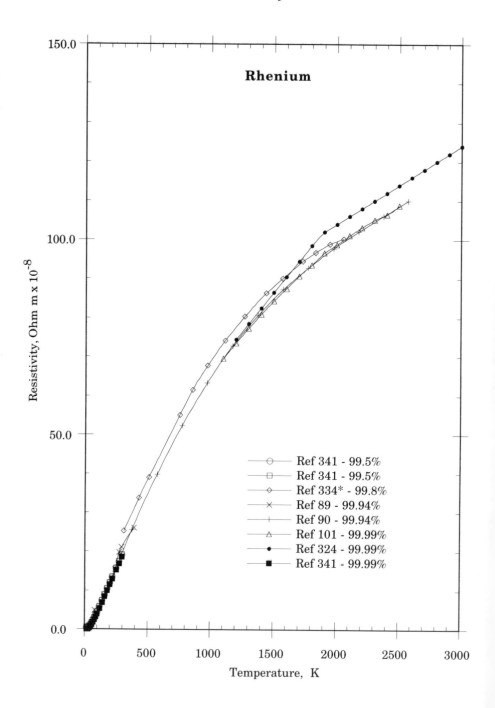

Rhenium

Resistivity, Ohm m x 10^{-8}

Temperature, K

Ref 341 - 99.5%
Ref 341 - 99.5%
Ref 334* - 99.8%
Ref 89 - 99.94%
Ref 90 - 99.94%
Ref 101 - 99.99%
Ref 324 - 99.99%
Ref 341 - 99.99%

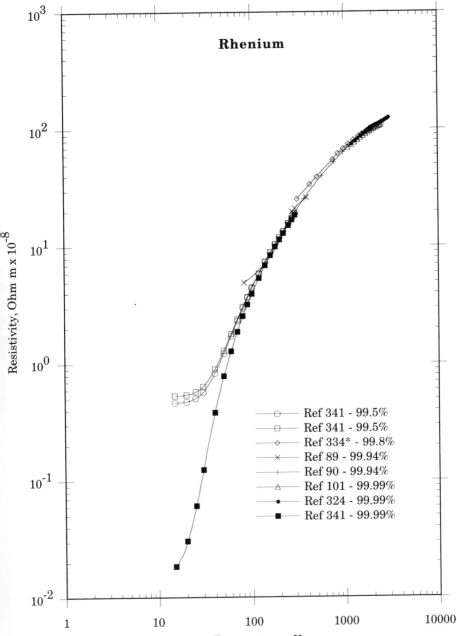

Rhenium

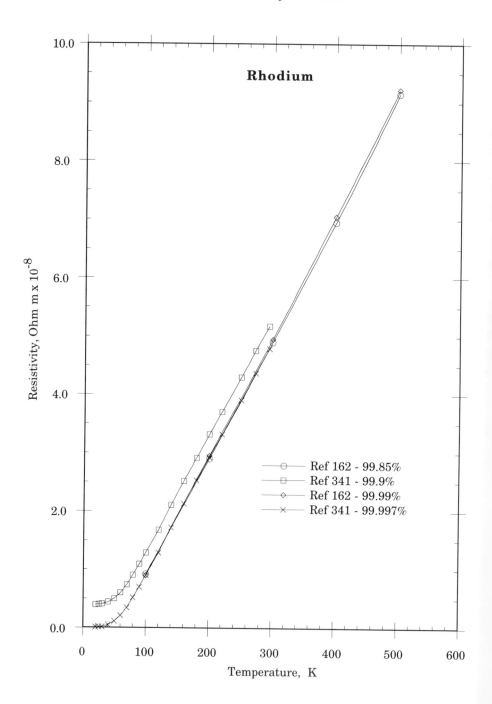

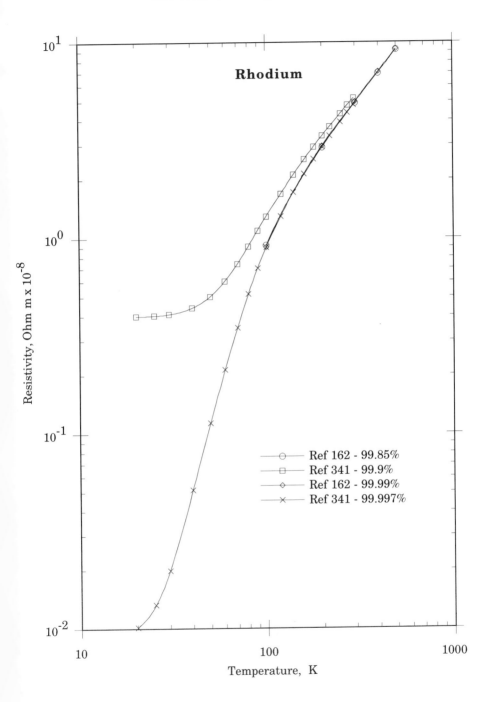

Rhodium

Resistivity, Ohm m x 10⁻⁸

Temperature, K

Ref 162 - 99.85%
Ref 341 - 99.9%
Ref 162 - 99.99%
Ref 341 - 99.997%

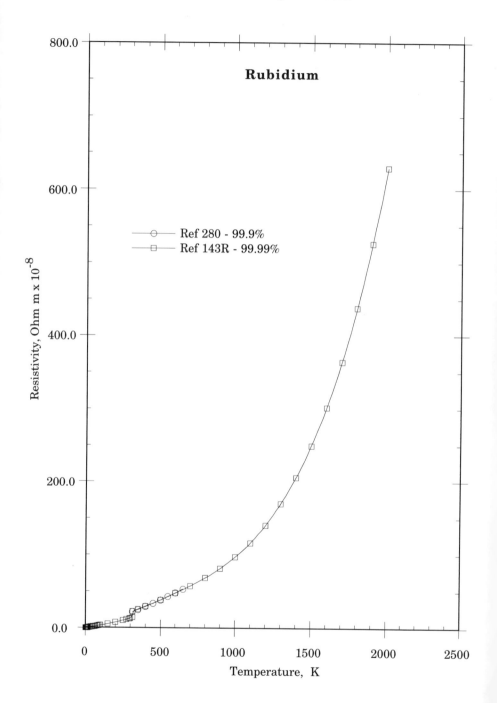

Rubidium

Ref 280 - 99.9%
Ref 143R - 99.99%

Resistivity, Ohm m x 10^{-8}

Temperature, K

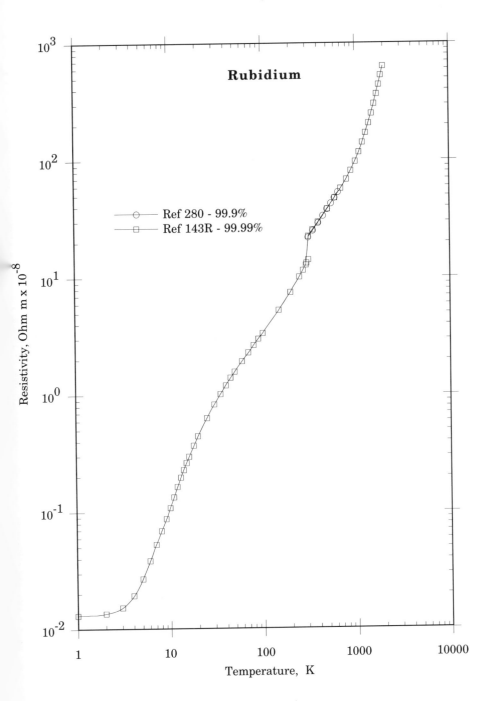

Rubidium

Ref 280 - 99.9%
Ref 143R - 99.99%

Resistivity, Ohm m x 10^{-8}

Temperature, K

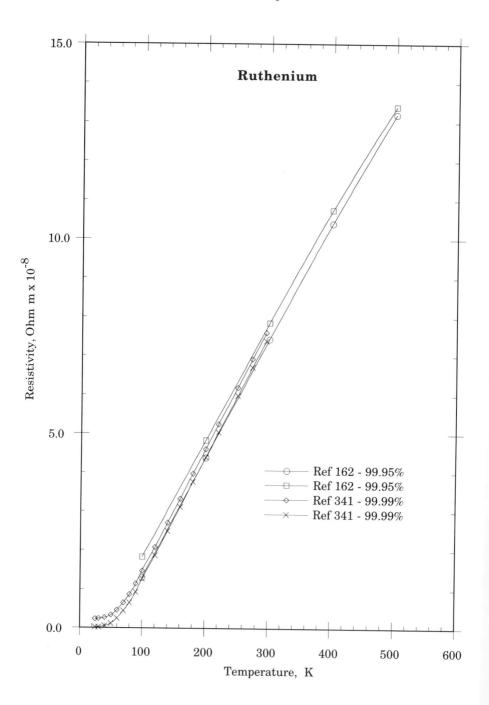

Ruthenium

Resistivity, Ohm m x 10⁻⁸

Temperature, K

Ref 162 - 99.95%
Ref 162 - 99.95%
Ref 341 - 99.99%
Ref 341 - 99.99%

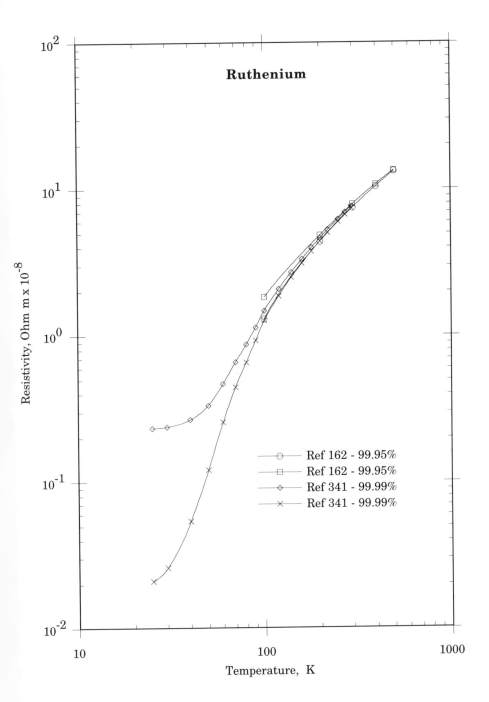

Ruthenium

Resistivity, Ohm m x 10^{-8}

Temperature, K

Ref 162 - 99.95%
Ref 162 - 99.95%
Ref 341 - 99.99%
Ref 341 - 99.99%

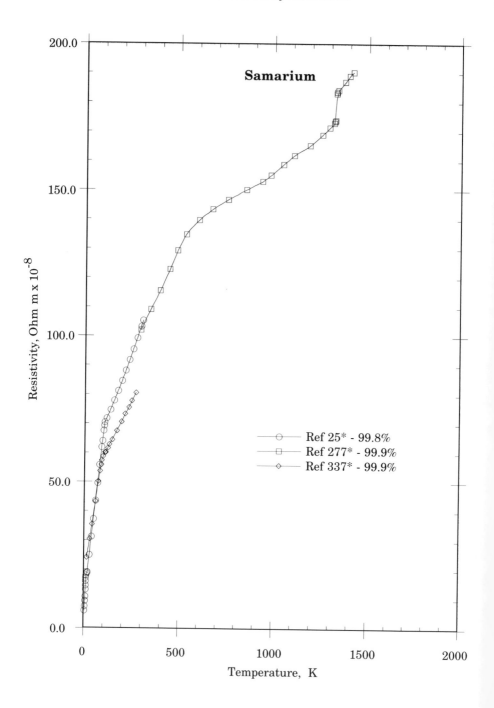

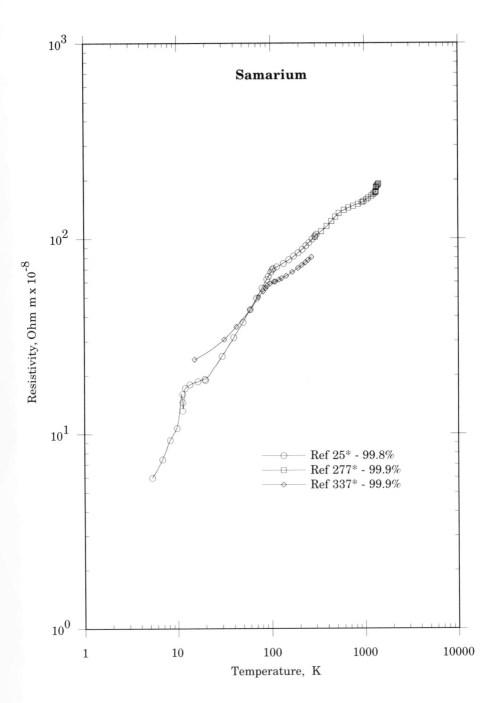

Samarium

Resistivity, Ohm m x 10⁻⁸

Temperature, K

Ref 25* - 99.8%
Ref 277* - 99.9%
Ref 337* - 99.9%

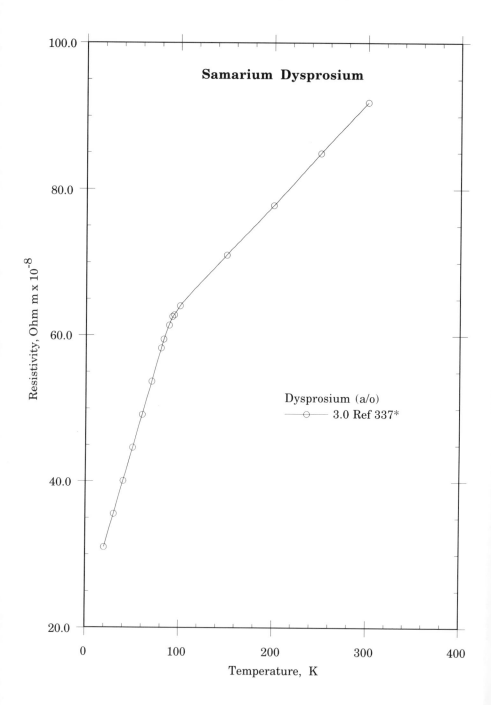

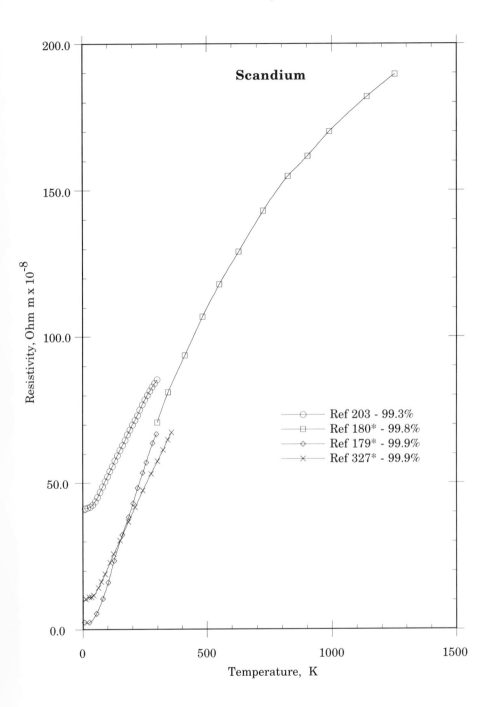

Scandium

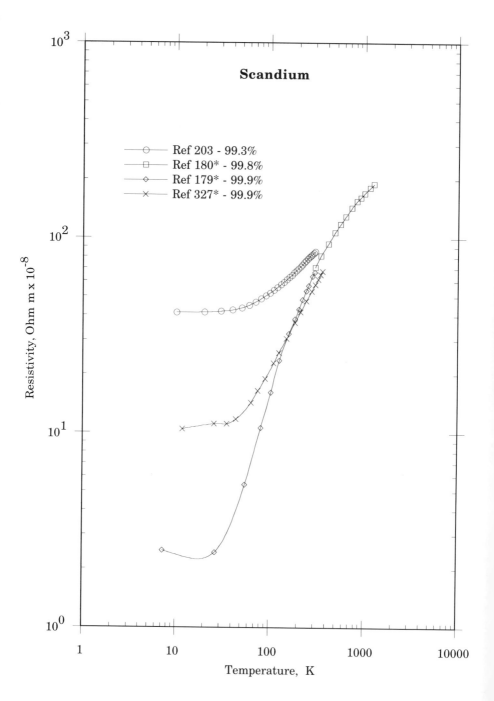

Scandium

Ref 203 - 99.3%
Ref 180* - 99.8%
Ref 179* - 99.9%
Ref 327* - 99.9%

Resistivity, Ohm m x 10^{-8}

Temperature, K

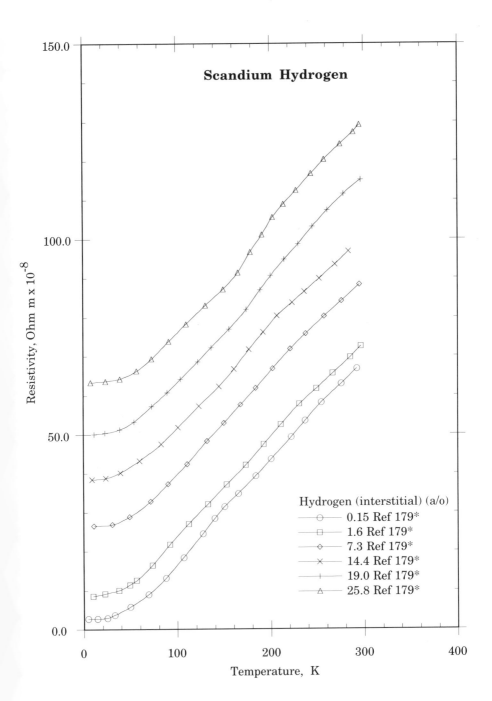

Scandium Hydrogen

Resistivity, Ohm m x 10⁻⁸

Temperature, K

Hydrogen (interstitial) (a/o)
- 0.15 Ref 179*
- 1.6 Ref 179*
- 7.3 Ref 179*
- 14.4 Ref 179*
- 19.0 Ref 179*
- 25.8 Ref 179*

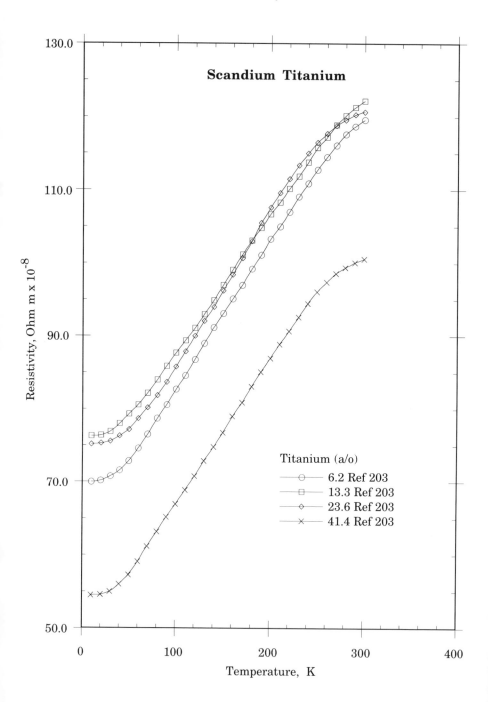

Scandium Titanium

Titanium (a/o)
- ——○—— 6.2 Ref 203
- ——□—— 13.3 Ref 203
- ——◇—— 23.6 Ref 203
- ——×—— 41.4 Ref 203

Resistivity, Ohm m x 10^{-8}

Temperature, K

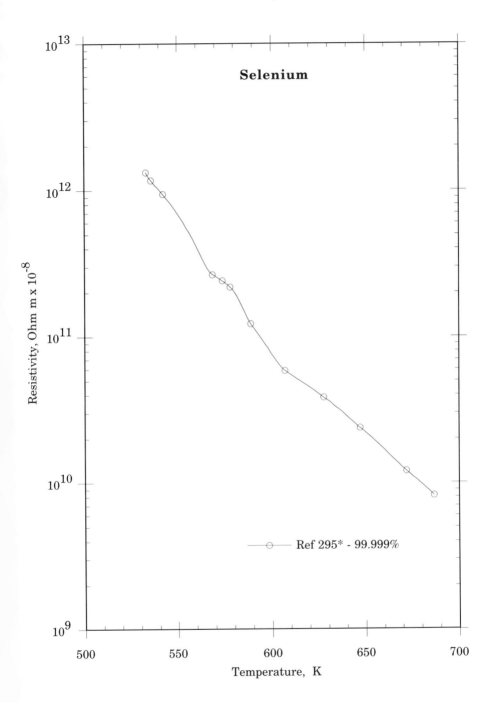

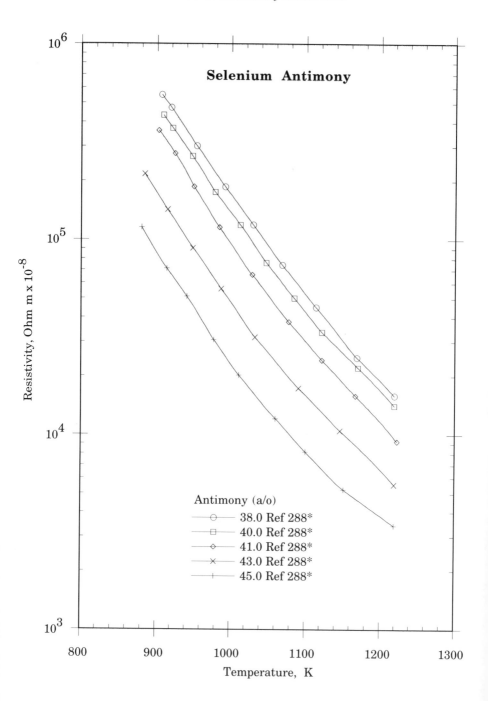

Selenium Antimony

Resistivity, Ohm m x 10^{-8}

Temperature, K

Antimony (a/o)
- ——⊖—— 38.0 Ref 288*
- ——□—— 40.0 Ref 288*
- ——◇—— 41.0 Ref 288*
- ——×—— 43.0 Ref 288*
- ——+—— 45.0 Ref 288*

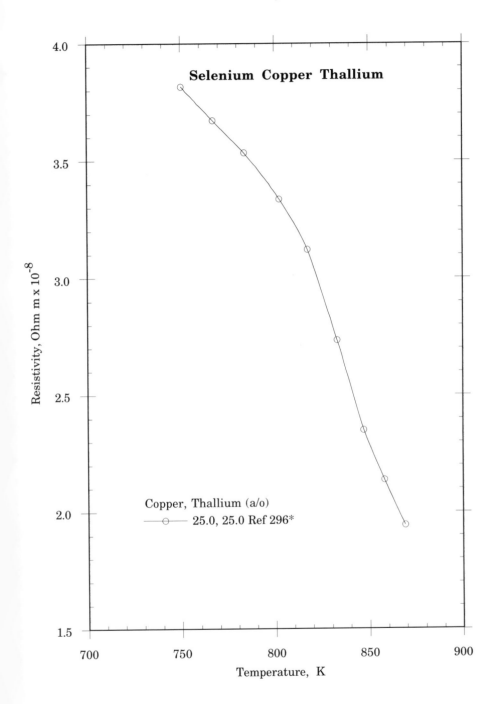

Selenium Copper Thallium

Copper, Thallium (a/o)
—⊙— 25.0, 25.0 Ref 296*

Resistivity, Ohm m x 10^{-8}

Temperature, K

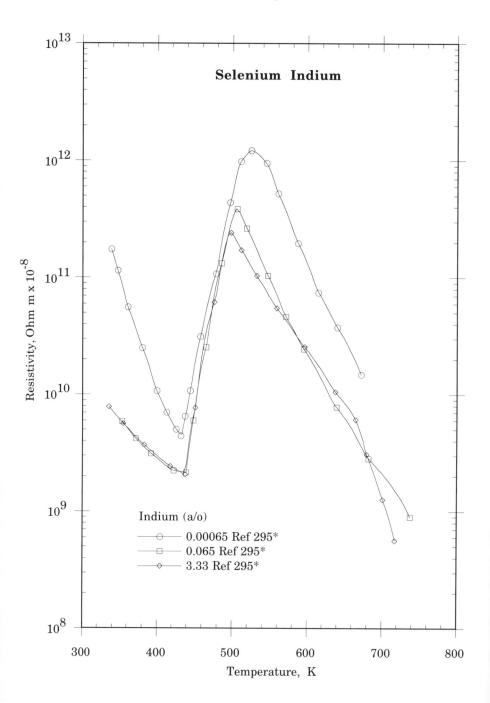

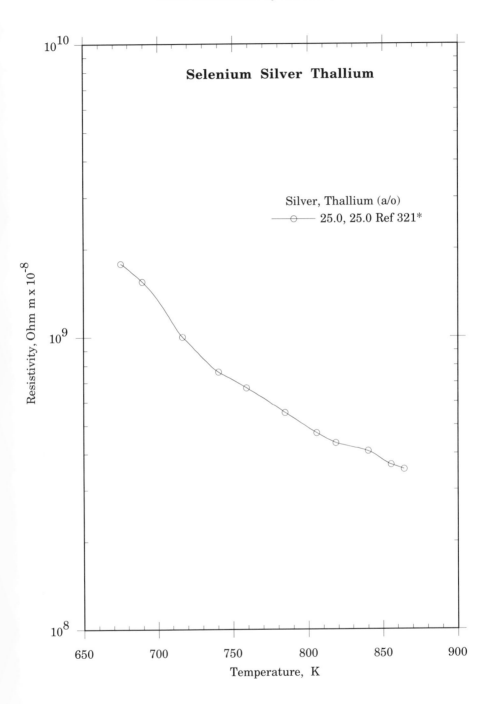

Selenium Silver Thallium

Silver, Thallium (a/o)
——⊙—— 25.0, 25.0 Ref 321*

Resistivity, Ohm m x 10⁻⁸

Temperature, K

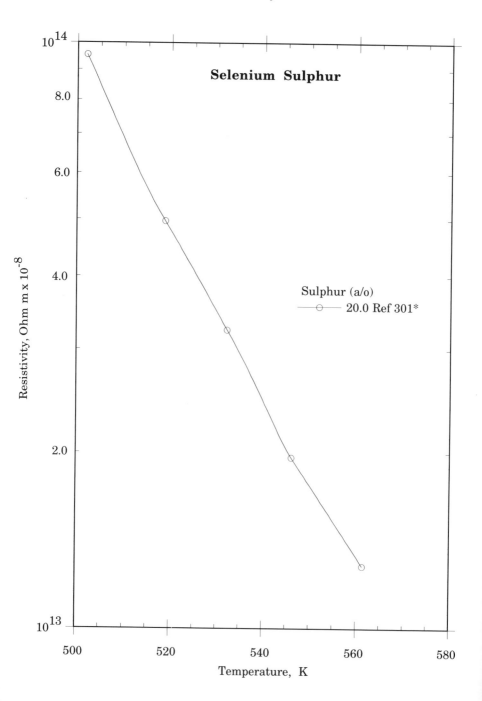

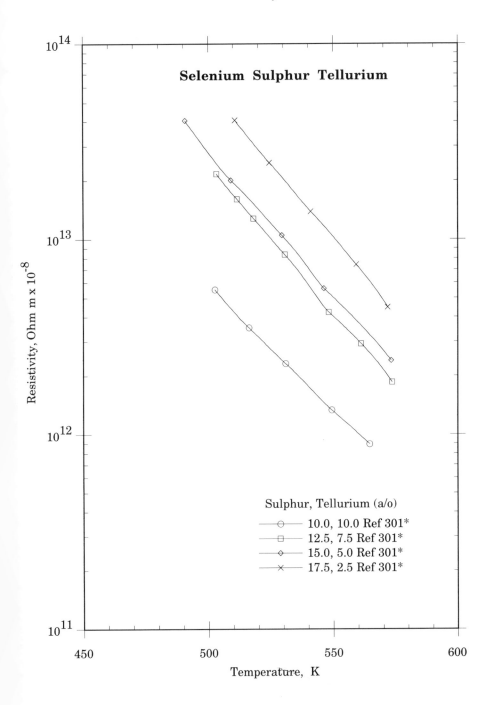

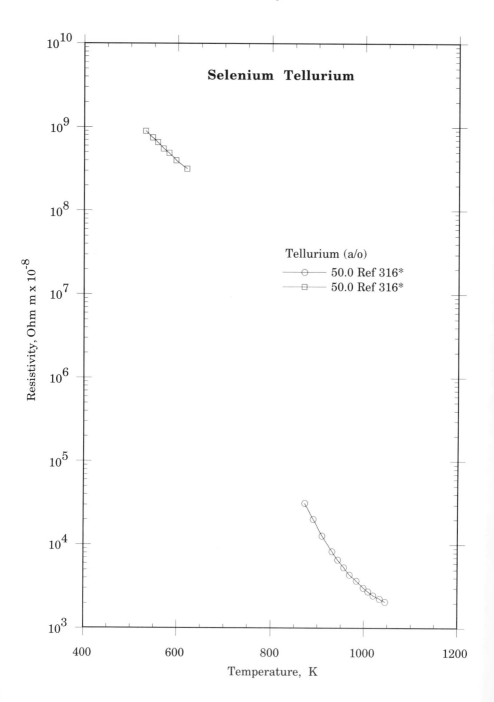

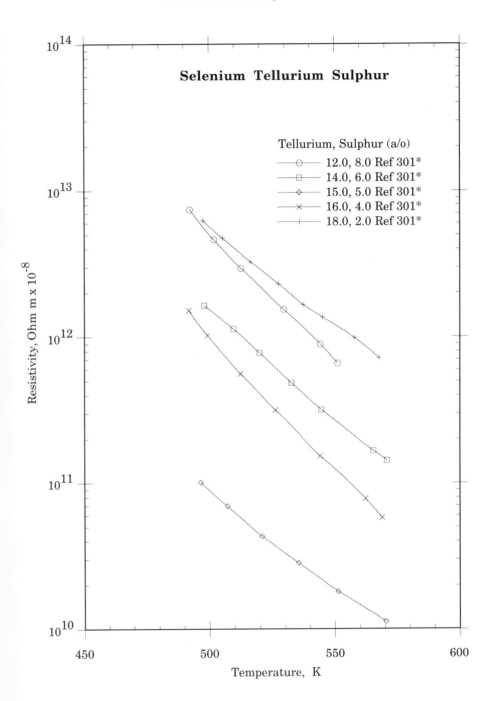

Selenium Tellurium Sulphur

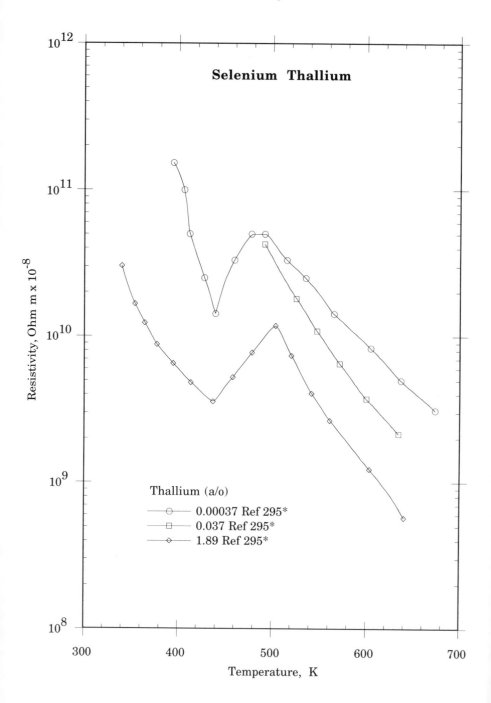

Selenium Thallium

Thallium (a/o)
- 0.00037 Ref 295*
- 0.037 Ref 295*
- 1.89 Ref 295*

Resistivity, Ohm m x 10^{-8}

Temperature, K

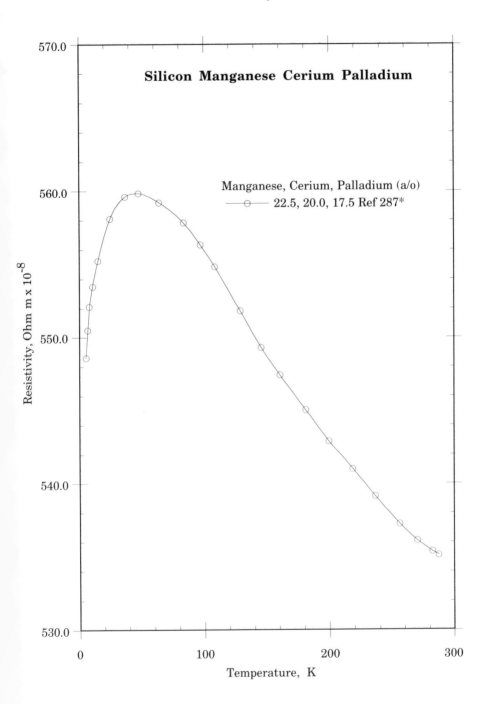

Silicon Manganese Cerium Palladium

Manganese, Cerium, Palladium (a/o)
——⊖—— 22.5, 20.0, 17.5 Ref 287*

Resistivity, Ohm m x 10^{-8}

Temperature, K

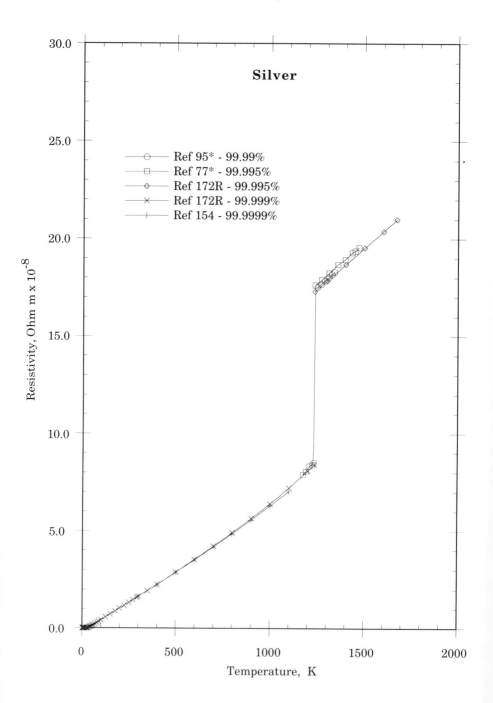

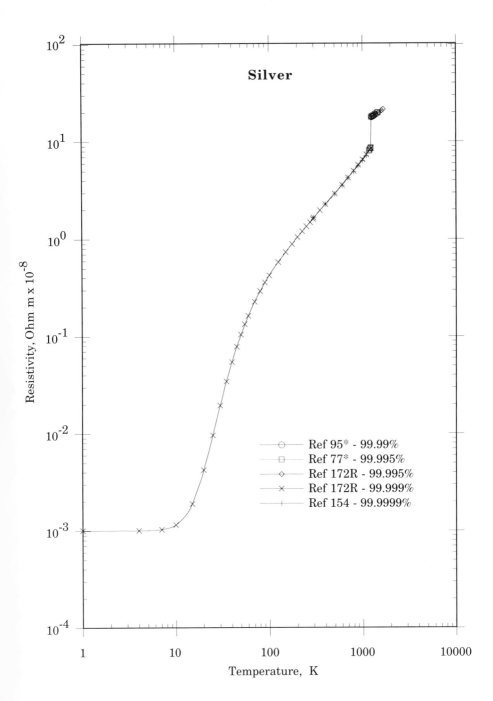

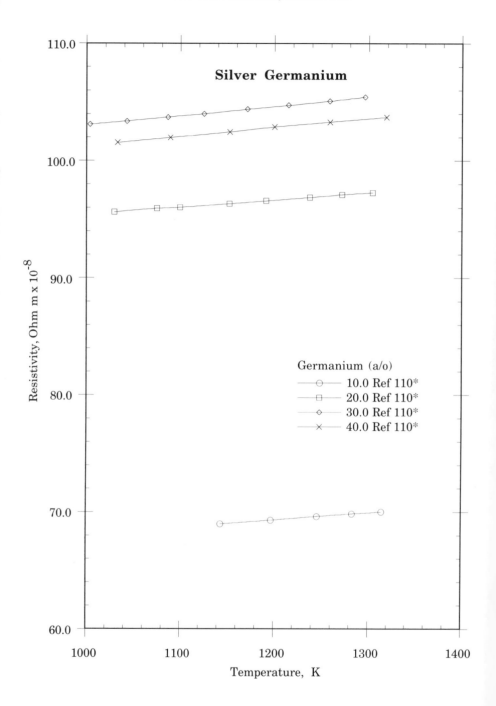

Silver Germanium

Germanium (a/o)
- ⊙ 10.0 Ref 110*
- ⊟ 20.0 Ref 110*
- ◇ 30.0 Ref 110*
- ✕ 40.0 Ref 110*

Resistivity, Ohm m x 10^{-8}

Temperature, K

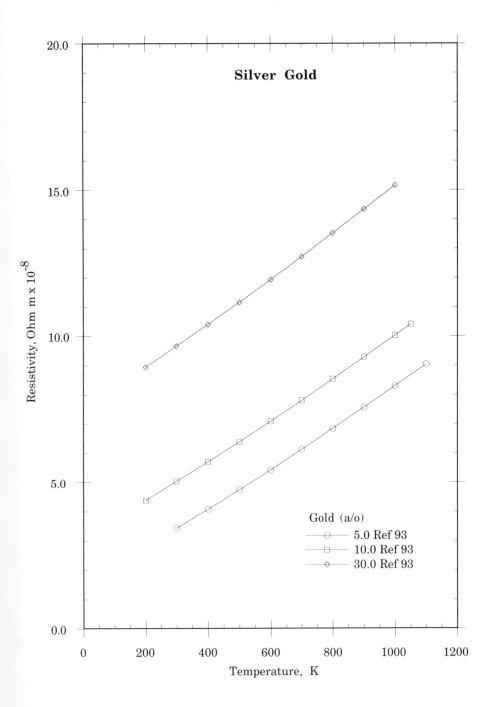

Silver Gold

Gold (a/o)
—⊙— 5.0 Ref 93
—▫— 10.0 Ref 93
—◇— 30.0 Ref 93

Resistivity, Ohm m x 10^{-8}

Temperature, K

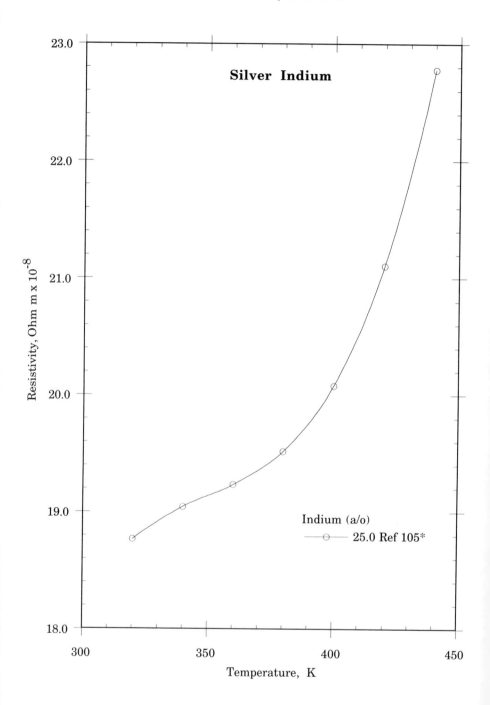

Silver Indium

Resistivity, Ohm m x 10^{-8}

Temperature, K

Indium (a/o)
25.0 Ref 105*

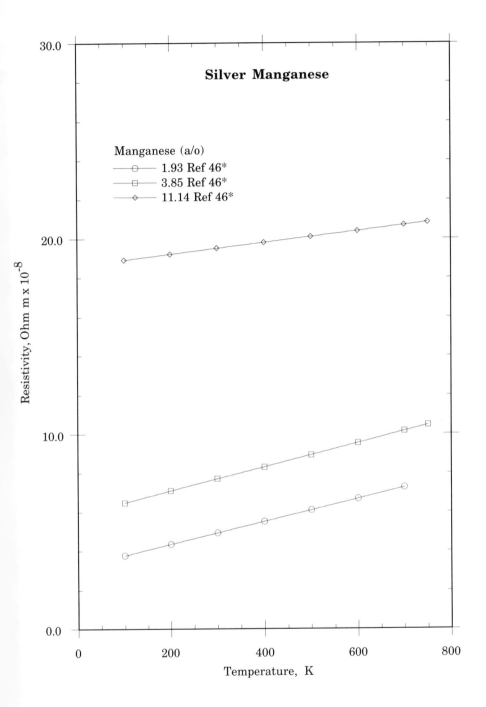

Silver Manganese

Manganese (a/o)
—⊖— 1.93 Ref 46*
—□— 3.85 Ref 46*
—◇— 11.14 Ref 46*

Resistivity, Ohm m x 10^{-8}

Temperature, K

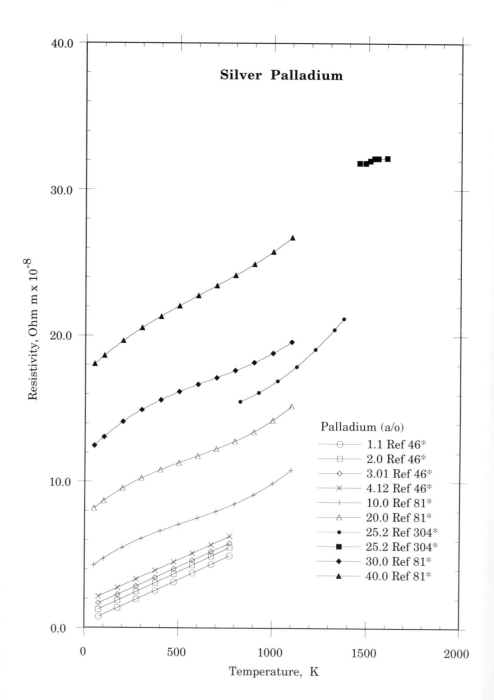

Silver Palladium

Resistivity, Ohm m x 10^{-8}

Temperature, K

Palladium (a/o)
- ⊖ 1.1 Ref 46*
- ⊟ 2.0 Ref 46*
- ◇ 3.01 Ref 46*
- ✕ 4.12 Ref 46*
- + 10.0 Ref 81*
- △ 20.0 Ref 81*
- ● 25.2 Ref 304*
- ■ 25.2 Ref 304*
- ◆ 30.0 Ref 81*
- ▲ 40.0 Ref 81*

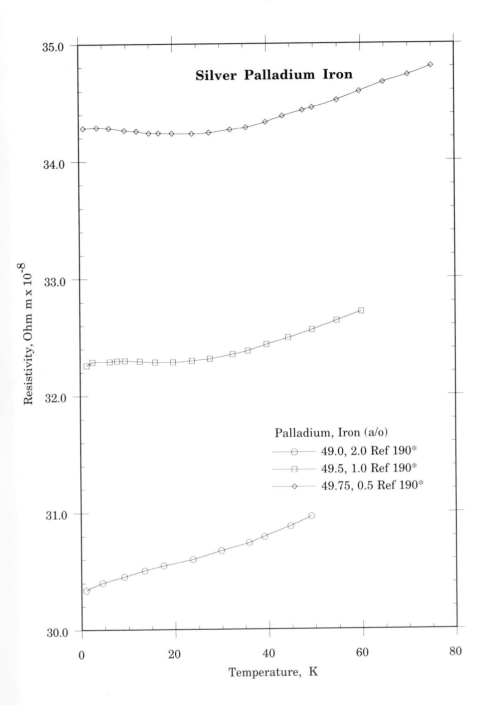

Silver Palladium Iron

Resistivity, Ohm m x 10^{-8}

Palladium, Iron (a/o)
—○— 49.0, 2.0 Ref 190*
—□— 49.5, 1.0 Ref 190*
—◇— 49.75, 0.5 Ref 190*

Temperature, K

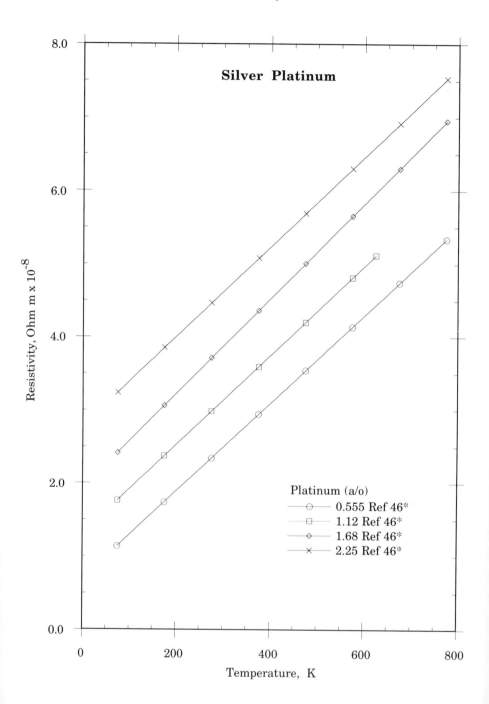

Silver Platinum

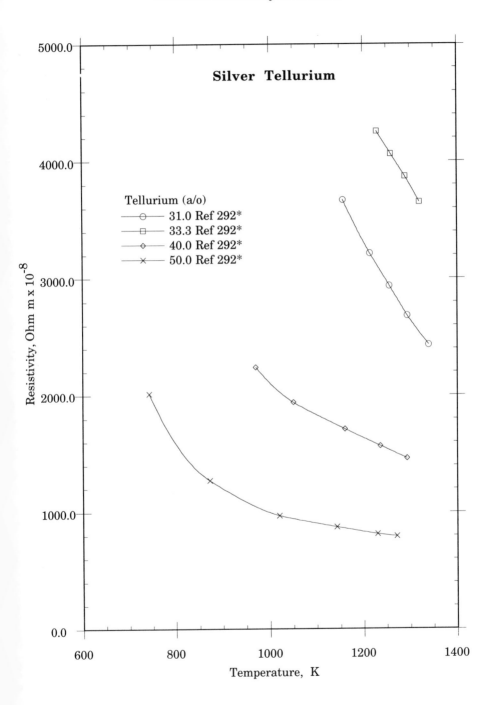

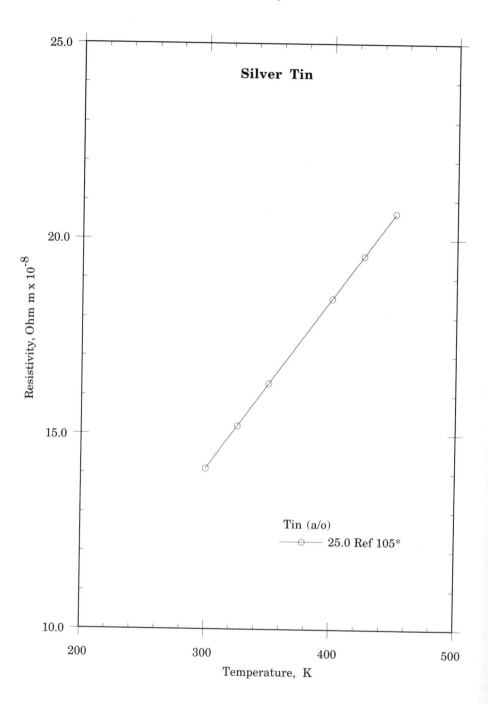

Silver Tin

Tin (a/o)
25.0 Ref 105*

Resistivity, Ohm m x 10^{-8}

Temperature, K

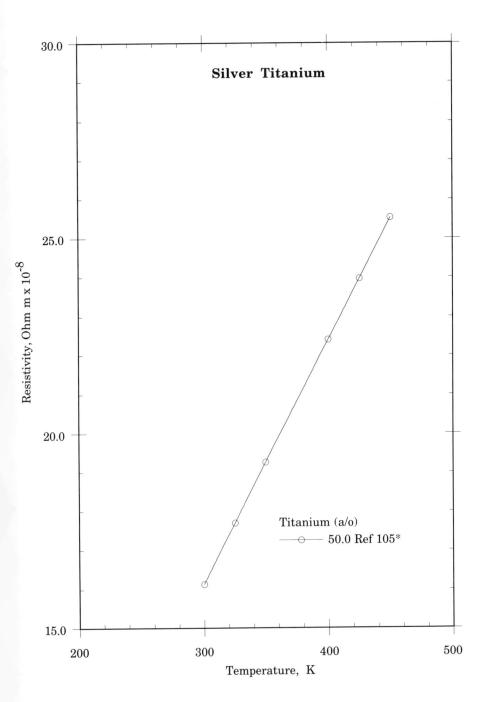

Silver Titanium

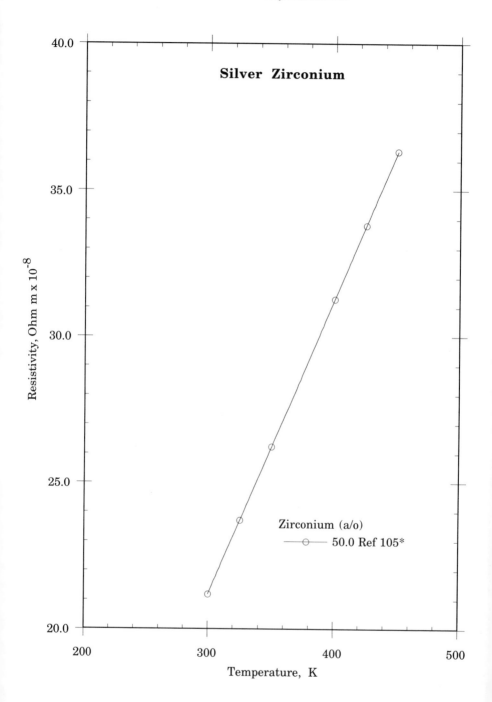

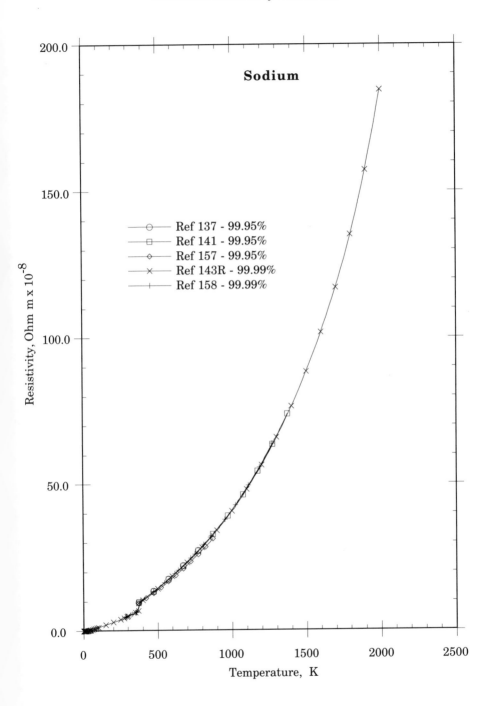

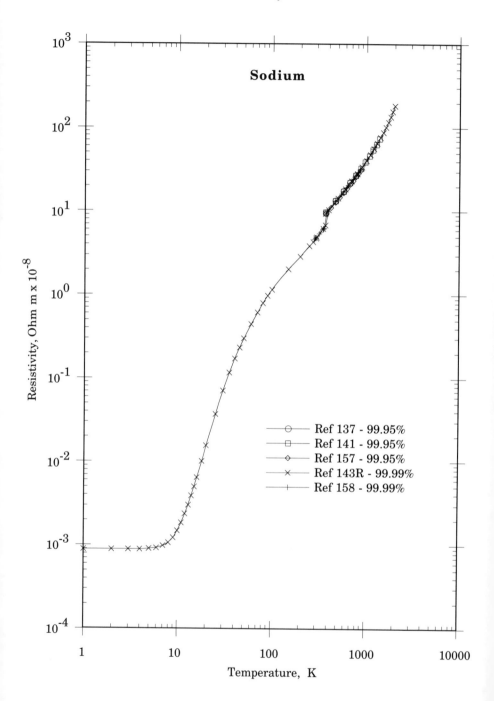

Sodium

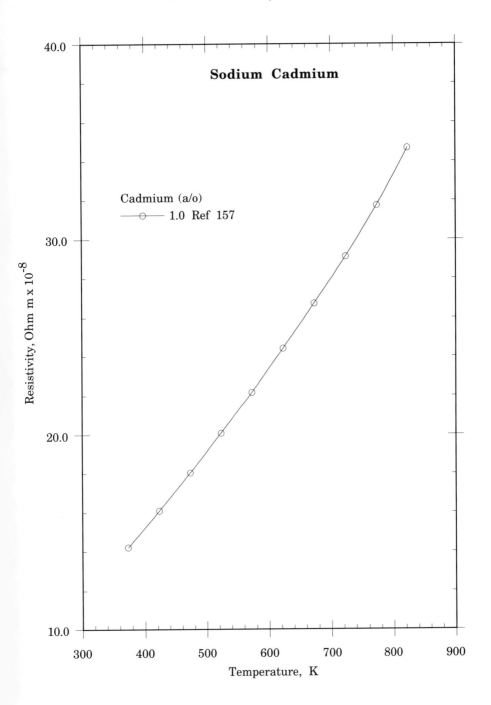

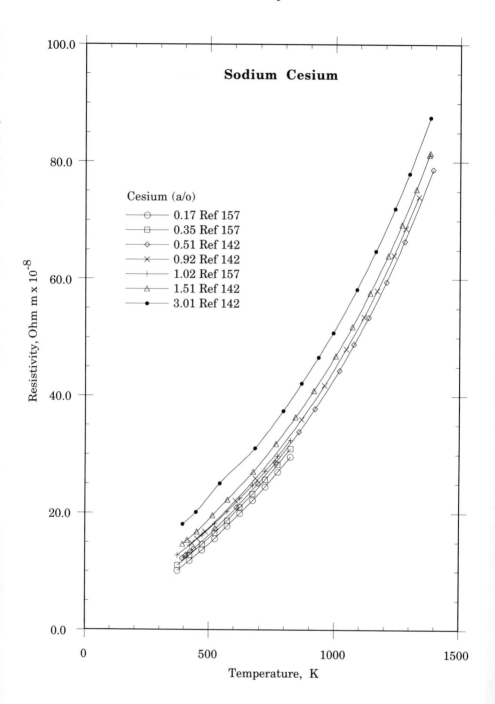

Sodium Cesium

Cesium (a/o)
— ⊖ — 0.17 Ref 157
— □ — 0.35 Ref 157
— ◇ — 0.51 Ref 142
— × — 0.92 Ref 142
— + — 1.02 Ref 157
— △ — 1.51 Ref 142
— • — 3.01 Ref 142

Resistivity, Ohm m x 10^{-8}

Temperature, K

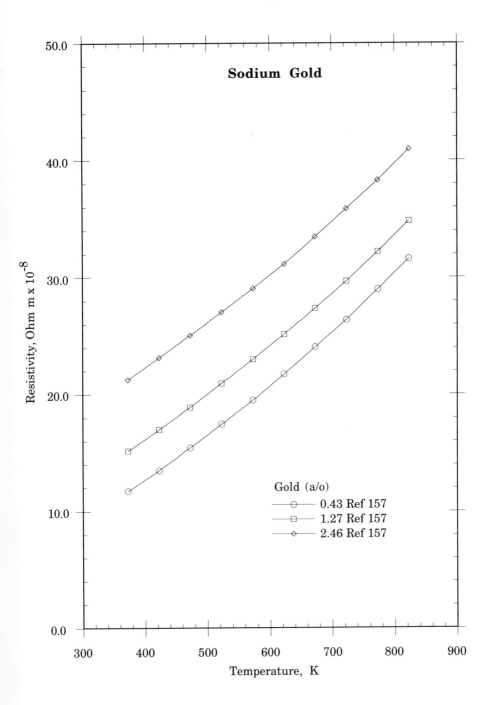

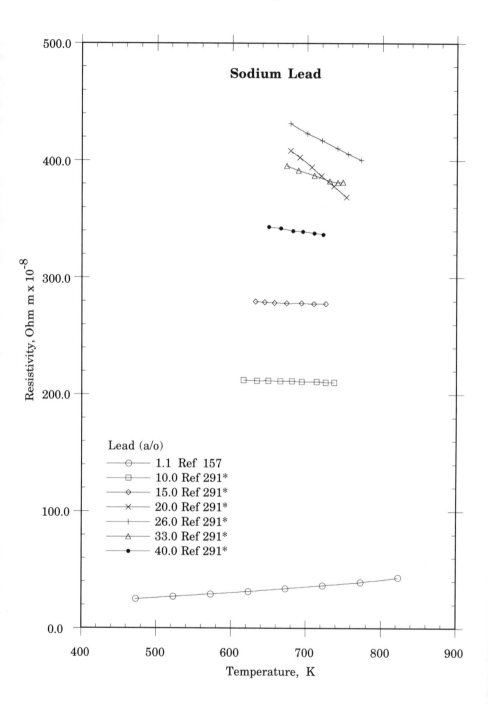

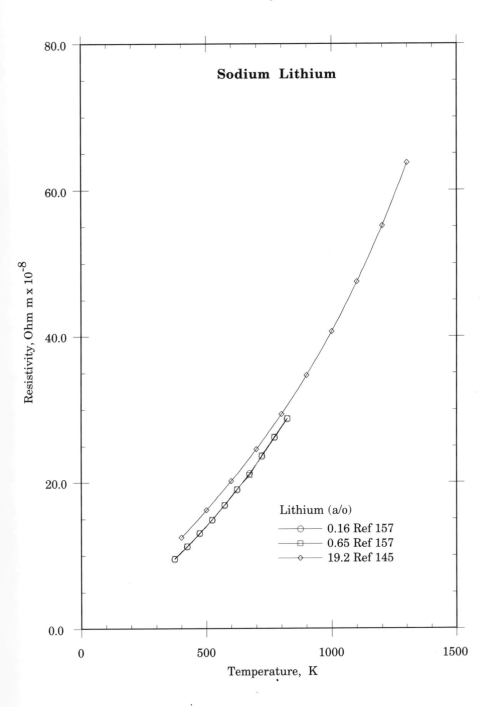

Sodium Lithium

Lithium (a/o)
——○—— 0.16 Ref 157
——□—— 0.65 Ref 157
——◇—— 19.2 Ref 145

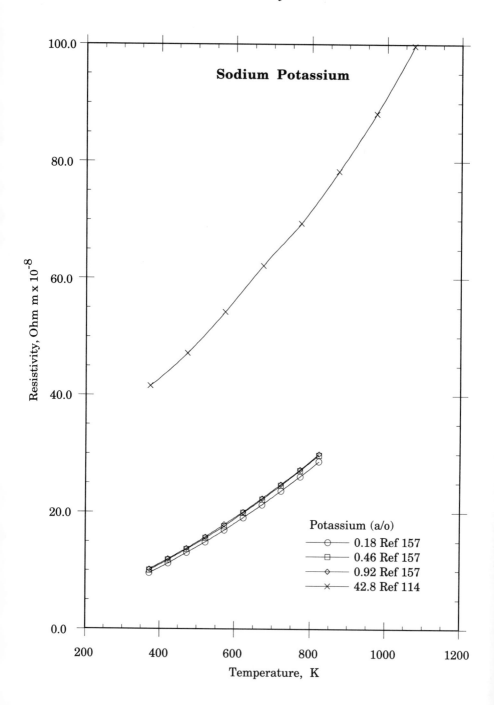

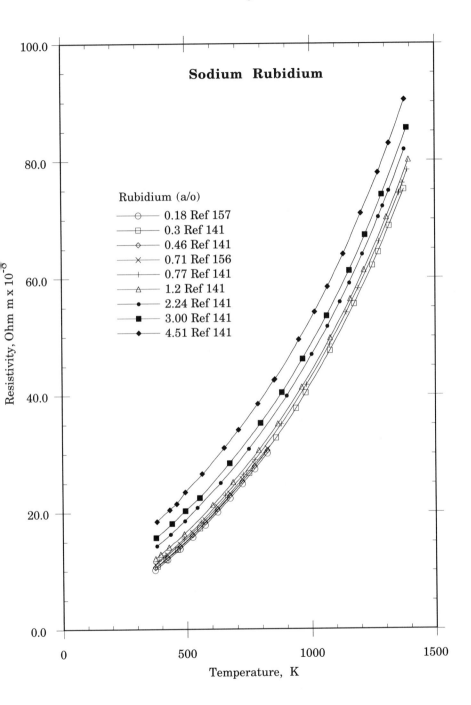

Sodium   Rubidium

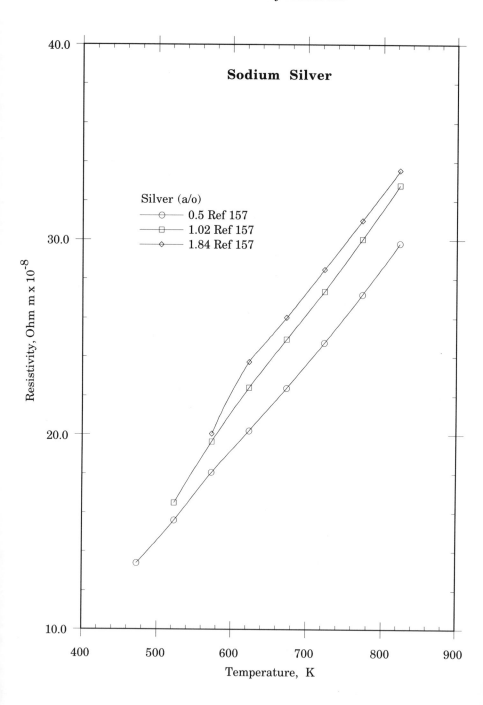

Sodium Silver

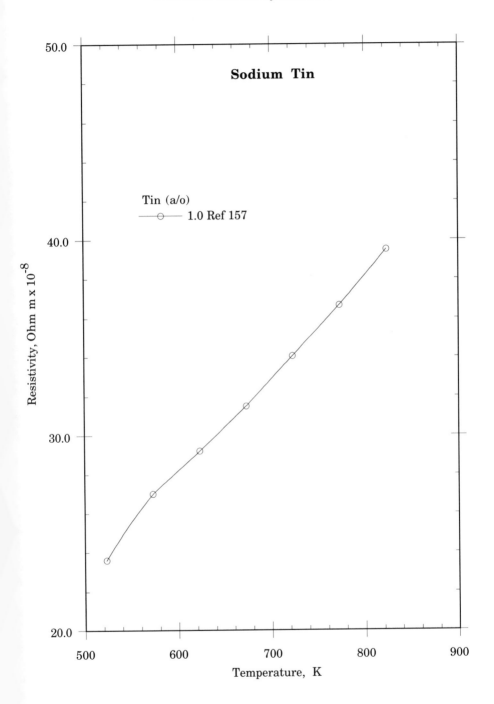

Sodium Tin

Tin (a/o)
⊕ 1.0 Ref 157

Resistivity, Ohm m x 10^{-8}

Temperature, K

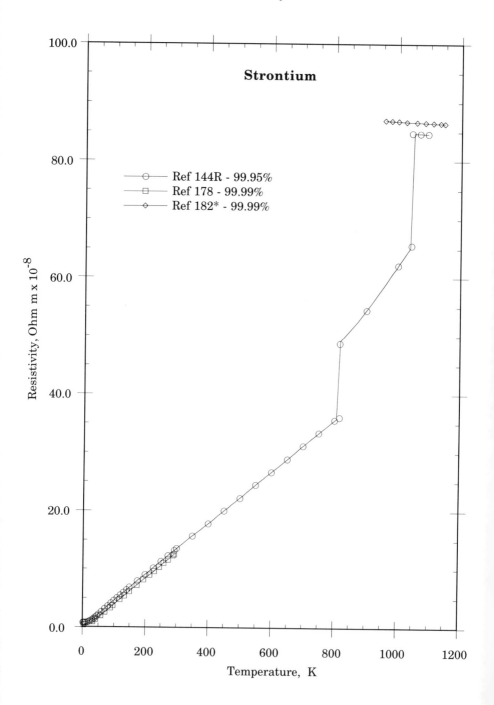

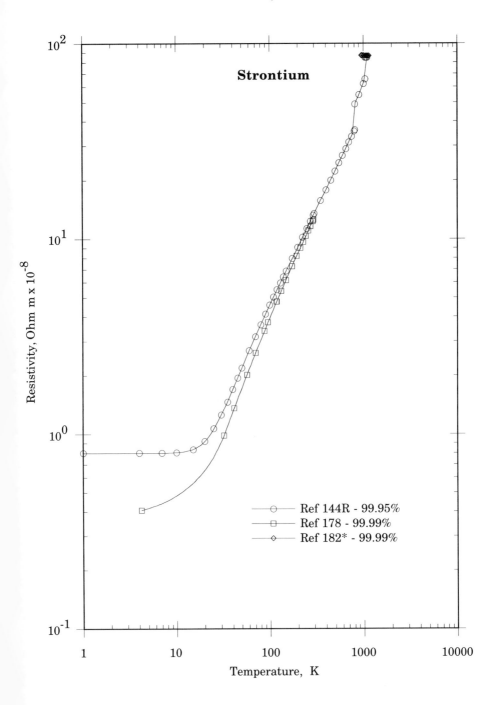

Strontium

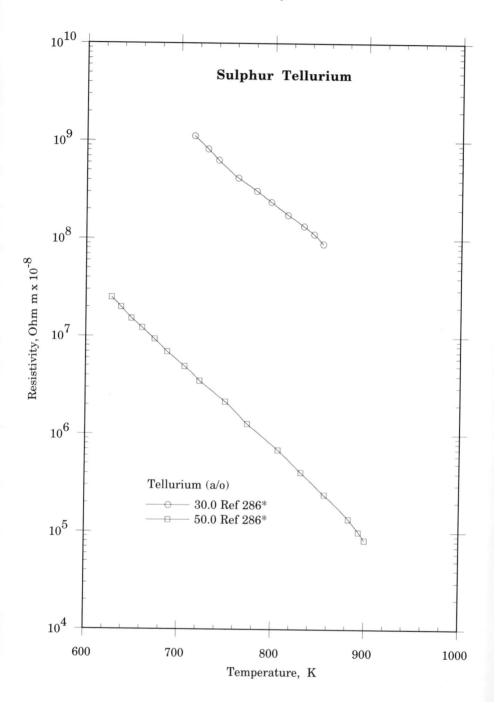

Sulphur Tellurium

Resistivity, Ohm m x 10^{-8}

Temperature, K

Tellurium (a/o)
○ 30.0 Ref 286*
□ 50.0 Ref 286*

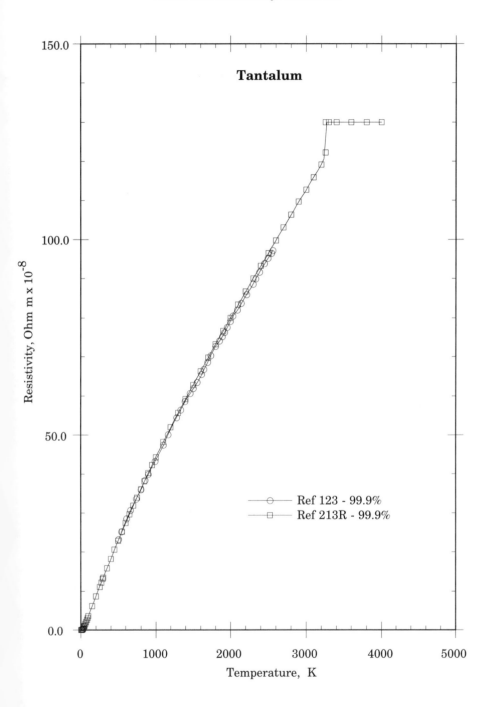

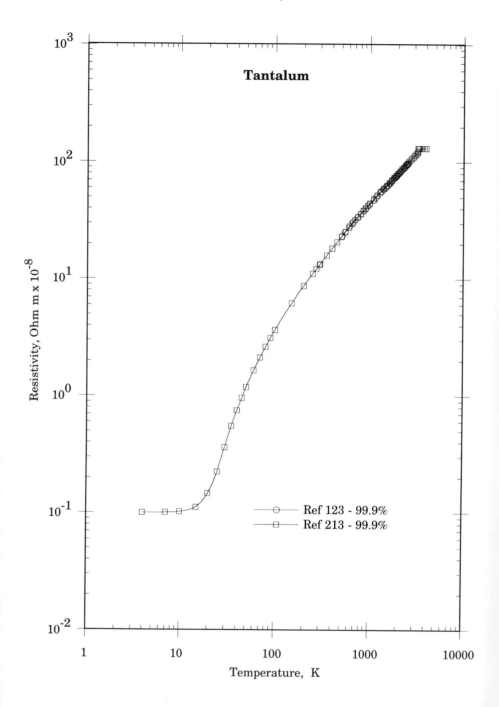

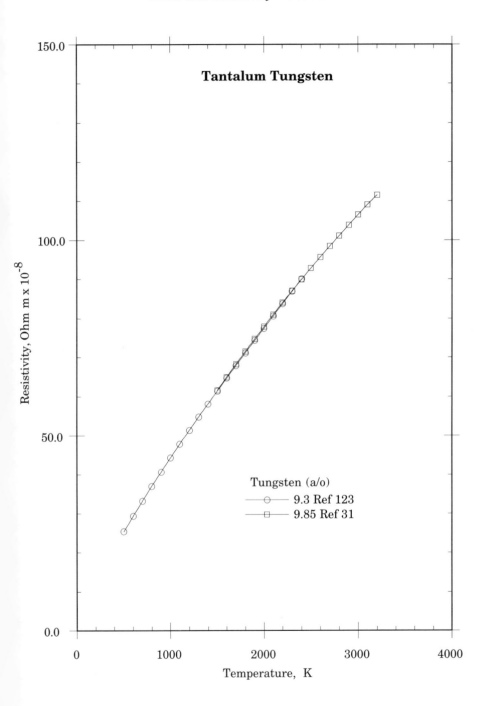

Tantalum Tungsten

Resistivity, Ohm m x 10^{-8}

Temperature, K

Tungsten (a/o)
—○— 9.3 Ref 123
—□— 9.85 Ref 31

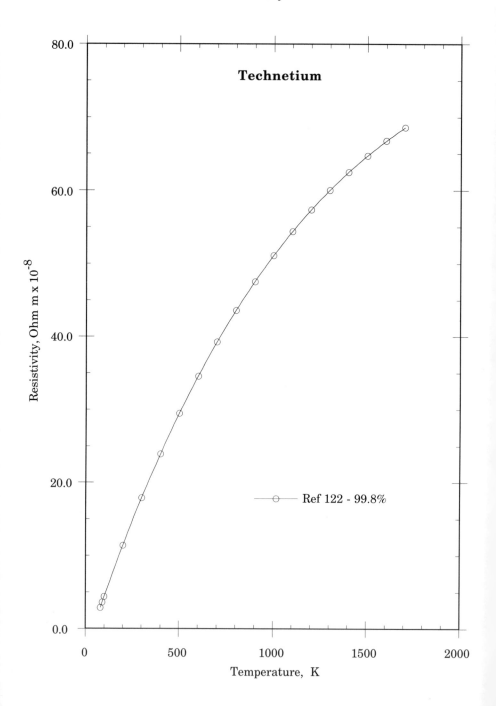

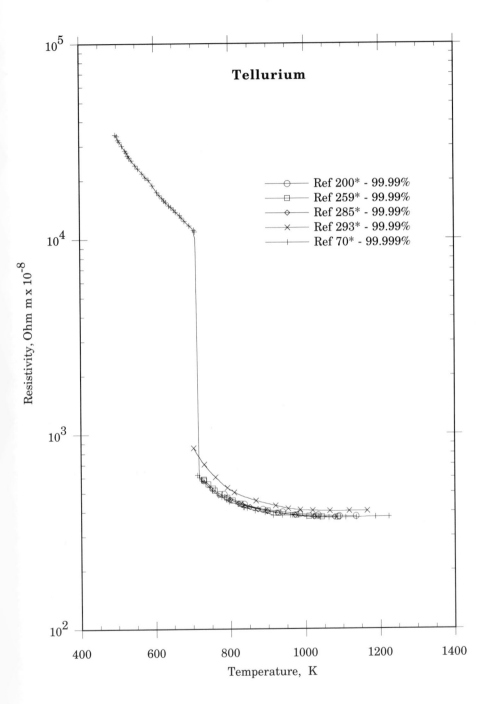

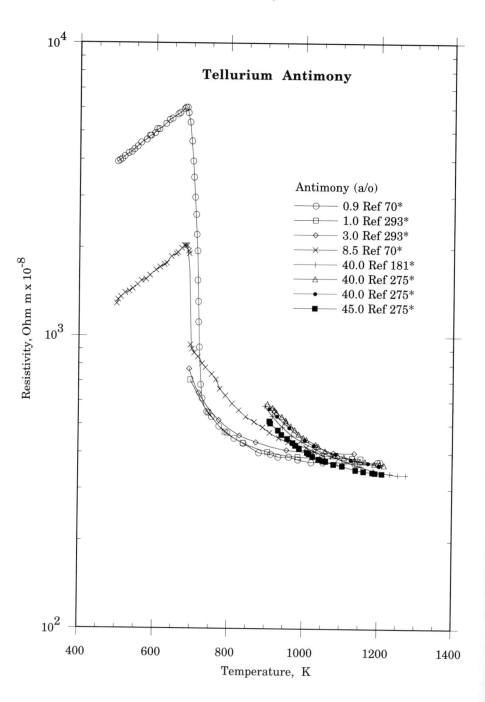

Tellurium Antimony

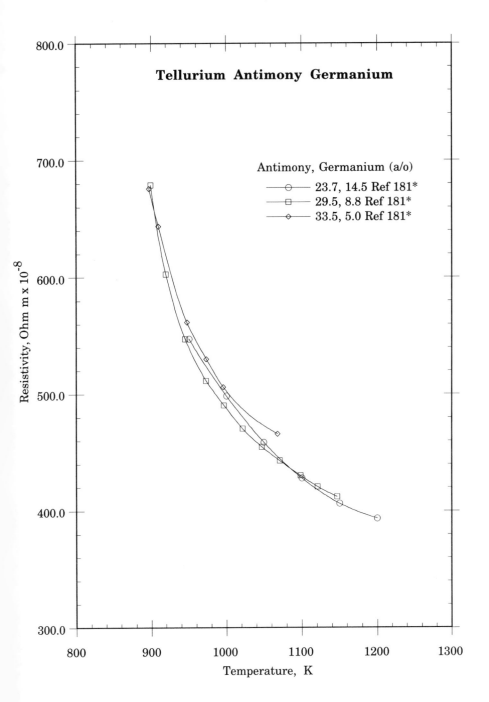

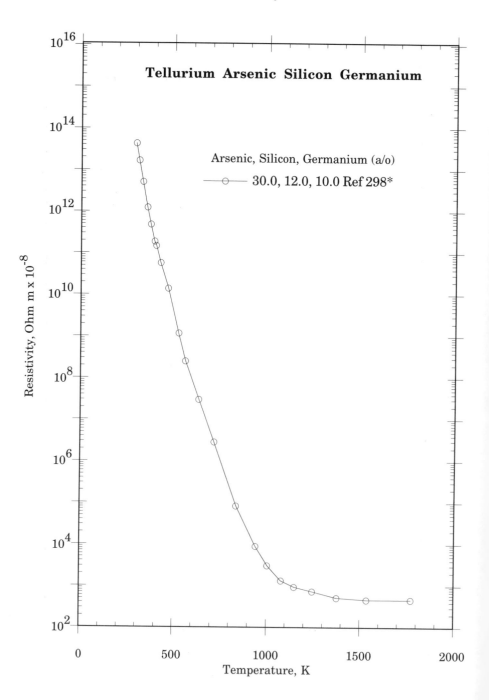

Tellurium Arsenic Silicon Germanium

Arsenic, Silicon, Germanium (a/o)

——⊝—— 30.0, 12.0, 10.0 Ref 298*

Resistivity, Ohm m x 10^{-8}

Temperature, K

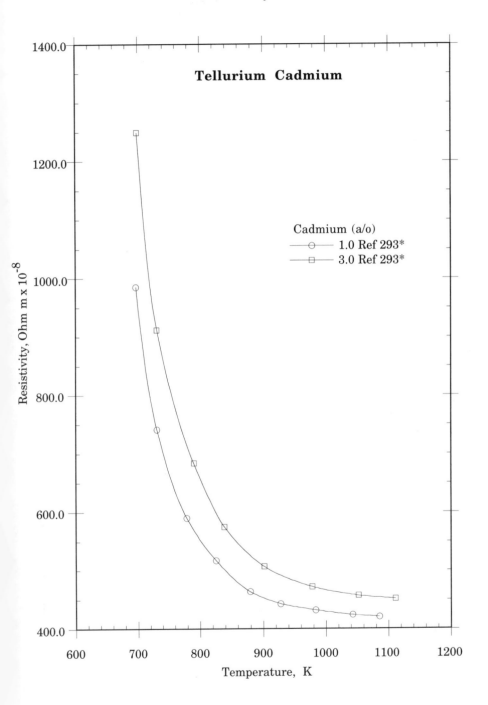

Tellurium Cadmium

Cadmium (a/o)
———o——— 1.0 Ref 293*
———□——— 3.0 Ref 293*

Resistivity, Ohm m x 10⁻⁸

Temperature, K

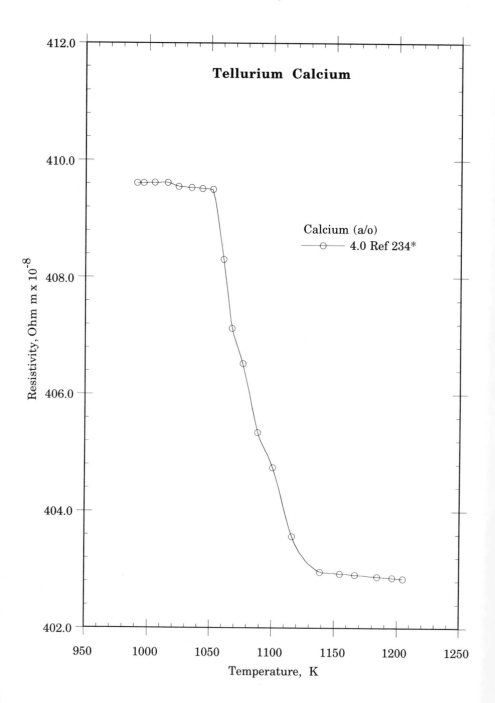

Tellurium Calcium

Calcium (a/o)
—⊖— 4.0 Ref 234*

Resistivity, Ohm m x 10⁻⁸

Temperature, K

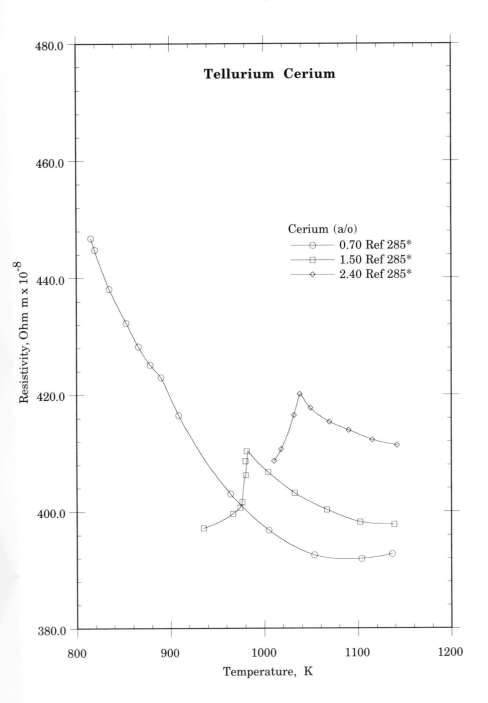

Tellurium Cerium

Cerium (a/o)
- 0.70 Ref 285*
- 1.50 Ref 285*
- 2.40 Ref 285*

Resistivity, Ohm m x 10^{-8}

Temperature, K

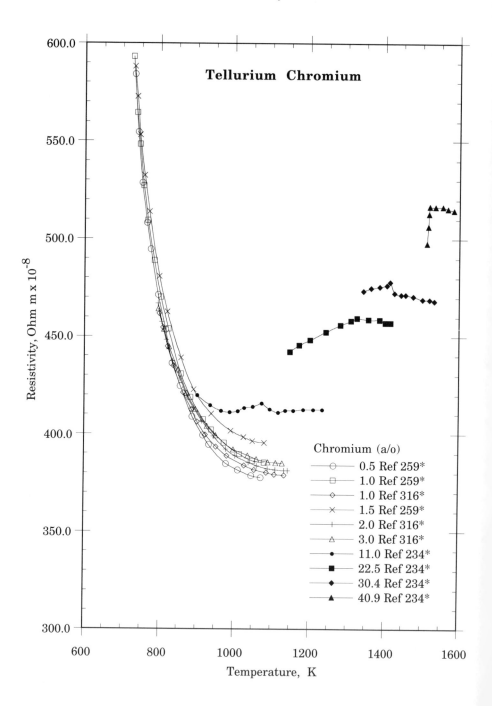

Tellurium Chromium

Resistivity, Ohm m x 10^{-8}

Temperature, K

Chromium (a/o)
- 0.5 Ref 259*
- 1.0 Ref 259*
- 1.0 Ref 316*
- 1.5 Ref 259*
- 2.0 Ref 316*
- 3.0 Ref 316*
- 11.0 Ref 234*
- 22.5 Ref 234*
- 30.4 Ref 234*
- 40.9 Ref 234*

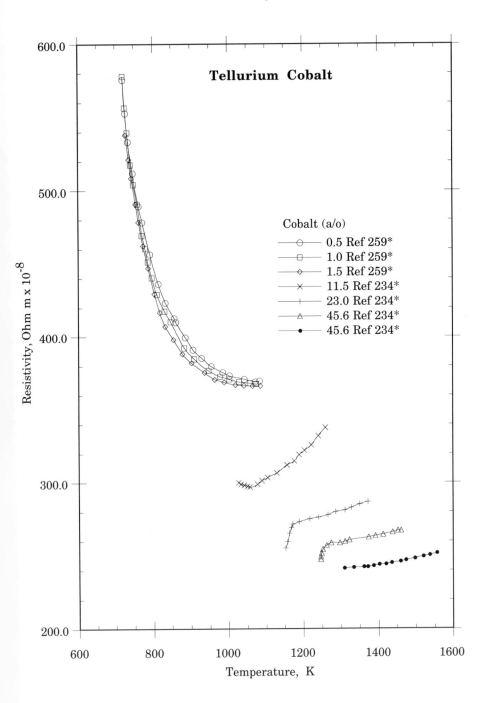

Tellurium Cobalt

Cobalt (a/o)
- ⊙ 0.5 Ref 259*
- □ 1.0 Ref 259*
- ◇ 1.5 Ref 259*
- × 11.5 Ref 234*
- + 23.0 Ref 234*
- △ 45.6 Ref 234*
- ● 45.6 Ref 234*

Resistivity, Ohm m x 10⁻⁸

Temperature, K

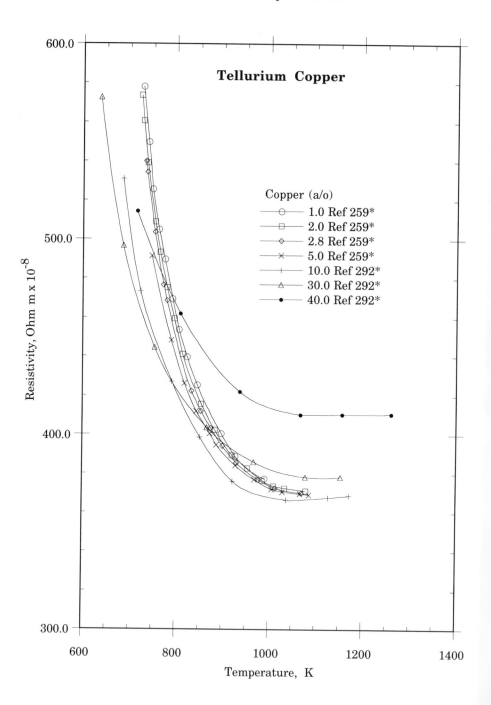

Tellurium Copper

Copper (a/o)
- ── ○ ── 1.0 Ref 259*
- ── □ ── 2.0 Ref 259*
- ── ◇ ── 2.8 Ref 259*
- ── ✕ ── 5.0 Ref 259*
- ── ＋ ── 10.0 Ref 292*
- ── △ ── 30.0 Ref 292*
- ── ● ── 40.0 Ref 292*

Resistivity, Ohm m x 10⁻⁸

Temperature, K

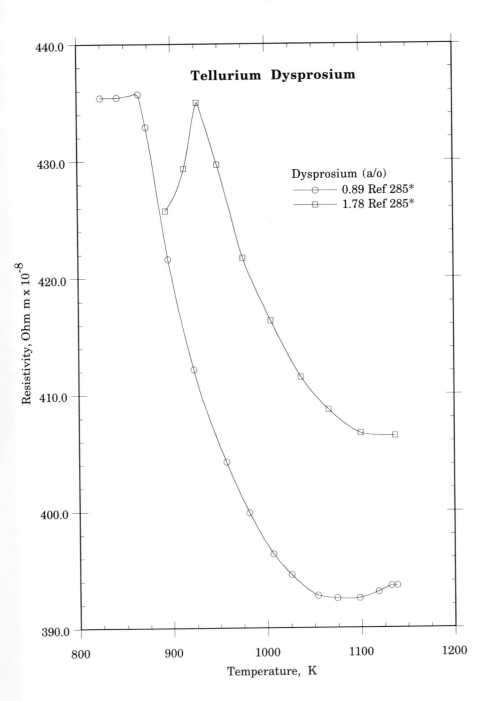

Tellurium Dysprosium

Dysprosium (a/o)
— ○ — 0.89 Ref 285*
— □ — 1.78 Ref 285*

Resistivity, Ohm m x 10^{-8}

Temperature, K

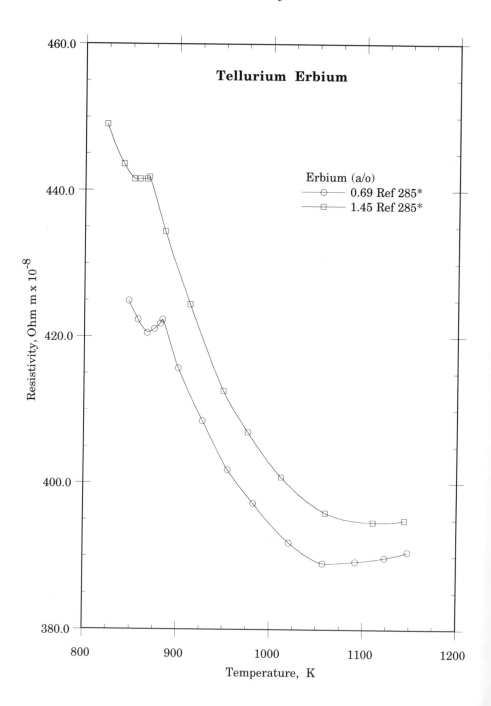

Tellurium Erbium

Erbium (a/o)
○ 0.69 Ref 285*
□ 1.45 Ref 285*

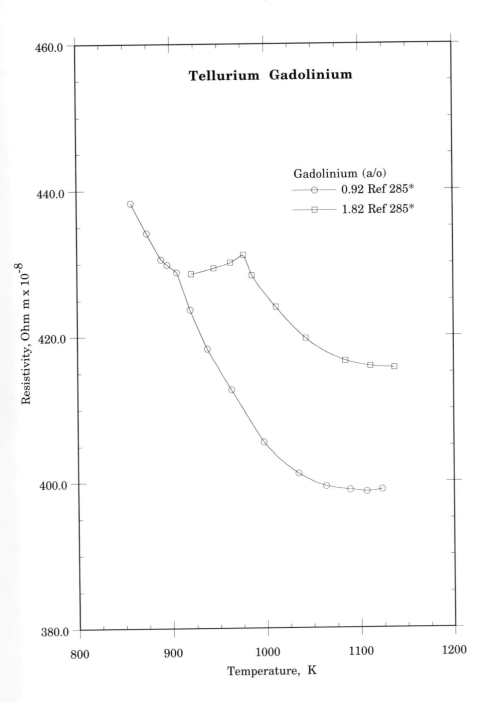

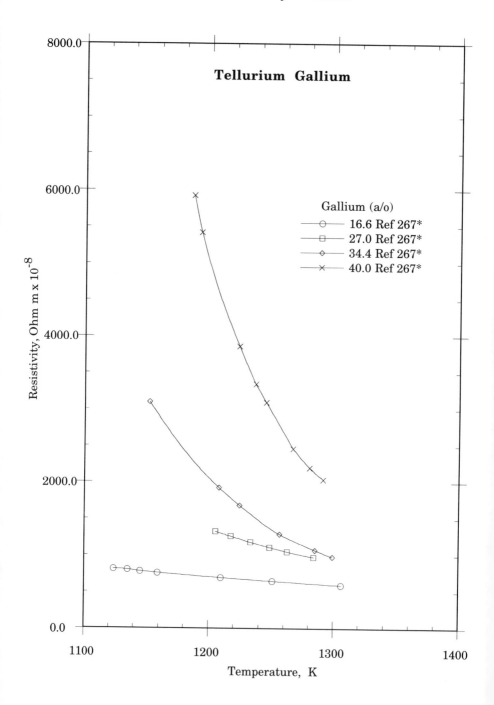

Tellurium Gallium

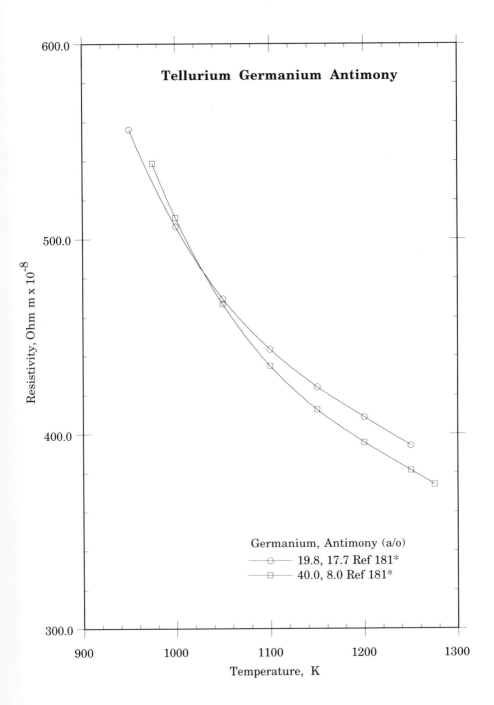

Tellurium Germanium Antimony

Resistivity, Ohm m x 10^{-8}

Temperature, K

Germanium, Antimony (a/o)
—⊙— 19.8, 17.7 Ref 181*
—☐— 40.0, 8.0 Ref 181*

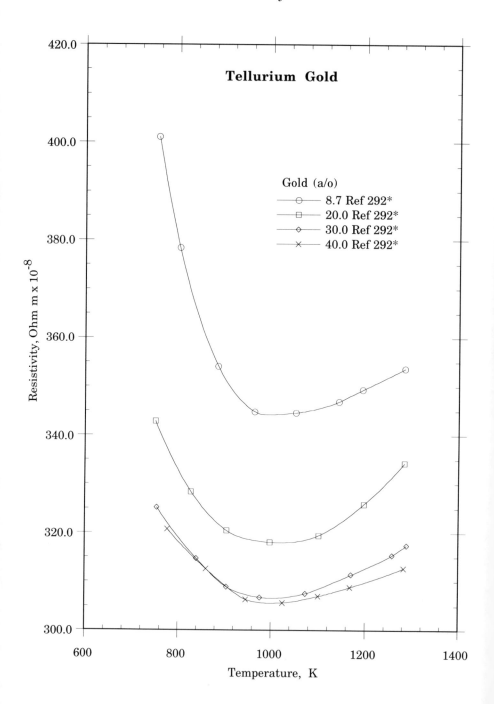

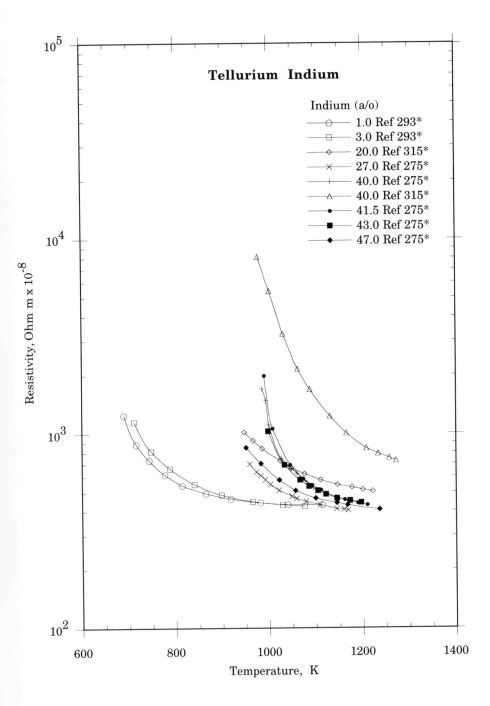

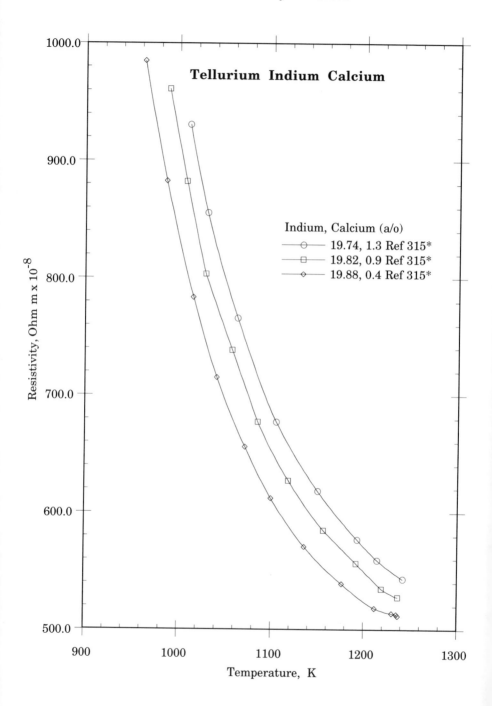

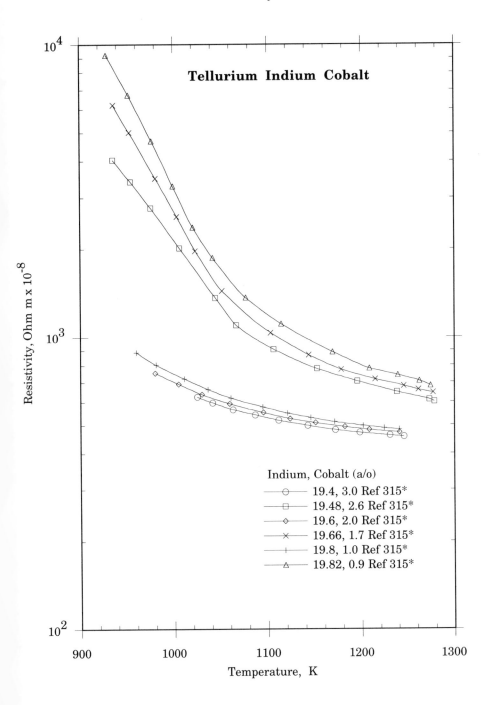

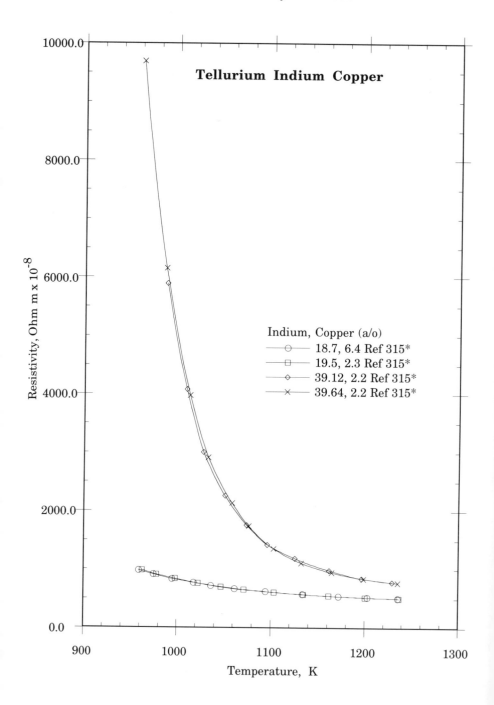

Tellurium Indium Copper

Indium, Copper (a/o)
——○—— 18.7, 6.4 Ref 315*
——□—— 19.5, 2.3 Ref 315*
——◇—— 39.12, 2.2 Ref 315*
——×—— 39.64, 2.2 Ref 315*

Resistivity, Ohm m x 10^{-8}

Temperature, K

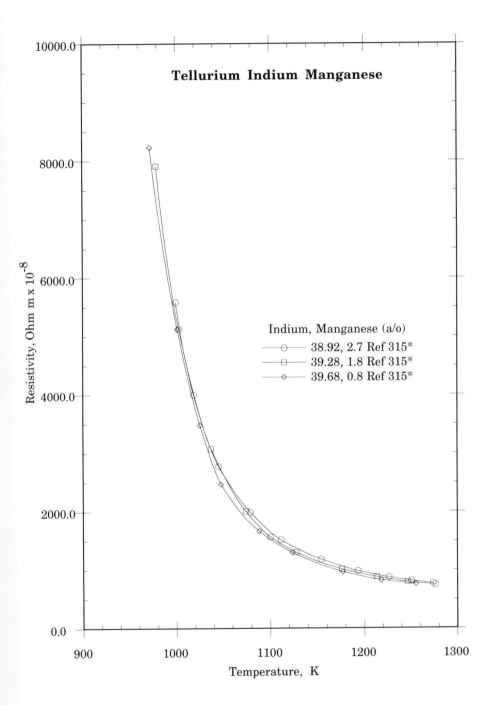

Tellurium Indium Manganese

Indium, Manganese (a/o)
— ○ — 38.92, 2.7 Ref 315*
— □ — 39.28, 1.8 Ref 315*
— ◇ — 39.68, 0.8 Ref 315*

Resistivity, Ohm m x 10^{-8}

Temperature, K

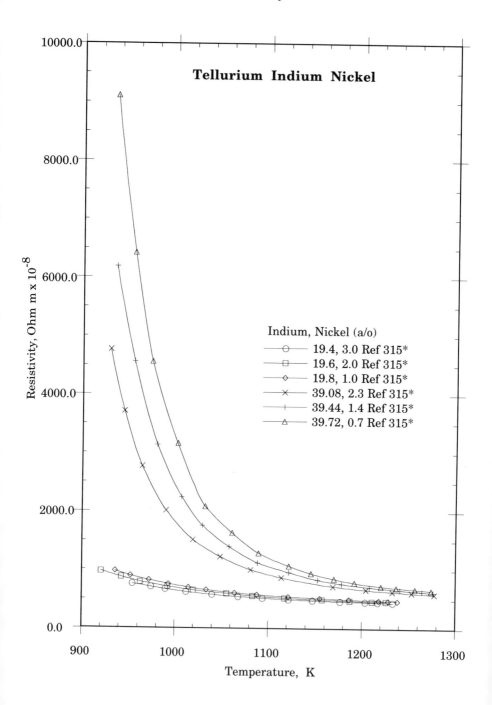

Tellurium Indium Nickel

Indium, Nickel (a/o)
— ⊖ — 19.4, 3.0 Ref 315*
— ☐ — 19.6, 2.0 Ref 315*
— ◇ — 19.8, 1.0 Ref 315*
— ✕ — 39.08, 2.3 Ref 315*
— + — 39.44, 1.4 Ref 315*
— △ — 39.72, 0.7 Ref 315*

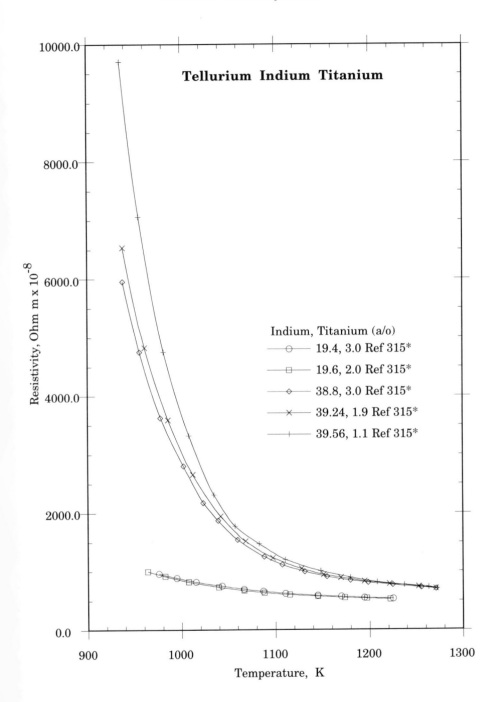

Tellurium Indium Titanium

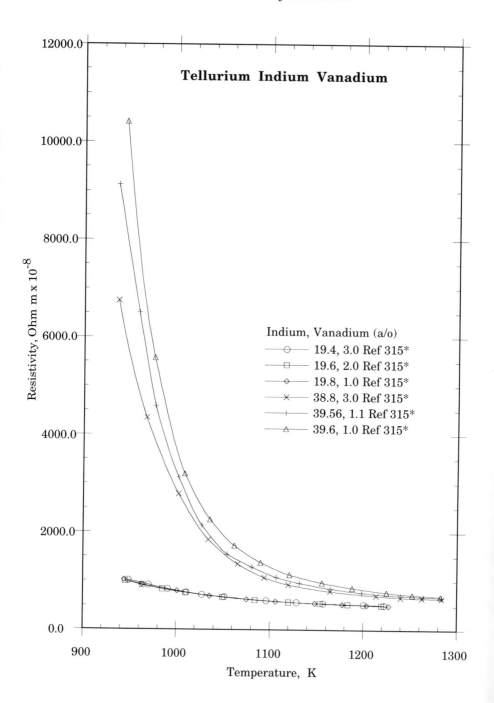

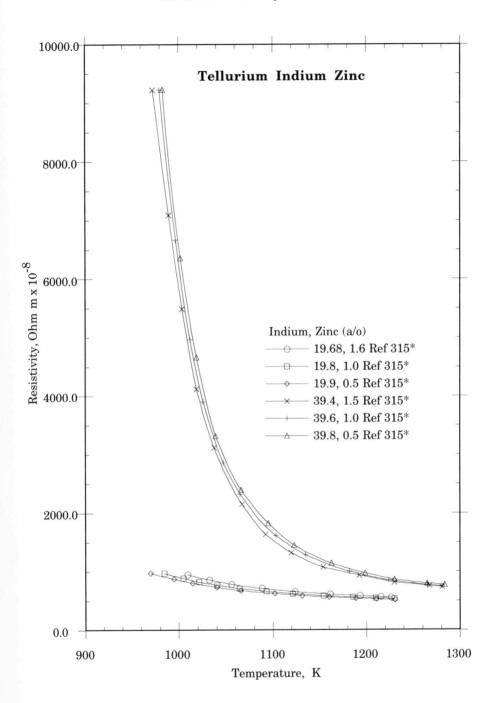

Tellurium Indium Zinc

Indium, Zinc (a/o)
- ——o—— 19.68, 1.6 Ref 315*
- ——□—— 19.8, 1.0 Ref 315*
- ——◇—— 19.9, 0.5 Ref 315*
- ——×—— 39.4, 1.5 Ref 315*
- ——+—— 39.6, 1.0 Ref 315*
- ——△—— 39.8, 0.5 Ref 315*

Resistivity, Ohm m x 10^{-8}

Temperature, K

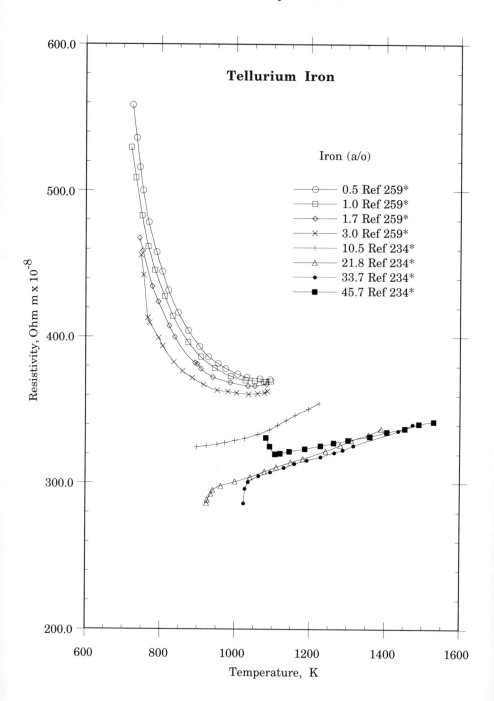

Tellurium Iron

Iron (a/o)

⊖	0.5 Ref 259*
▢	1.0 Ref 259*
◇	1.7 Ref 259*
×	3.0 Ref 259*
+	10.5 Ref 234*
△	21.8 Ref 234*
●	33.7 Ref 234*
■	45.7 Ref 234*

Resistivity, Ohm m x 10⁻⁸

Temperature, K

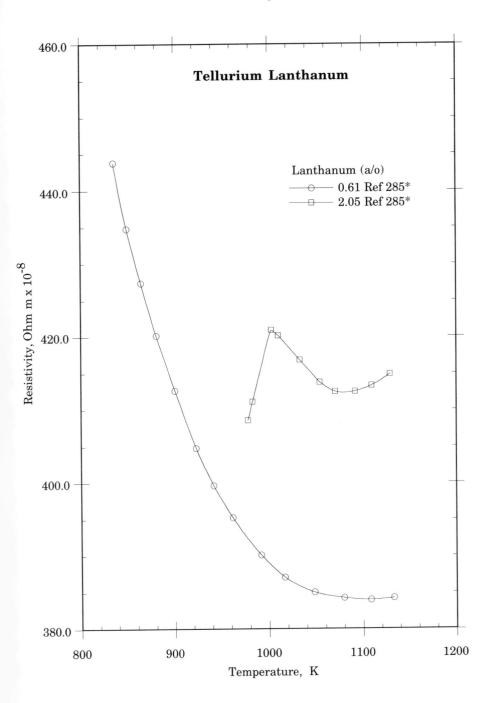

Tellurium Lanthanum

Lanthanum (a/o)
— ○ — 0.61 Ref 285*
— □ — 2.05 Ref 285*

Resistivity, Ohm m x 10^{-8}

Temperature, K

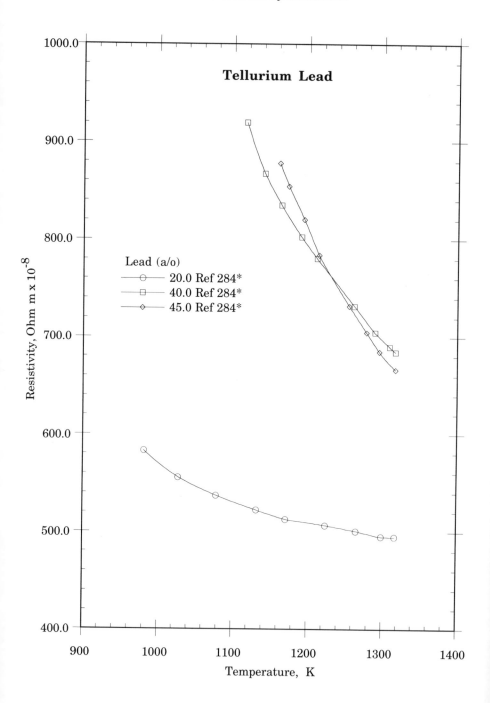

Tellurium Lead

Lead (a/o)
— ⊖ — 20.0 Ref 284*
— ☐ — 40.0 Ref 284*
— ◇ — 45.0 Ref 284*

Resistivity, Ohm m x 10⁻⁸

Temperature, K

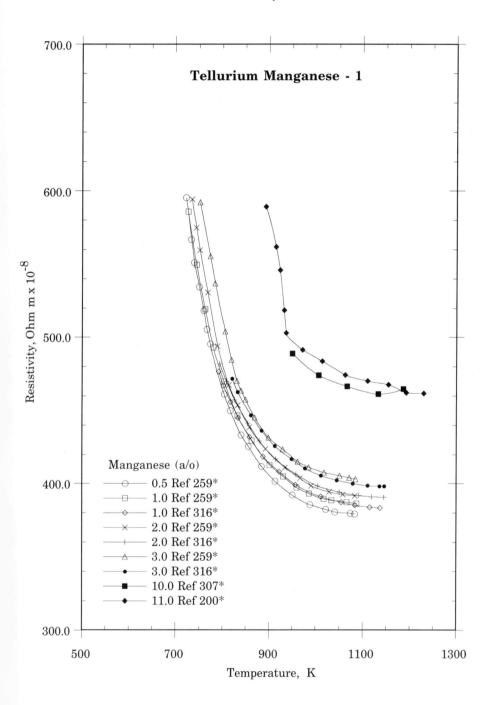

Tellurium Manganese - 1

Resistivity, Ohm m x 10^{-8}

Manganese (a/o)
— ⊖ — 0.5 Ref 259*
— ☐ — 1.0 Ref 259*
— ◇ — 1.0 Ref 316*
— ✕ — 2.0 Ref 259*
— + — 2.0 Ref 316*
— △ — 3.0 Ref 259*
— ● — 3.0 Ref 316*
— ■ — 10.0 Ref 307*
— ◆ — 11.0 Ref 200*

Temperature, K

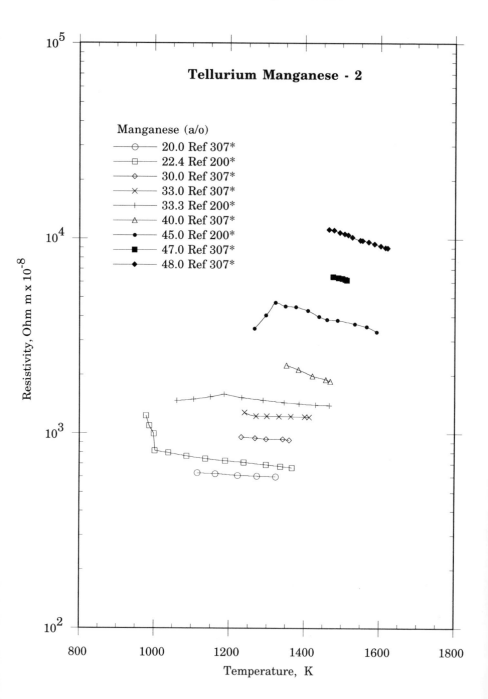

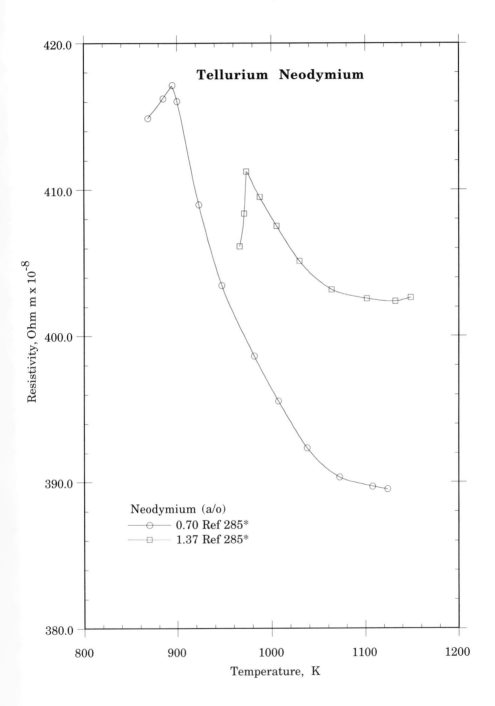

Tellurium Neodymium

Resistivity, Ohm m x 10^{-8}

Temperature, K

Neodymium (a/o)
— ⊙ — 0.70 Ref 285*
— □ — 1.37 Ref 285*

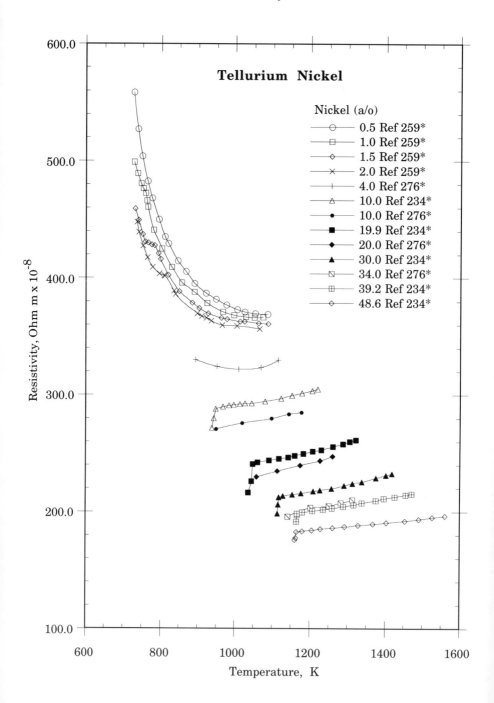

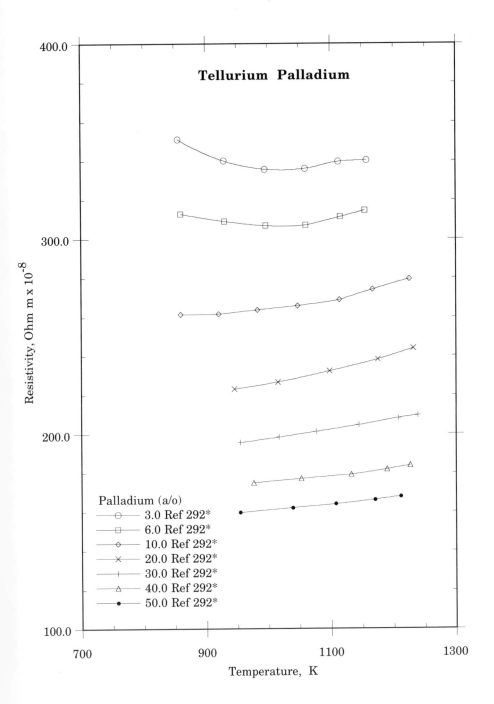

Tellurium Palladium

Resistivity, Ohm m x 10^{-8}

Temperature, K

Palladium (a/o)
- ○ 3.0 Ref 292*
- □ 6.0 Ref 292*
- ◇ 10.0 Ref 292*
- × 20.0 Ref 292*
- + 30.0 Ref 292*
- △ 40.0 Ref 292*
- ● 50.0 Ref 292*

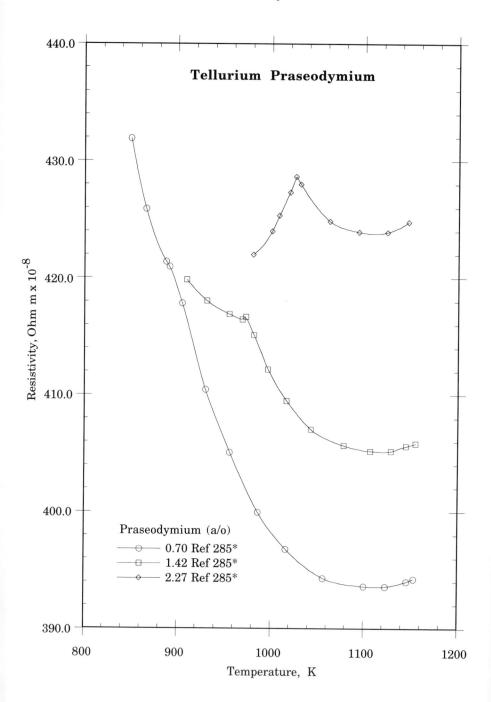

Tellurium Praseodymium

Resistivity, Ohm m x 10^{-8}

Temperature, K

Praseodymium (a/o)
○ 0.70 Ref 285*
□ 1.42 Ref 285*
◇ 2.27 Ref 285*

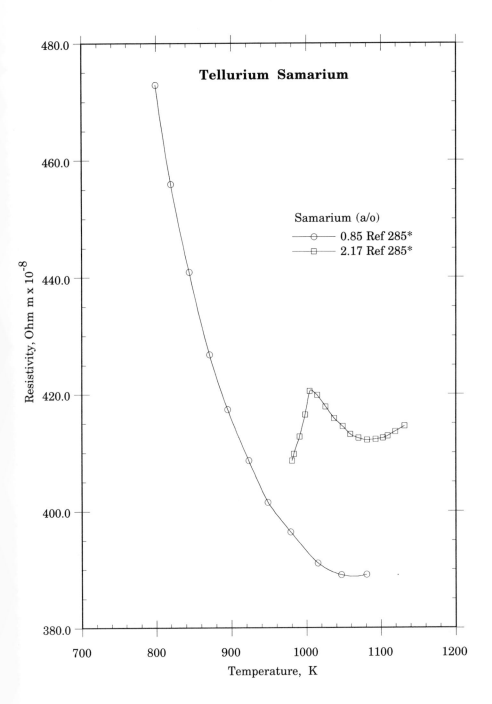

Tellurium Samarium

Samarium (a/o)
— ⊙ — 0.85 Ref 285*
— □ — 2.17 Ref 285*

Resistivity, Ohm m x 10⁻⁸

Temperature, K

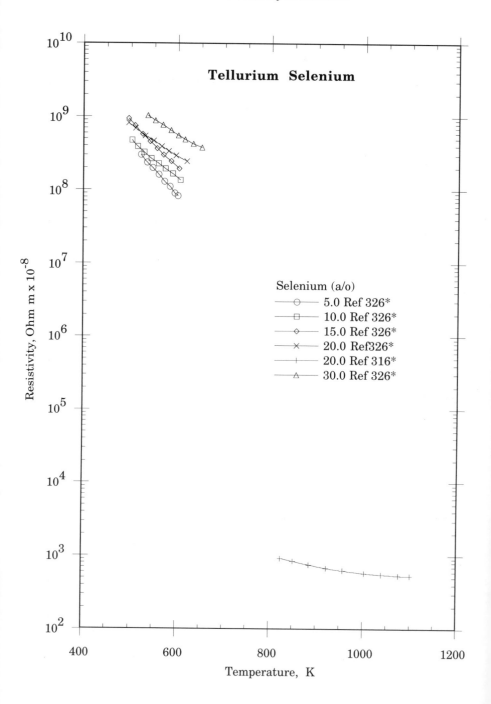

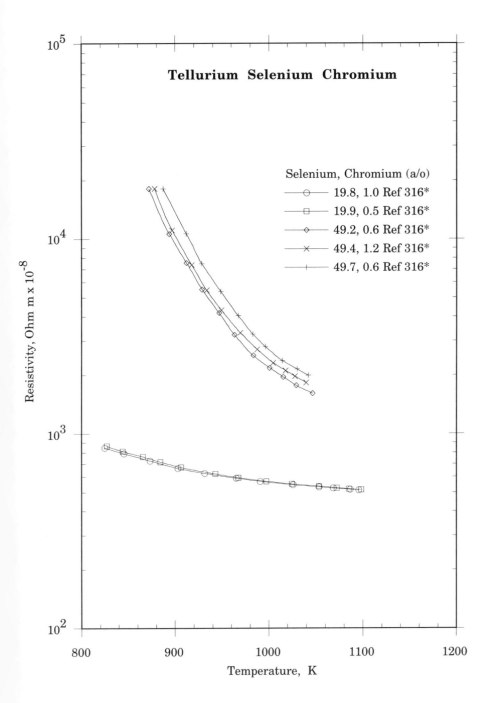

Tellurium Selenium Chromium

Selenium, Chromium (a/o)
— ○ — 19.8, 1.0 Ref 316*
— □ — 19.9, 0.5 Ref 316*
— ◇ — 49.2, 0.6 Ref 316*
— × — 49.4, 1.2 Ref 316*
— + — 49.7, 0.6 Ref 316*

Resistivity, Ohm m x 10^{-8}

Temperature, K

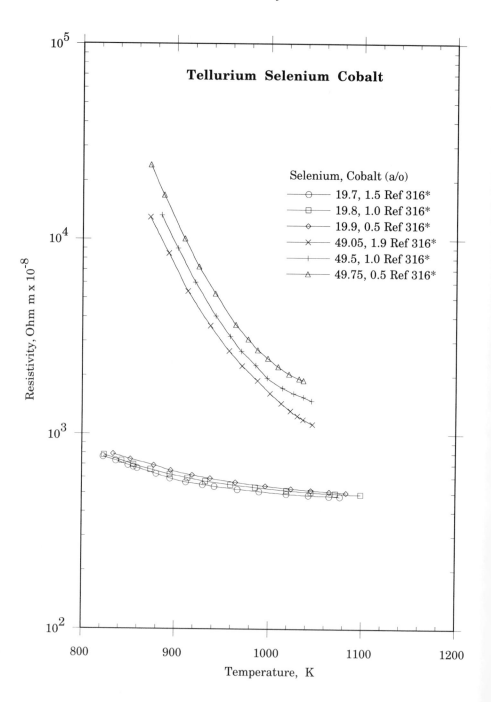

Tellurium Selenium Cobalt

Selenium, Cobalt (a/o)
- 19.7, 1.5 Ref 316*
- 19.8, 1.0 Ref 316*
- 19.9, 0.5 Ref 316*
- 49.05, 1.9 Ref 316*
- 49.5, 1.0 Ref 316*
- 49.75, 0.5 Ref 316*

Resistivity, Ohm m x 10^{-8}

Temperature, K

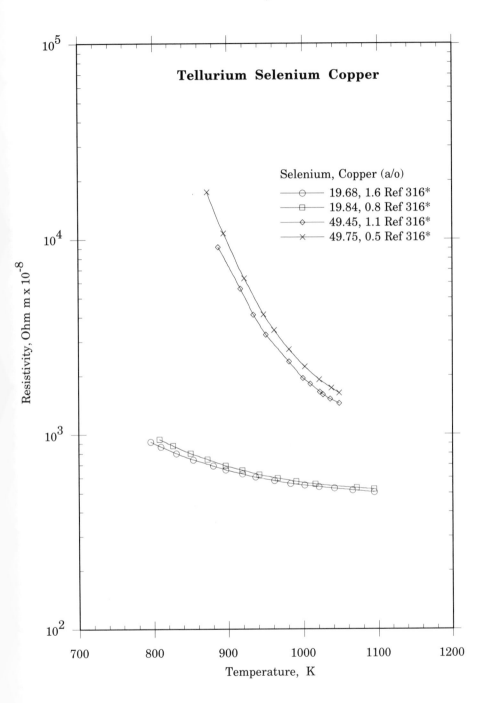

Tellurium Selenium Copper

Selenium, Copper (a/o)
- ⊖ 19.68, 1.6 Ref 316*
- □ 19.84, 0.8 Ref 316*
- ◇ 49.45, 1.1 Ref 316*
- ✕ 49.75, 0.5 Ref 316*

Resistivity, Ohm m x 10^{-8}

Temperature, K

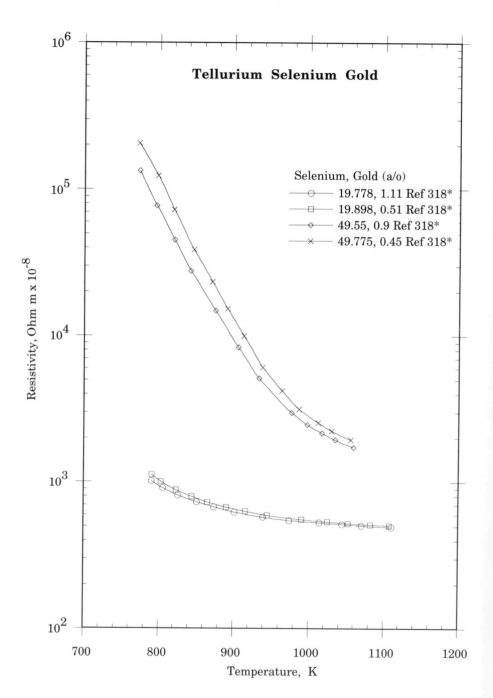

Tellurium Selenium Gold

Selenium, Gold (a/o)
— ⊙ — 19.778, 1.11 Ref 318*
— ☐ — 19.898, 0.51 Ref 318*
— ◇ — 49.55, 0.9 Ref 318*
— ✕ — 49.775, 0.45 Ref 318*

Resistivity, Ohm m x 10⁻⁸

Temperature, K

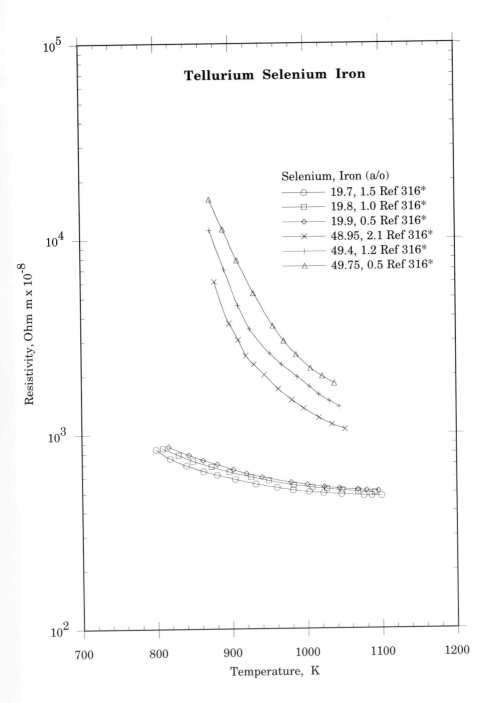

Tellurium Selenium Iron

Selenium, Iron (a/o)
— ⊖ — 19.7, 1.5 Ref 316*
— ☐ — 19.8, 1.0 Ref 316*
— ◇ — 19.9, 0.5 Ref 316*
— ✕ — 48.95, 2.1 Ref 316*
— + — 49.4, 1.2 Ref 316*
— △ — 49.75, 0.5 Ref 316*

Resistivity, Ohm m x 10⁻⁸

Temperature, K

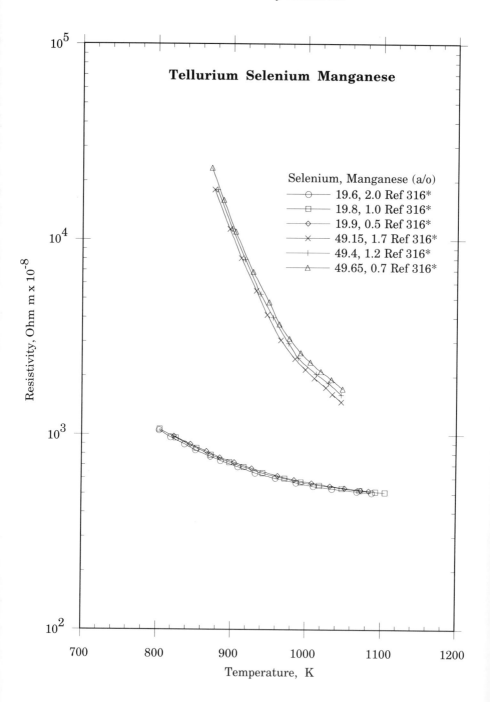

Tellurium Selenium Manganese

Selenium, Manganese (a/o)
———⊖——— 19.6, 2.0 Ref 316*
———☐——— 19.8, 1.0 Ref 316*
———◇——— 19.9, 0.5 Ref 316*
———✕——— 49.15, 1.7 Ref 316*
———+——— 49.4, 1.2 Ref 316*
———△——— 49.65, 0.7 Ref 316*

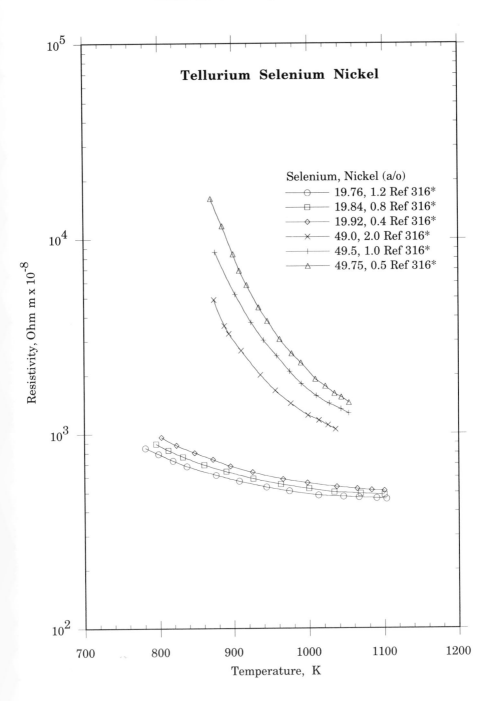

Tellurium Selenium Nickel

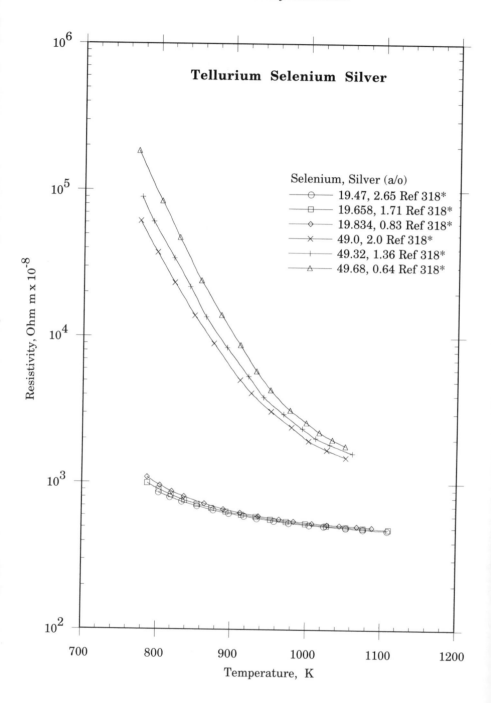

Tellurium Selenium Silver

Selenium, Silver (a/o)
- ⊖ 19.47, 2.65 Ref 318*
- ⊟ 19.658, 1.71 Ref 318*
- ◇ 19.834, 0.83 Ref 318*
- ✕ 49.0, 2.0 Ref 318*
- ⊹ 49.32, 1.36 Ref 318*
- △ 49.68, 0.64 Ref 318*

Resistivity, Ohm m x 10^{-8}

Temperature, K

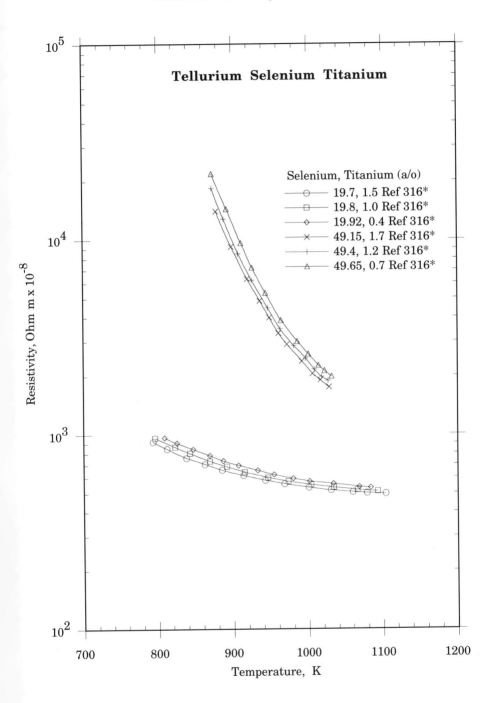

Tellurium Selenium Titanium

Selenium, Titanium (a/o)
- ─○─ 19.7, 1.5 Ref 316*
- ─□─ 19.8, 1.0 Ref 316*
- ─◇─ 19.92, 0.4 Ref 316*
- ─×─ 49.15, 1.7 Ref 316*
- ─+─ 49.4, 1.2 Ref 316*
- ─△─ 49.65, 0.7 Ref 316*

Resistivity, Ohm m x 10^{-8}

Temperature, K

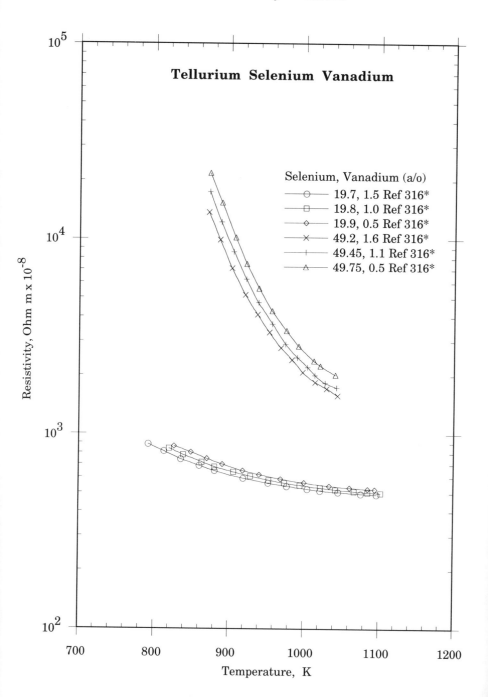

Tellurium Selenium Vanadium

Selenium, Vanadium (a/o)
- ⊙ 19.7, 1.5 Ref 316*
- □ 19.8, 1.0 Ref 316*
- ◇ 19.9, 0.5 Ref 316*
- × 49.2, 1.6 Ref 316*
- + 49.45, 1.1 Ref 316*
- △ 49.75, 0.5 Ref 316*

Resistivity, Ohm m x 10^{-8}

Temperature, K

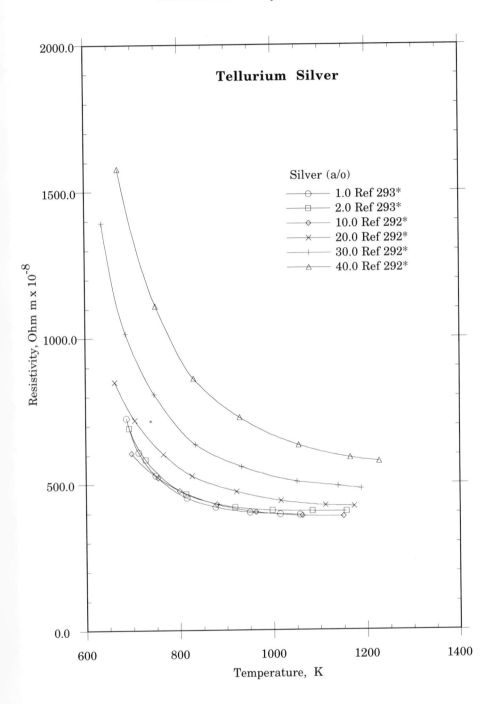

Tellurium Silver

Silver (a/o)
- ─○─ 1.0 Ref 293*
- ─□─ 2.0 Ref 293*
- ─◇─ 10.0 Ref 292*
- ─×─ 20.0 Ref 292*
- ─+─ 30.0 Ref 292*
- ─△─ 40.0 Ref 292*

Resistivity, Ohm m x 10^{-8}

Temperature, K

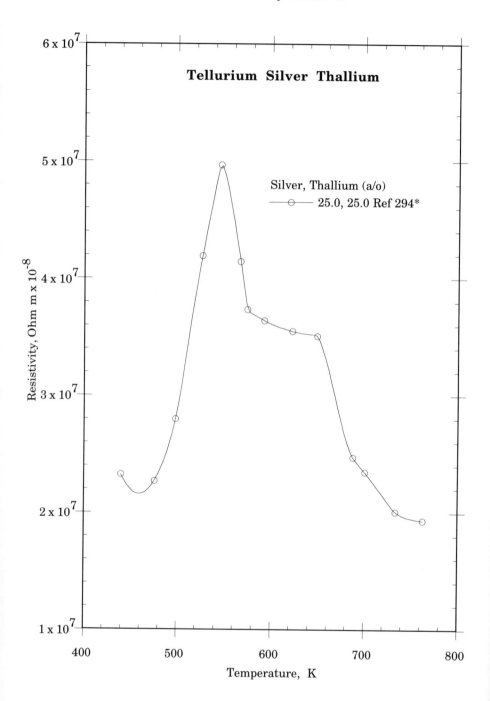

Tellurium Silver Thallium

Silver, Thallium (a/o)
——○—— 25.0, 25.0 Ref 294*

Resistivity, Ohm m x 10^{-8}

Temperature, K

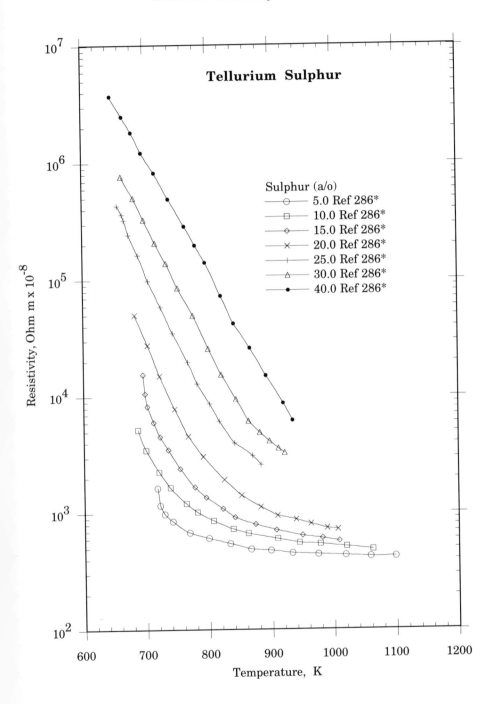

Tellurium Sulphur

Sulphur (a/o)
- ○ — 5.0 Ref 286*
- □ — 10.0 Ref 286*
- ◇ — 15.0 Ref 286*
- × — 20.0 Ref 286*
- + — 25.0 Ref 286*
- △ — 30.0 Ref 286*
- ● — 40.0 Ref 286*

Resistivity, Ohm m x 10^{-8}

Temperature, K

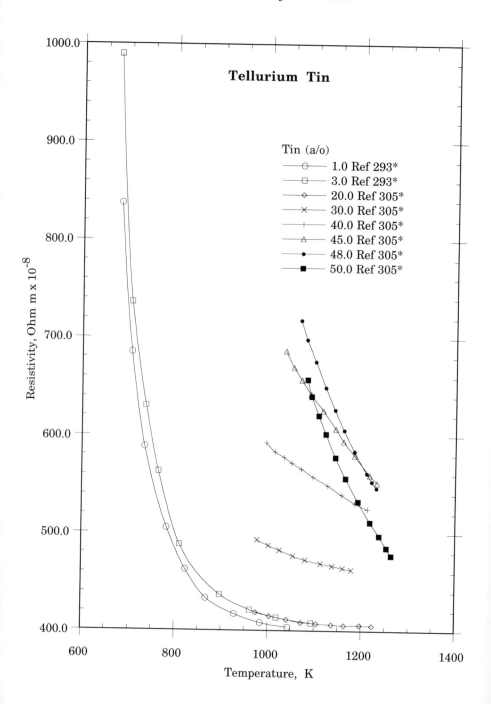

Tellurium Tin

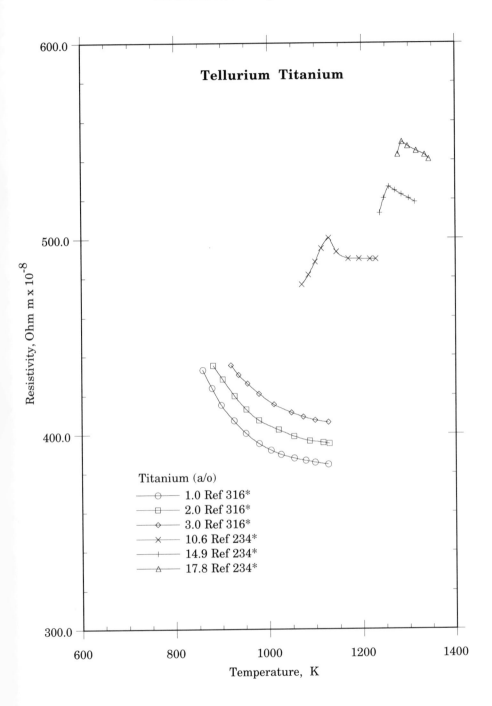

Tellurium Titanium

Resistivity, Ohm m x 10^{-8}

Temperature, K

Titanium (a/o)
- ——o—— 1.0 Ref 316*
- ——□—— 2.0 Ref 316*
- ——◇—— 3.0 Ref 316*
- ——×—— 10.6 Ref 234*
- ——+—— 14.9 Ref 234*
- ——△—— 17.8 Ref 234*

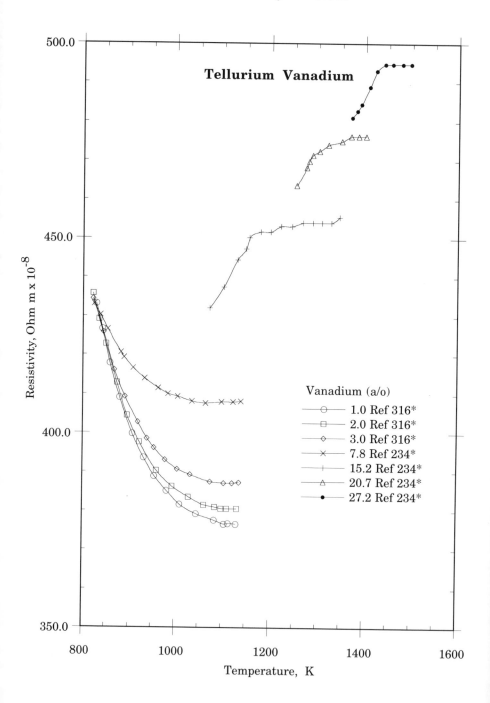

Tellurium Vanadium

Resistivity, Ohm m x 10⁻⁸

Temperature, K

Vanadium (a/o)
- ⊖— 1.0 Ref 316*
- ⊟— 2.0 Ref 316*
- ◇— 3.0 Ref 316*
- ✕— 7.8 Ref 234*
- —+— 15.2 Ref 234*
- △— 20.7 Ref 234*
- ●— 27.2 Ref 234*

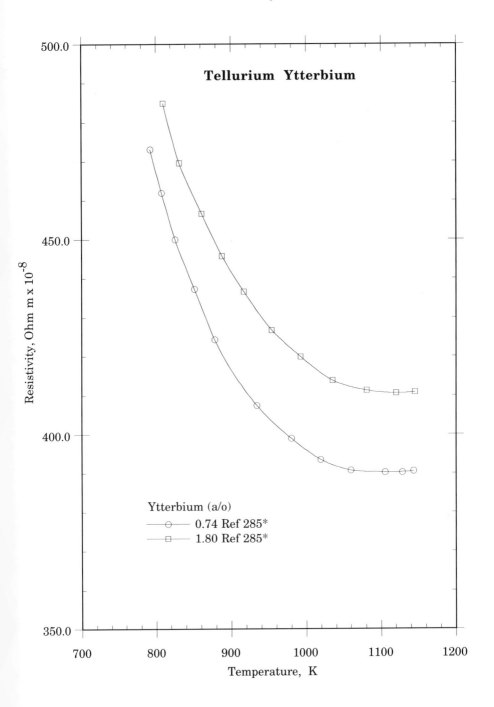

Tellurium Ytterbium

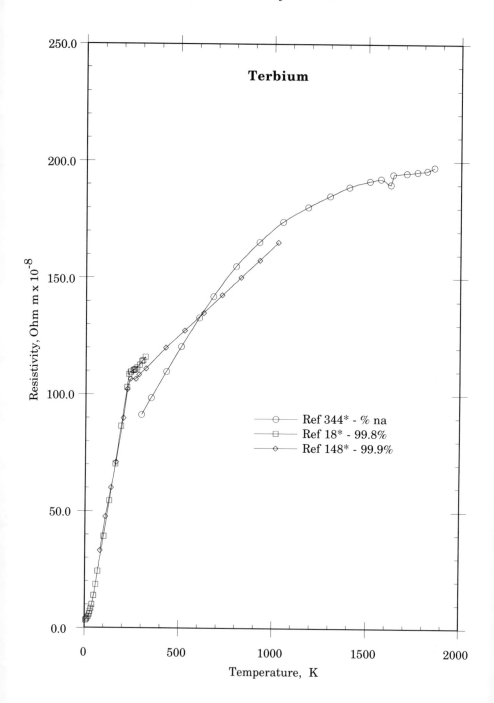

Terbium

Resistivity, Ohm m x 10^{-8}

Temperature, K

Ref 344* - % na
Ref 18* - 99.8%
Ref 148* - 99.9%

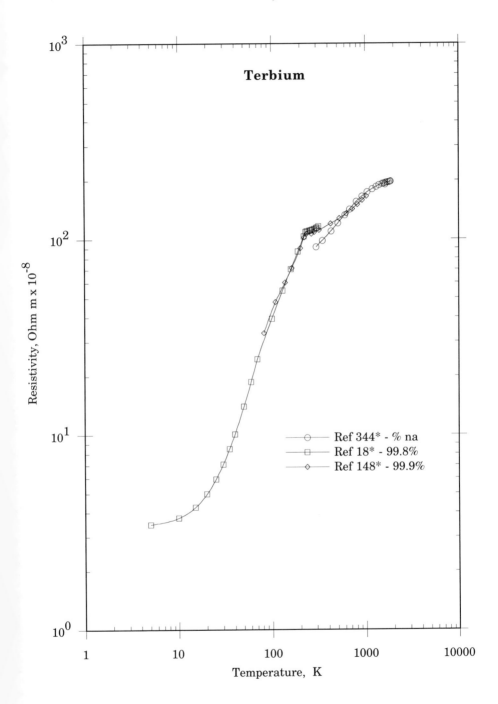

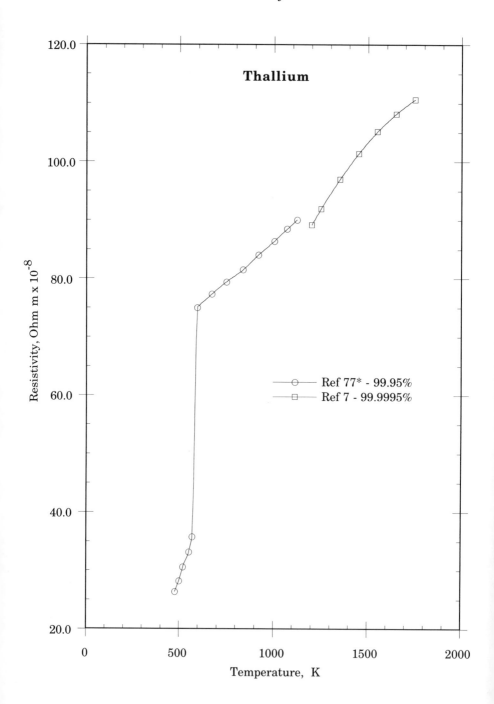

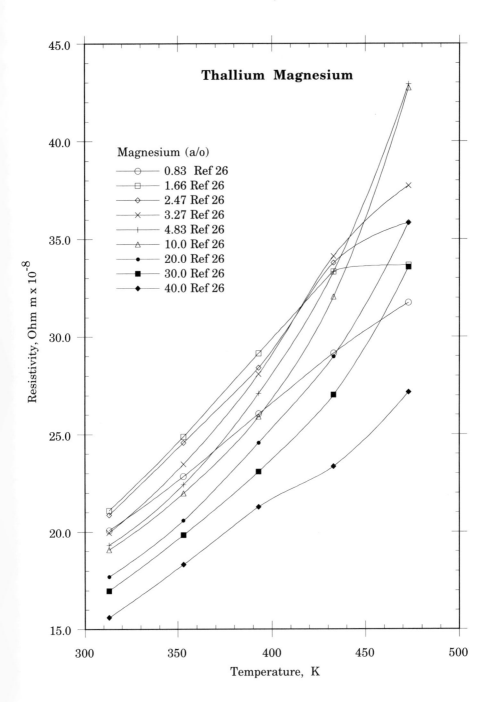

Thallium Magnesium

Magnesium (a/o)
- —○— 0.83 Ref 26
- —□— 1.66 Ref 26
- —◇— 2.47 Ref 26
- —×— 3.27 Ref 26
- —+— 4.83 Ref 26
- —△— 10.0 Ref 26
- —●— 20.0 Ref 26
- —■— 30.0 Ref 26
- —◆— 40.0 Ref 26

Resistivity, Ohm m x 10^{-8}

Temperature, K

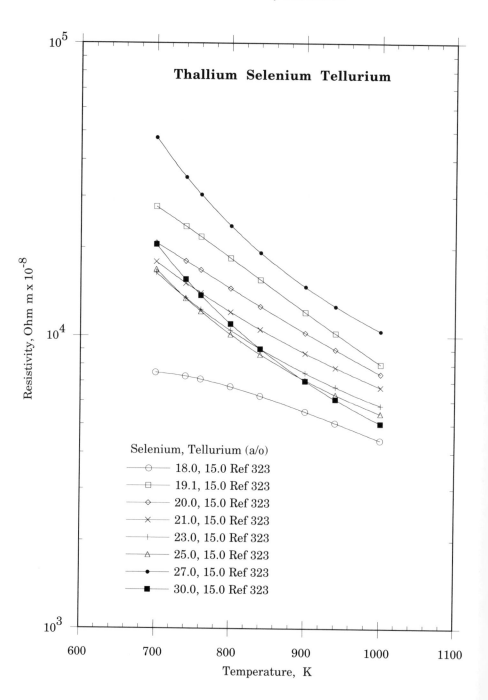

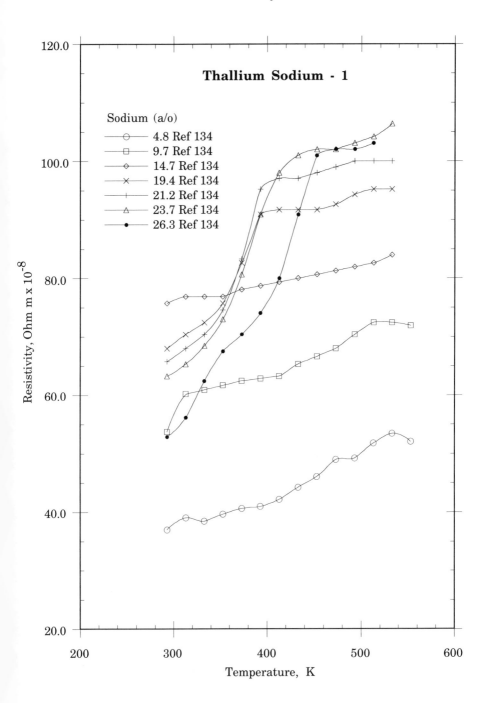

Thallium Sodium - 1

Sodium (a/o)
- ○ 4.8 Ref 134
- □ 9.7 Ref 134
- ◇ 14.7 Ref 134
- × 19.4 Ref 134
- + 21.2 Ref 134
- △ 23.7 Ref 134
- ● 26.3 Ref 134

Resistivity, Ohm m x 10^{-8}

Temperature, K

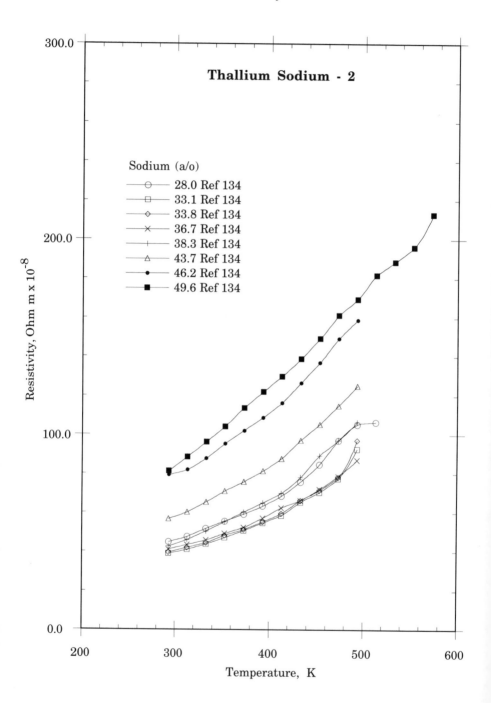

Thallium Sodium - 2

Sodium (a/o)
- 28.0 Ref 134
- 33.1 Ref 134
- 33.8 Ref 134
- 36.7 Ref 134
- 38.3 Ref 134
- 43.7 Ref 134
- 46.2 Ref 134
- 49.6 Ref 134

Resistivity, Ohm m x 10^{-8}

Temperature, K

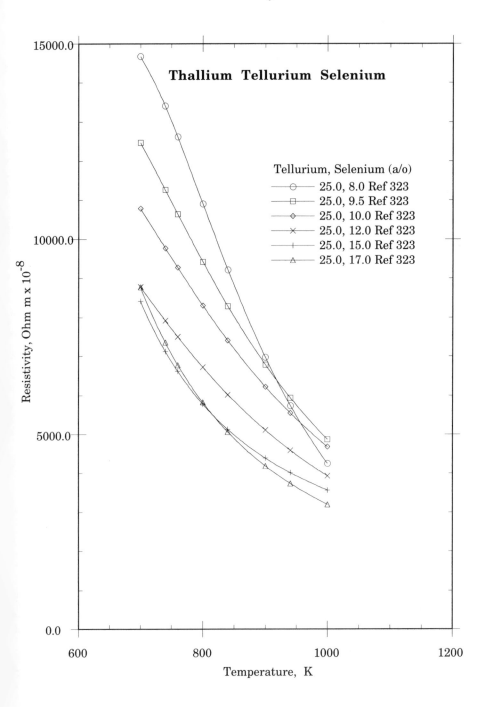

Thallium Tellurium Selenium

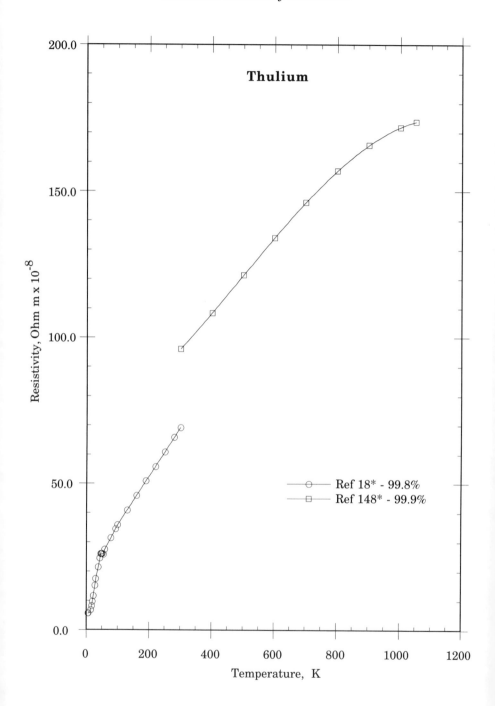

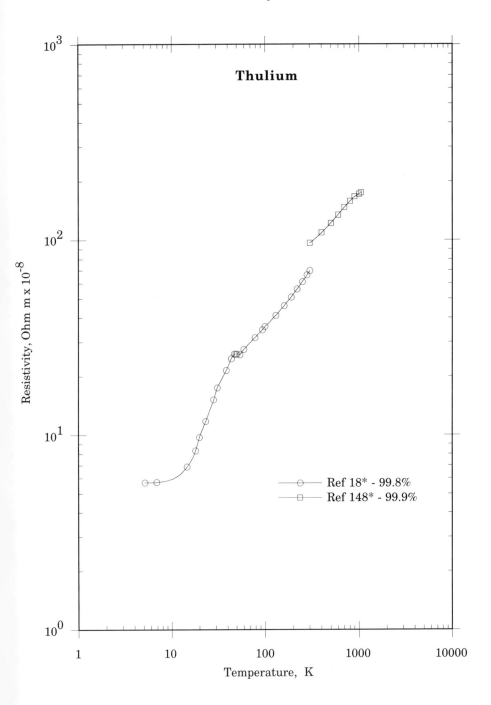

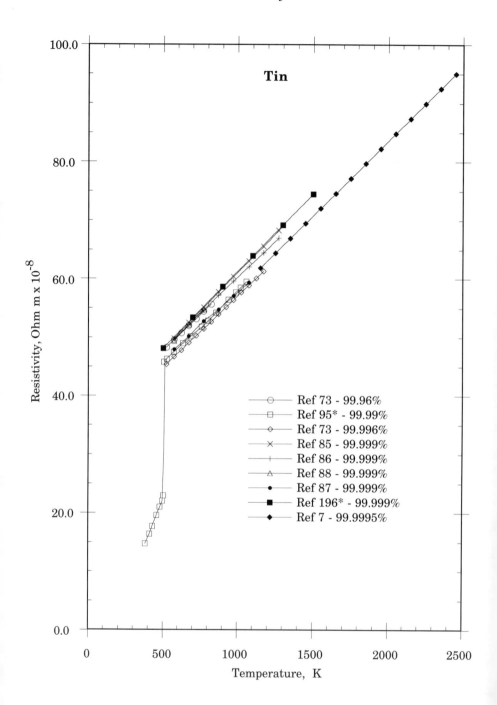

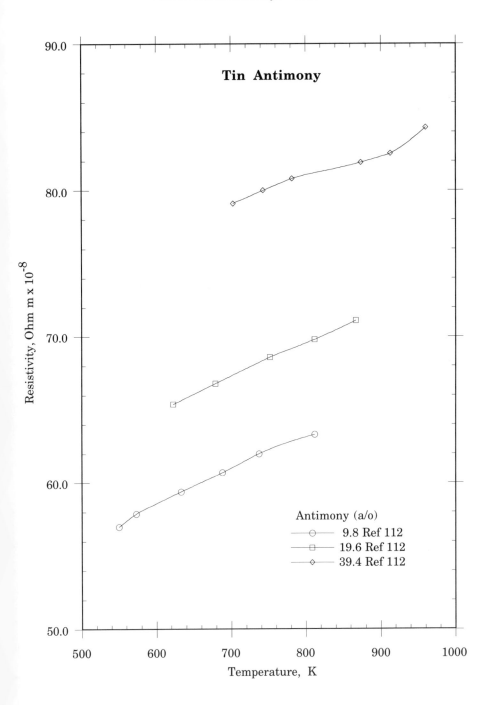

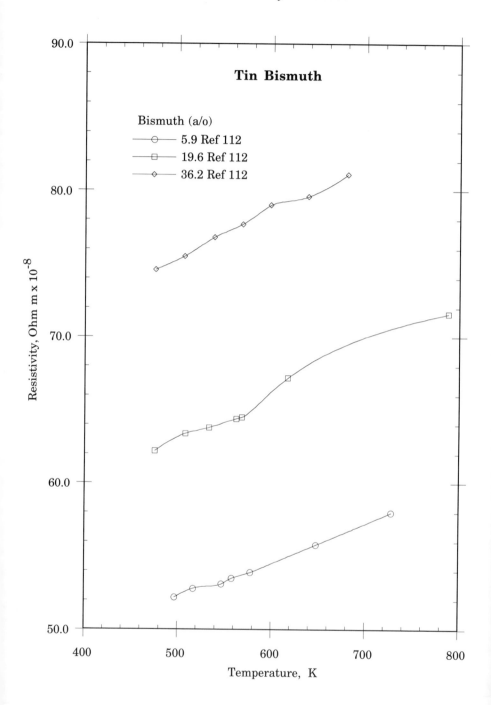

Tin Bismuth

Bismuth (a/o)
- ⊖ 5.9 Ref 112
- ☐ 19.6 Ref 112
- ◇ 36.2 Ref 112

Resistivity, Ohm m x 10^{-8}

Temperature, K

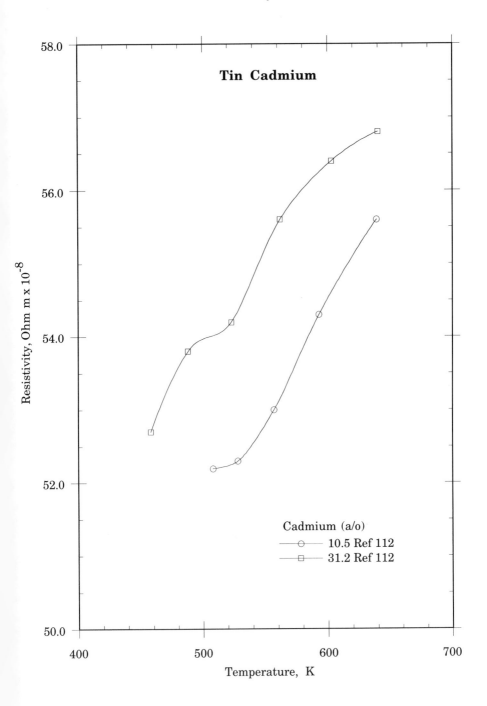

Tin Cadmium

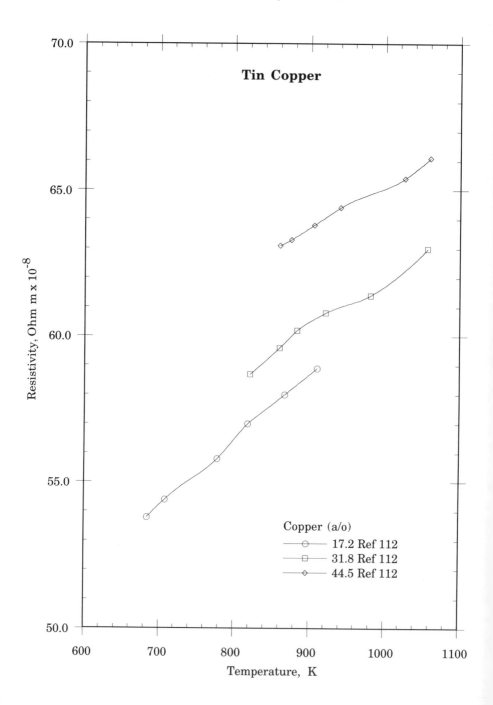

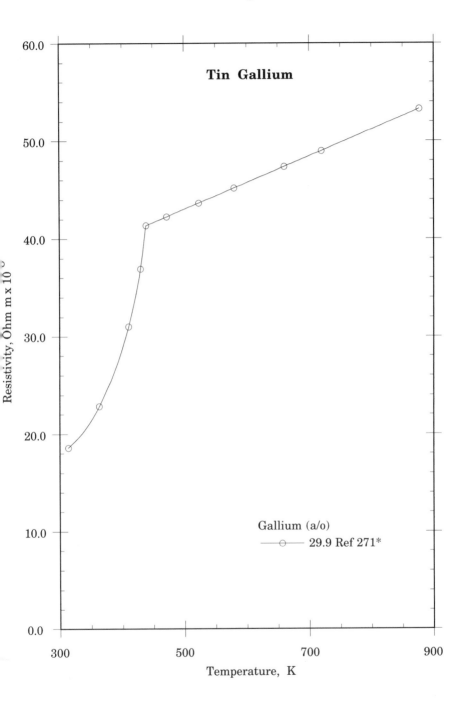

Tin Gallium

Gallium (a/o)
—⊙— 29.9 Ref 271*

Resistivity, Ohm m x 10⁻⁸

Temperature, K

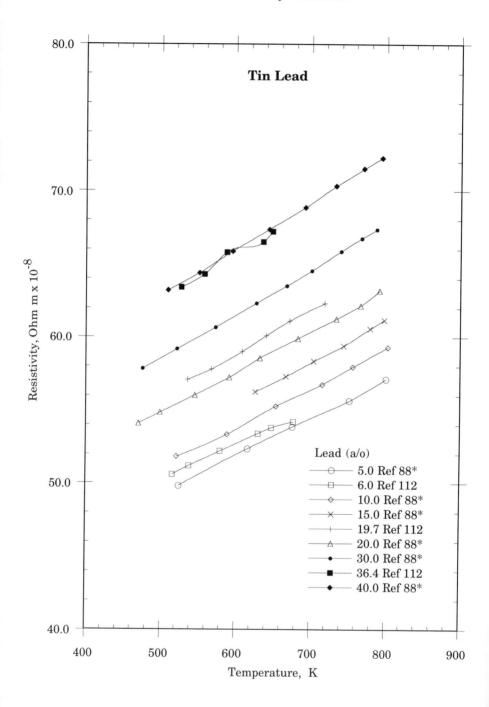

Tin Lead

Lead (a/o)
- ○ 5.0 Ref 88*
- □ 6.0 Ref 112
- ◇ 10.0 Ref 88*
- × 15.0 Ref 88*
- + 19.7 Ref 112
- △ 20.0 Ref 88*
- ● 30.0 Ref 88*
- ■ 36.4 Ref 112
- ◆ 40.0 Ref 88*

Resistivity, Ohm m x 10^{-8}

Temperature, K

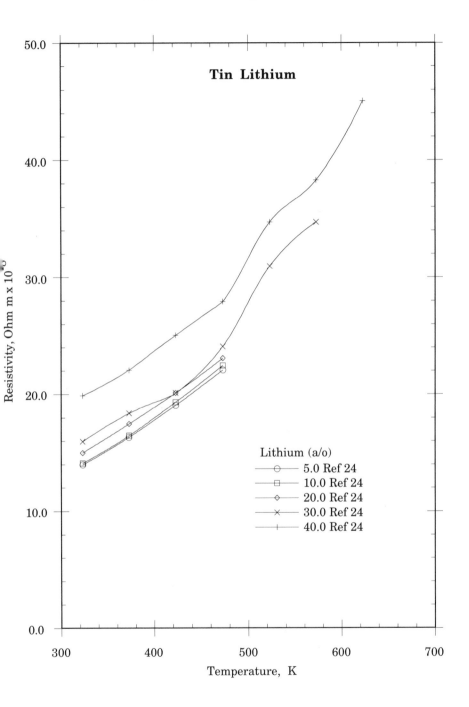

Tin Lithium

Resistivity, Ohm m x 10^{-8}

Temperature, K

Lithium (a/o)
—o— 5.0 Ref 24
—□— 10.0 Ref 24
—◇— 20.0 Ref 24
—×— 30.0 Ref 24
—+— 40.0 Ref 24

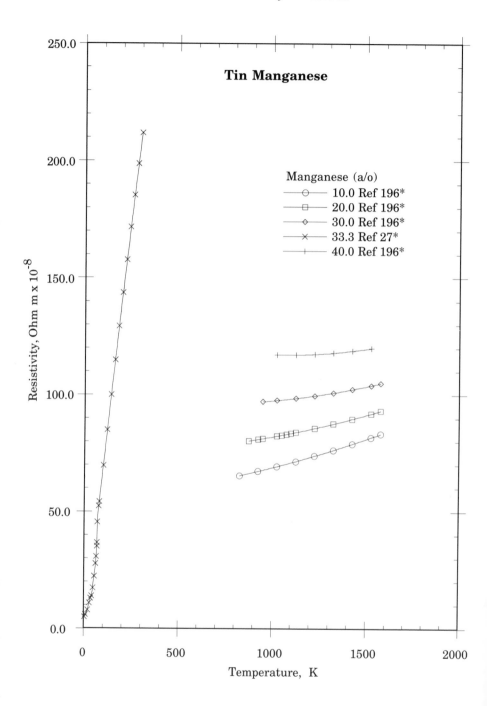

Tin Manganese

Manganese (a/o)
- 10.0 Ref 196*
- 20.0 Ref 196*
- 30.0 Ref 196*
- 33.3 Ref 27*
- 40.0 Ref 196*

Resistivity, Ohm m x 10⁻⁸

Temperature, K

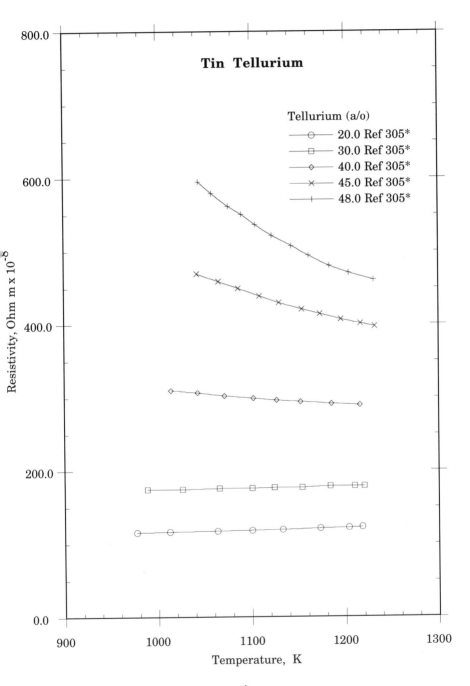

Tin Tellurium

Tellurium (a/o)
- ─○─ 20.0 Ref 305*
- ─□─ 30.0 Ref 305*
- ─◇─ 40.0 Ref 305*
- ─×─ 45.0 Ref 305*
- ─+─ 48.0 Ref 305*

Resistivity, Ohm m x 10^{-8}

Temperature, K

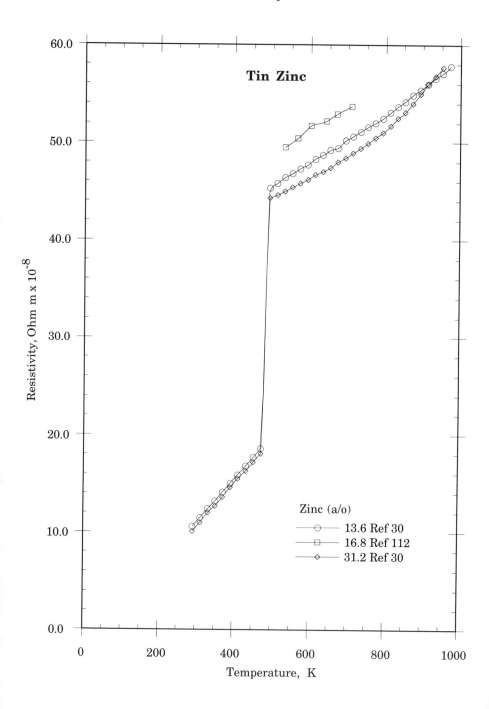

Tin Zinc

Resistivity, Ohm m x 10^{-8}

Temperature, K

Zinc (a/o)
—⊙— 13.6 Ref 30
—□— 16.8 Ref 112
—◇— 31.2 Ref 30

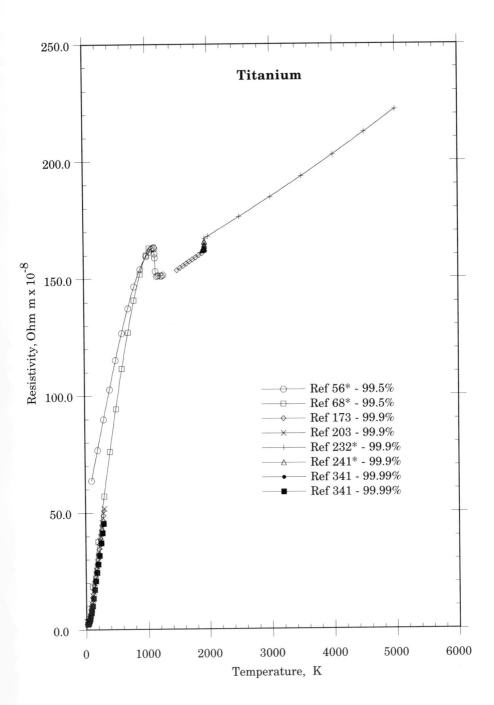

Titanium

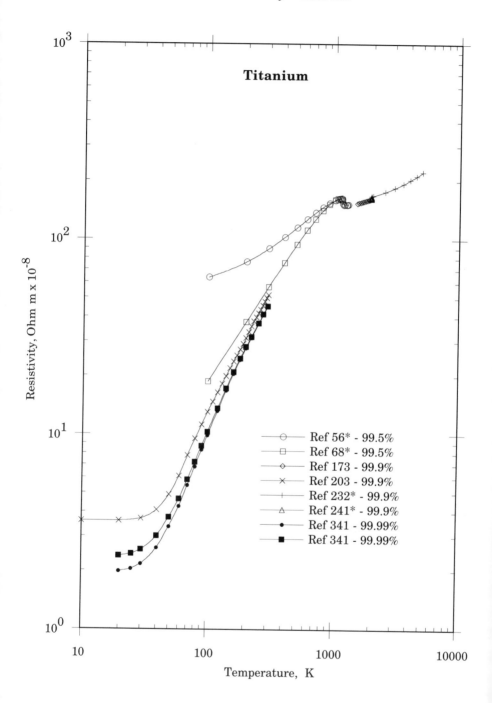

Titanium

Resistivity, Ohm m x 10^{-8}

Temperature, K

Ref 56* - 99.5%
Ref 68* - 99.5%
Ref 173 - 99.9%
Ref 203 - 99.9%
Ref 232* - 99.9%
Ref 241* - 99.9%
Ref 341 - 99.99%
Ref 341 - 99.99%

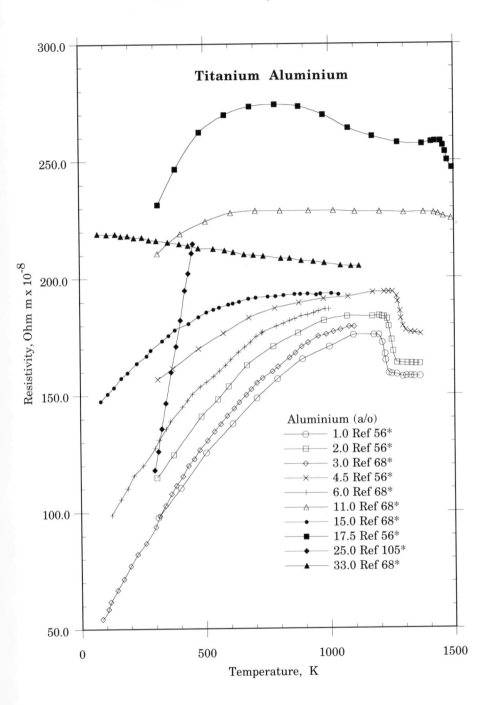

Titanium Aluminium

Aluminium (a/o)
- 1.0 Ref 56*
- 2.0 Ref 56*
- 3.0 Ref 68*
- 4.5 Ref 56*
- 6.0 Ref 68*
- 11.0 Ref 68*
- 15.0 Ref 68*
- 17.5 Ref 56*
- 25.0 Ref 105*
- 33.0 Ref 68*

Resistivity, Ohm m x 10^{-8}

Temperature, K

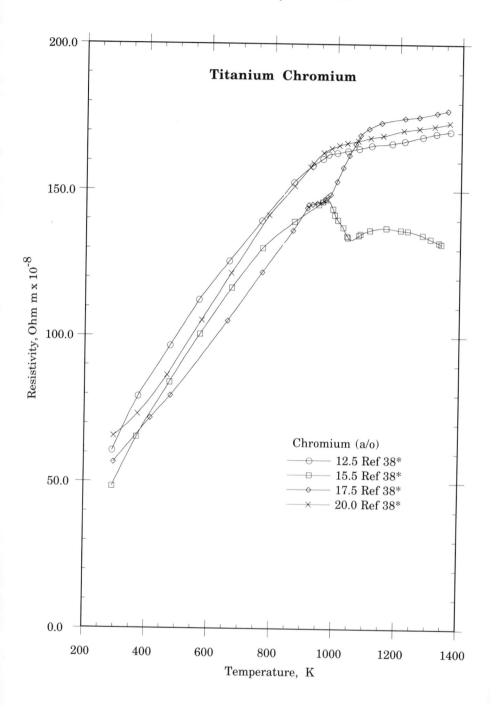

Titanium Chromium

Resistivity, Ohm m x 10⁻⁸

Temperature, K

Chromium (a/o)
- 12.5 Ref 38*
- 15.5 Ref 38*
- 17.5 Ref 38*
- 20.0 Ref 38*

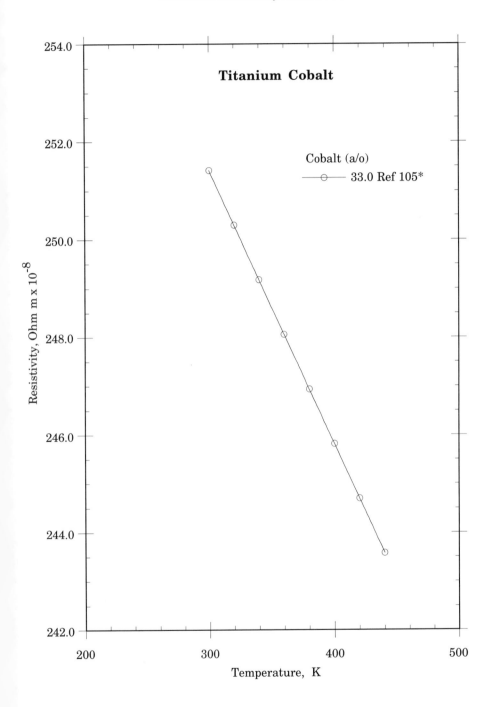

Titanium Cobalt

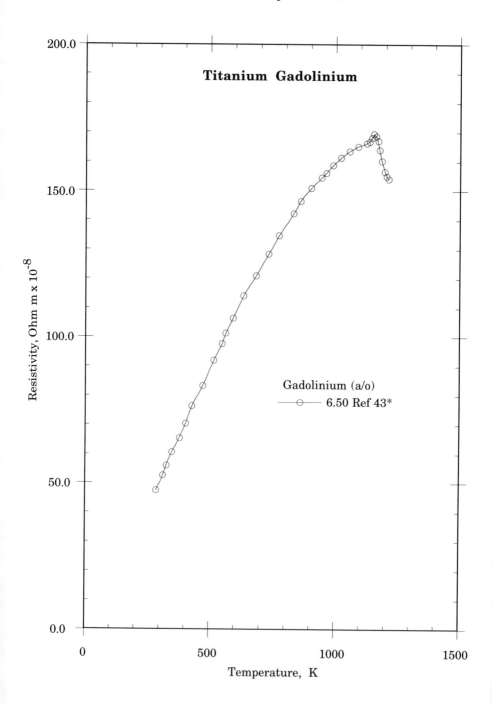

Titanium Gadolinium

Gadolinium (a/o)
6.50 Ref 43*

Resistivity, Ohm m x 10^{-8}

Temperature, K

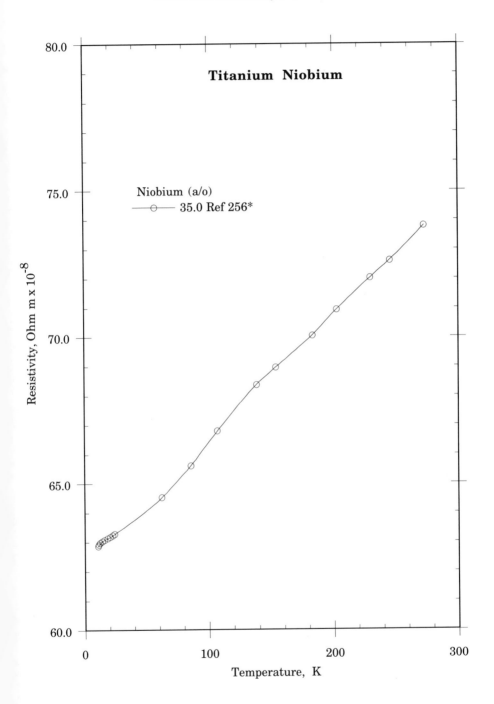

Titanium Niobium

Niobium (a/o)
——○—— 35.0 Ref 256*

Resistivity, Ohm m x 10^{-8}

Temperature, K

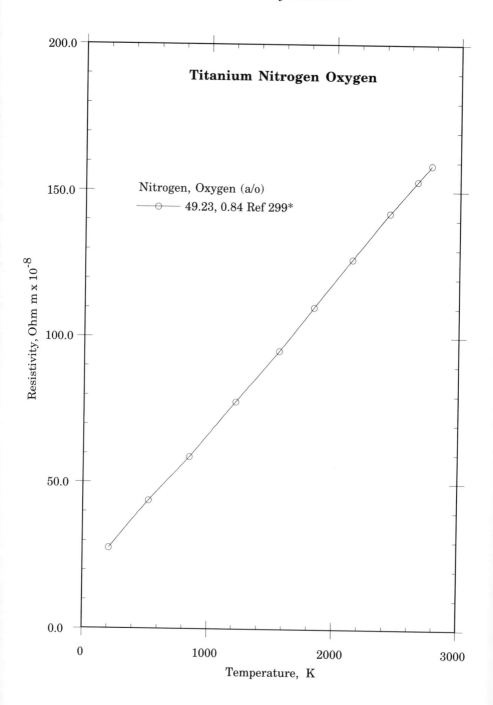

Titanium Nitrogen Oxygen

Nitrogen, Oxygen (a/o)
—o— 49.23, 0.84 Ref 299*

Resistivity, Ohm m x 10^{-8}

Temperature, K

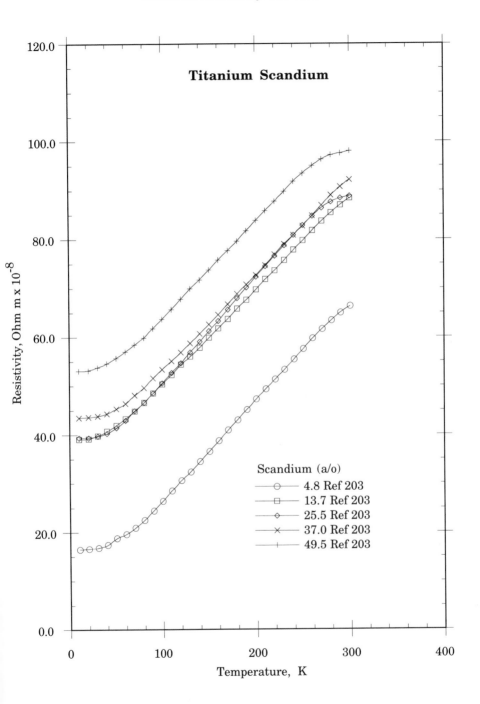

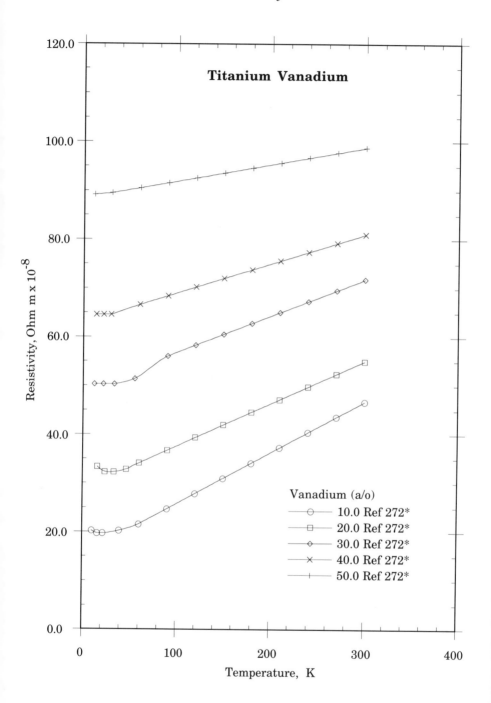

Titanium Vanadium

Resistivity, Ohm m x 10⁻⁸

Temperature, K

Vanadium (a/o)
- —○— 10.0 Ref 272*
- —□— 20.0 Ref 272*
- —◇— 30.0 Ref 272*
- —×— 40.0 Ref 272*
- —+— 50.0 Ref 272*

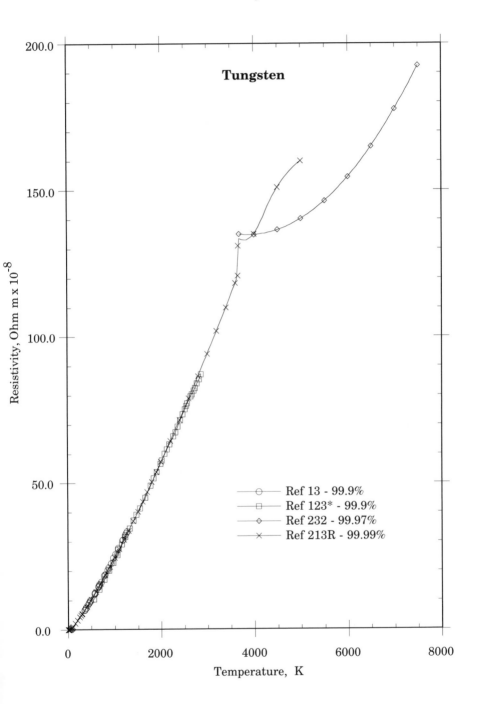

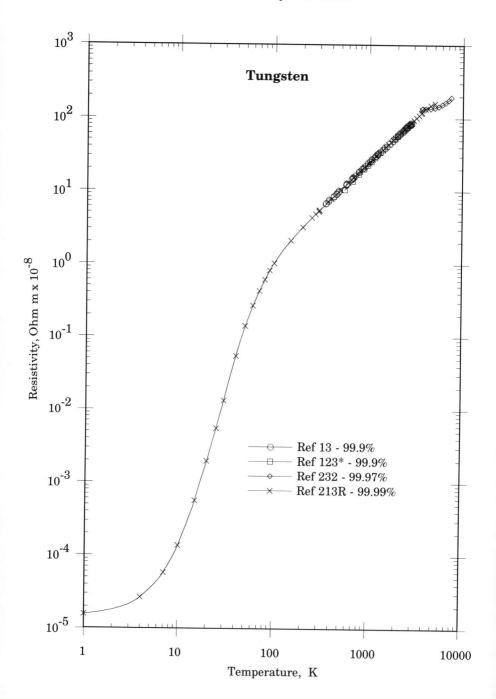

Tungsten

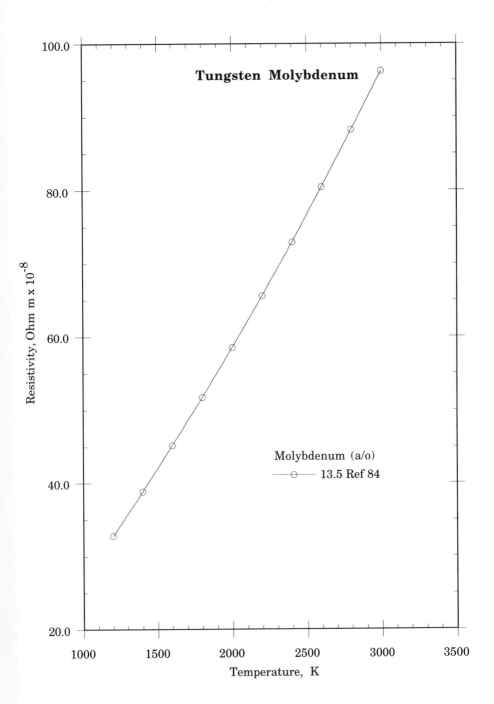

Tungsten Molybdenum

Molybdenum (a/o)
—○— 13.5 Ref 84

Resistivity, Ohm m x 10^{-8}

Temperature, K

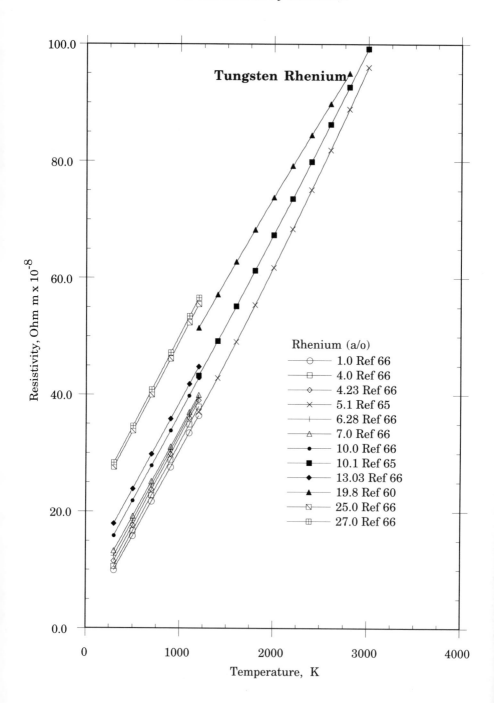

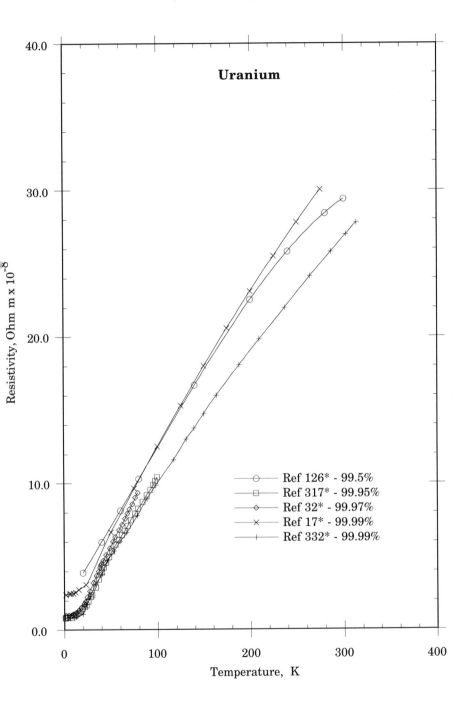

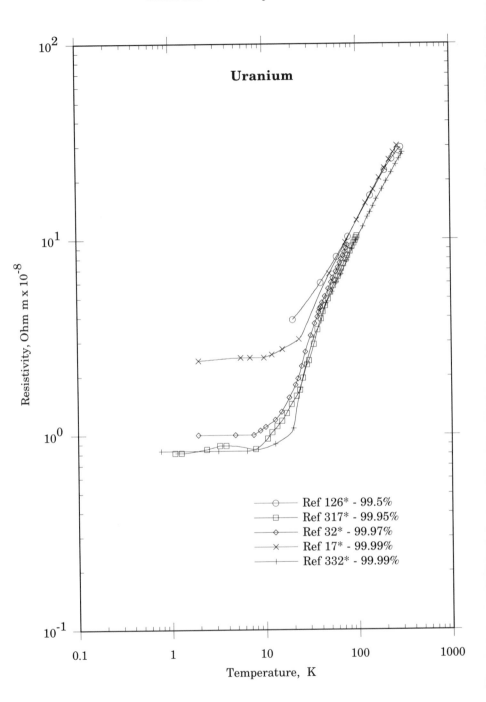

Uranium

Ref 126* - 99.5%
Ref 317* - 99.95%
Ref 32* - 99.97%
Ref 17* - 99.99%
Ref 332* - 99.99%

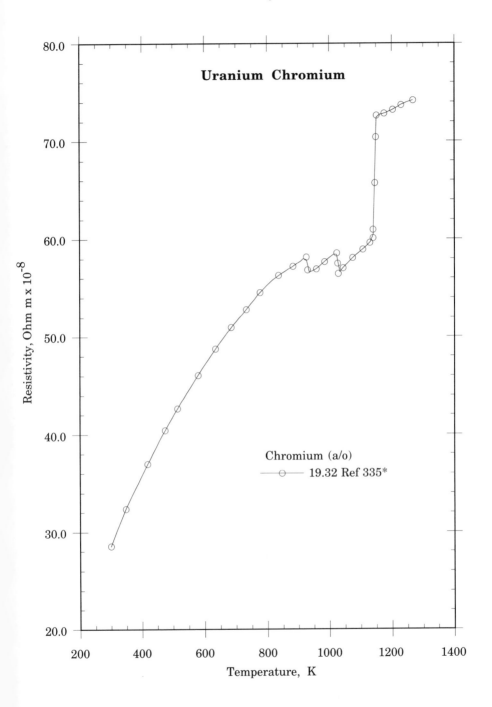

Uranium Chromium

Resistivity, Ohm m x 10^{-8}

Temperature, K

Chromium (a/o)
—⊙— 19.32 Ref 335*

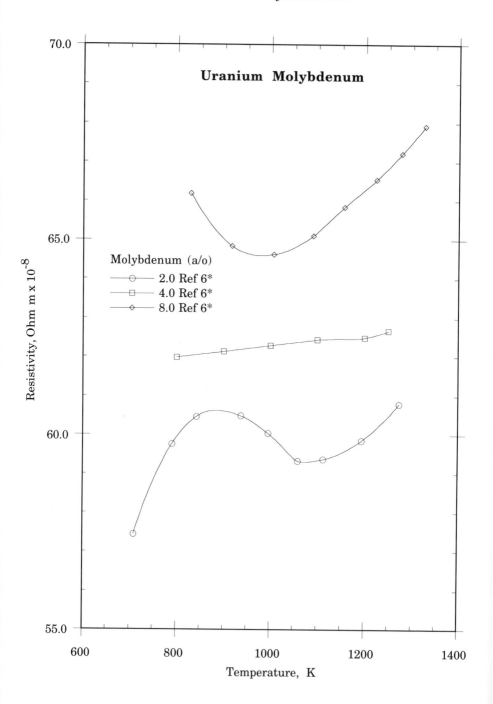

Uranium Molybdenum

Molybdenum (a/o)
— ⊖ — 2.0 Ref 6*
— □ — 4.0 Ref 6*
— ◇ — 8.0 Ref 6*

Resistivity, Ohm m x 10⁻⁸

Temperature, K

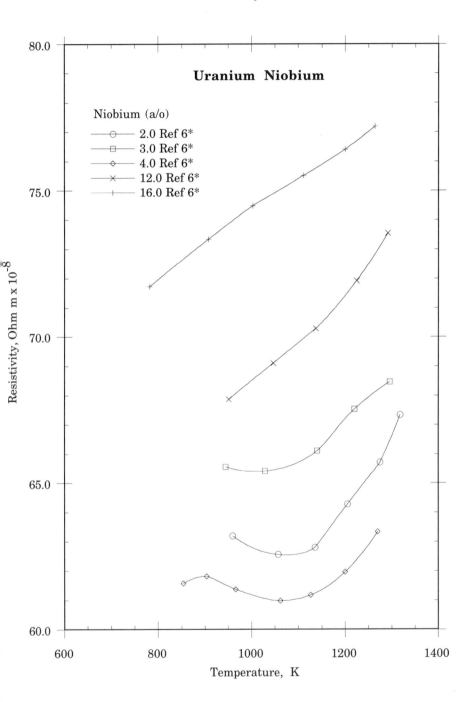

Uranium Niobium

Niobium (a/o)
- —⊙— 2.0 Ref 6*
- —□— 3.0 Ref 6*
- —◇— 4.0 Ref 6*
- —✕— 12.0 Ref 6*
- —+— 16.0 Ref 6*

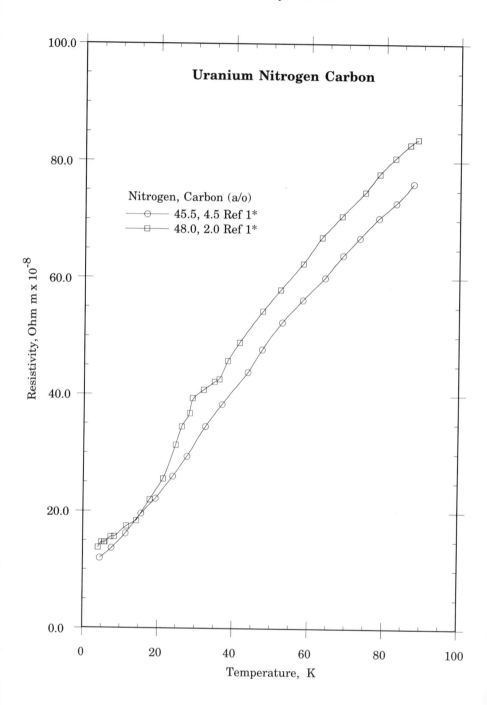

Uranium Nitrogen Carbon

Nitrogen, Carbon (a/o)
- ─○─ 45.5, 4.5 Ref 1*
- ─□─ 48.0, 2.0 Ref 1*

Resistivity, Ohm m x 10⁻⁸

Temperature, K

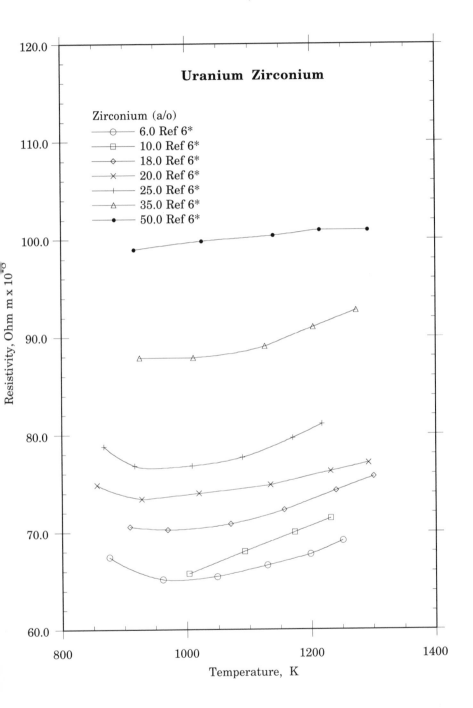

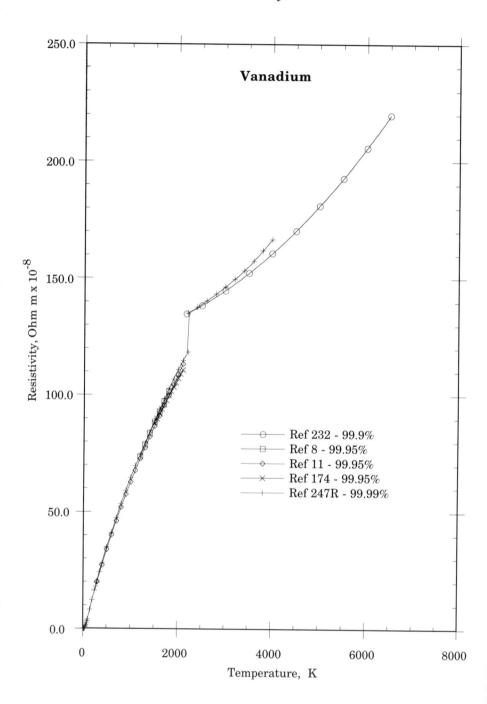

Vanadium

Resistivity, Ohm m x 10^{-8}

Temperature, K

Ref 232 - 99.9%
Ref 8 - 99.95%
Ref 11 - 99.95%
Ref 174 - 99.95%
Ref 247R - 99.99%

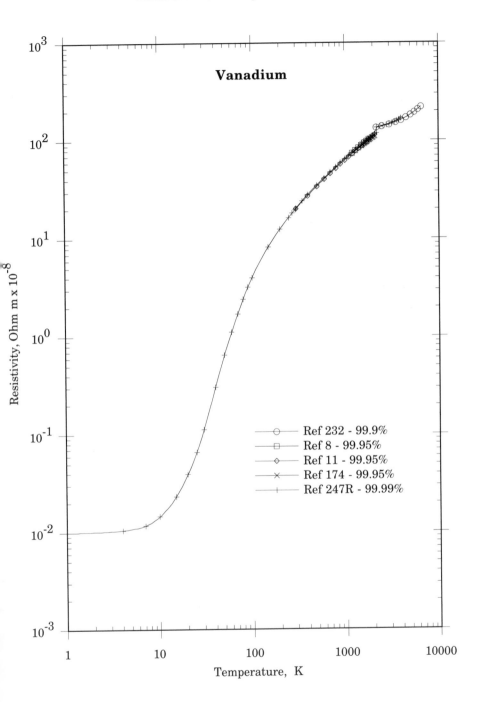

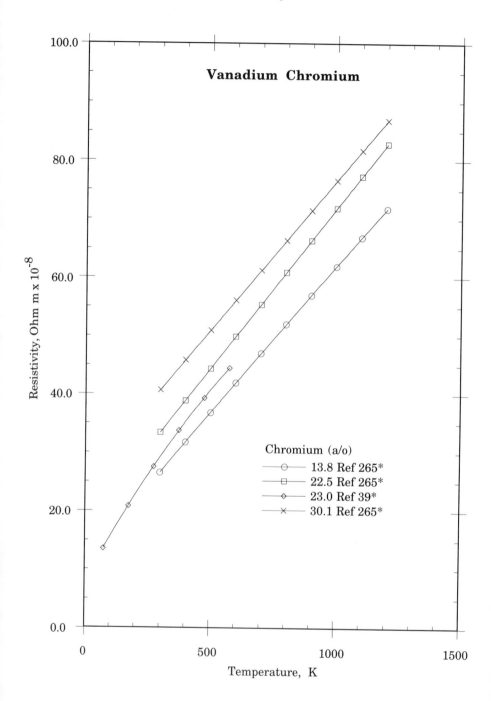

Vanadium Chromium

Chromium (a/o)
— ○ — 13.8 Ref 265*
— ☐ — 22.5 Ref 265*
— ◇ — 23.0 Ref 39*
— ✕ — 30.1 Ref 265*

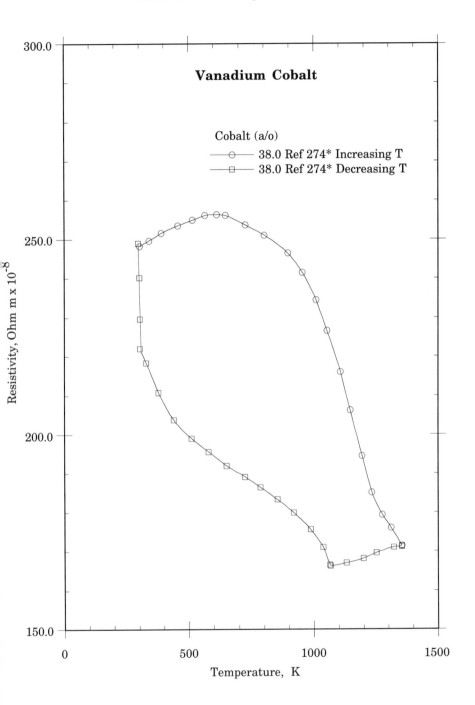

Vanadium Cobalt

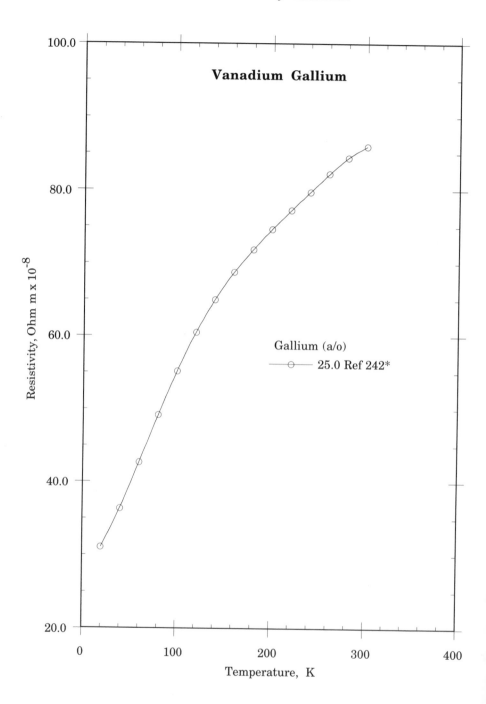

Vanadium Gallium

Gallium (a/o)
—○— 25.0 Ref 242*

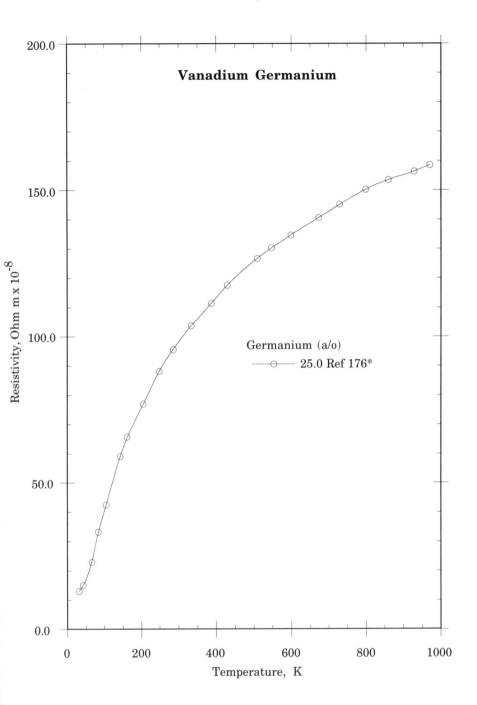

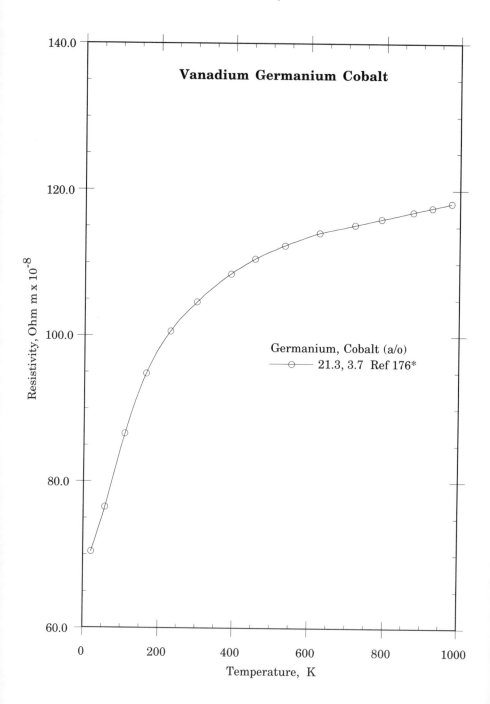

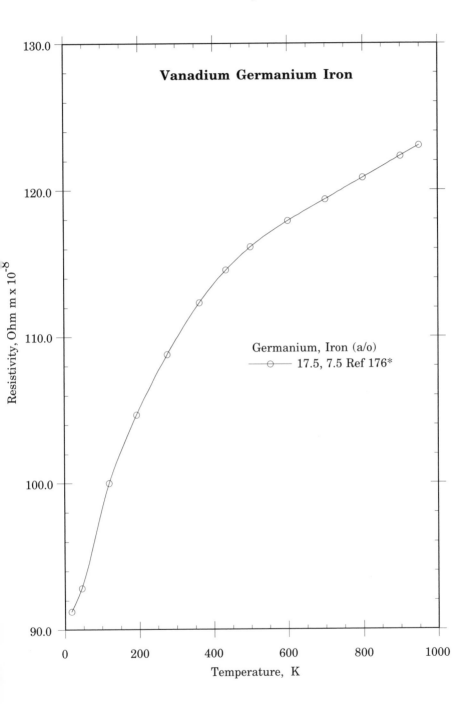

Vanadium Germanium Iron

Resistivity, Ohm m x 10^{-8}

Temperature, K

Germanium, Iron (a/o)
— ⊝ — 17.5, 7.5 Ref 176*

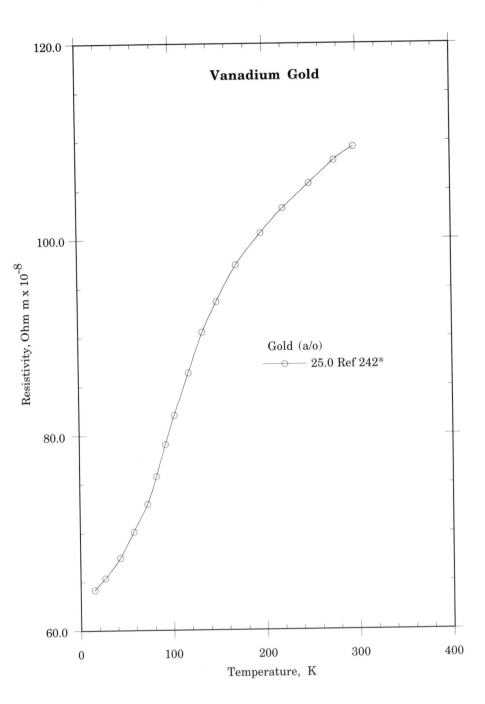

Vanadium Gold

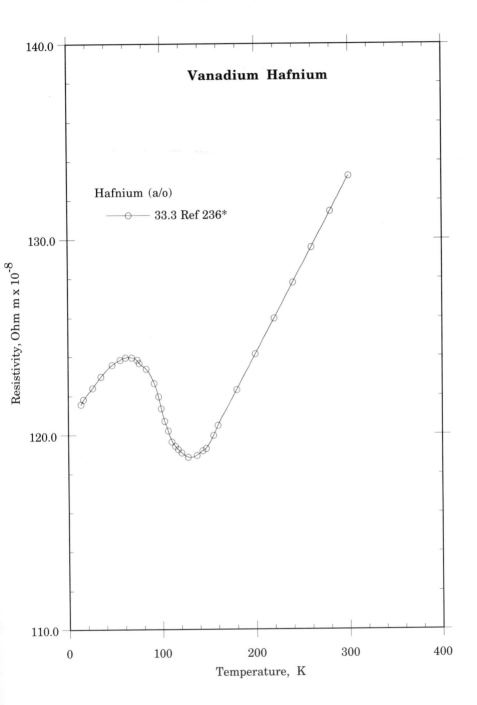

Vanadium Hafnium

Hafnium (a/o)

⊙ 33.3 Ref 236*

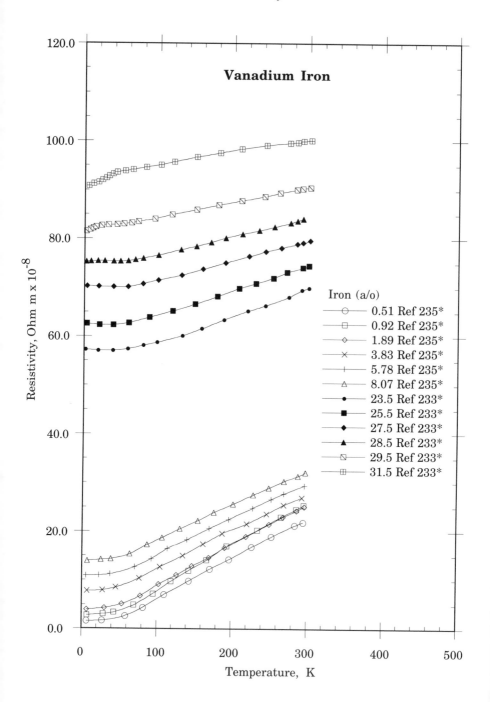

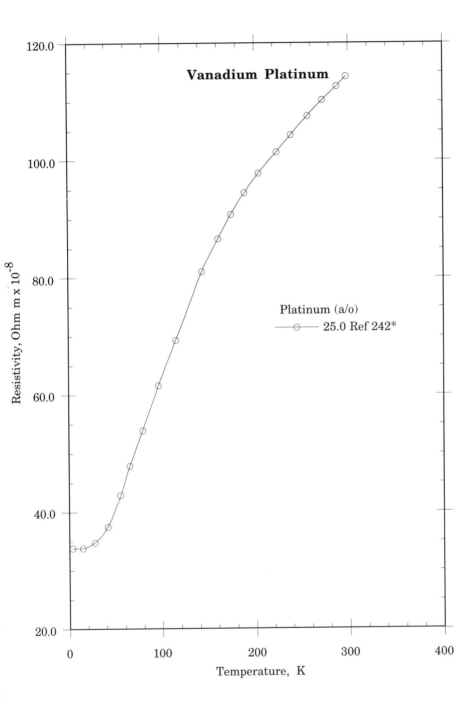

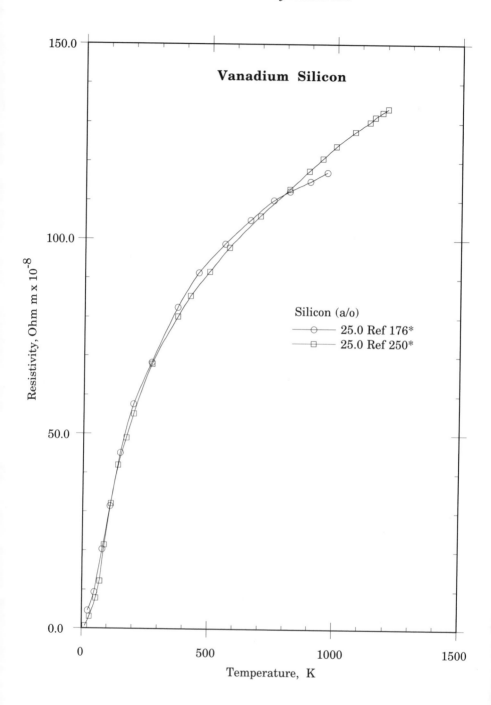

Vanadium Silicon

Silicon (a/o)
⊙ 25.0 Ref 176*
□ 25.0 Ref 250*

Resistivity, Ohm m x 10⁻⁸

Temperature, K

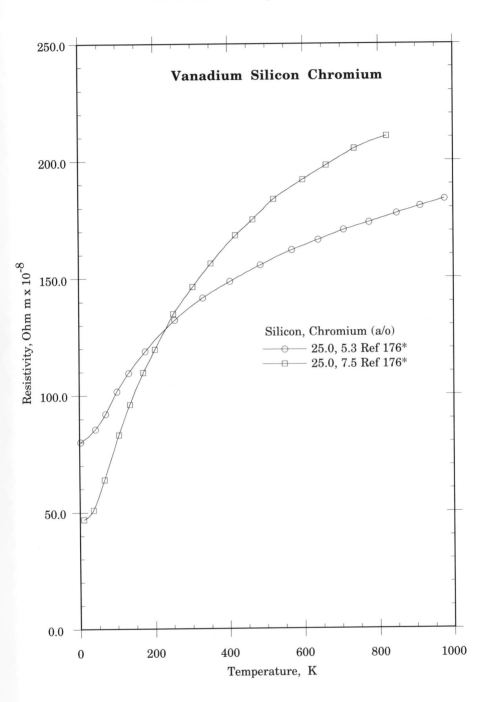

Vanadium Silicon Chromium

Silicon, Chromium (a/o)
⊖ 25.0, 5.3 Ref 176*
□ 25.0, 7.5 Ref 176*

Resistivity, Ohm m x 10^{-8}

Temperature, K

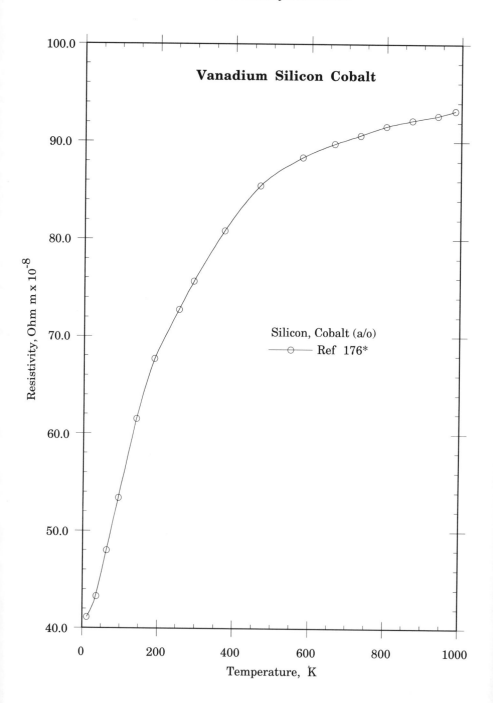

Vanadium Silicon Cobalt

Silicon, Cobalt (a/o)
—⊖— Ref 176*

Resistivity, Ohm m x 10⁻⁸

Temperature, K̇

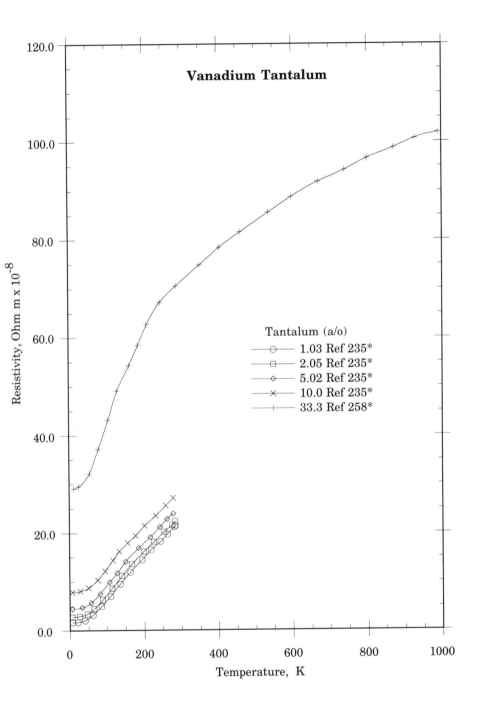

Vanadium Tantalum

Resistivity, Ohm m x 10^{-8}

Temperature, K

Tantalum (a/o)
- 1.03 Ref 235*
- 2.05 Ref 235*
- 5.02 Ref 235*
- 10.0 Ref 235*
- 33.3 Ref 258*

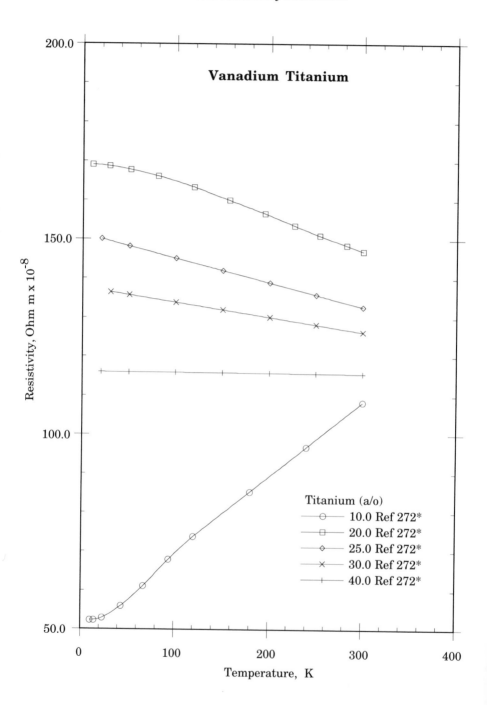

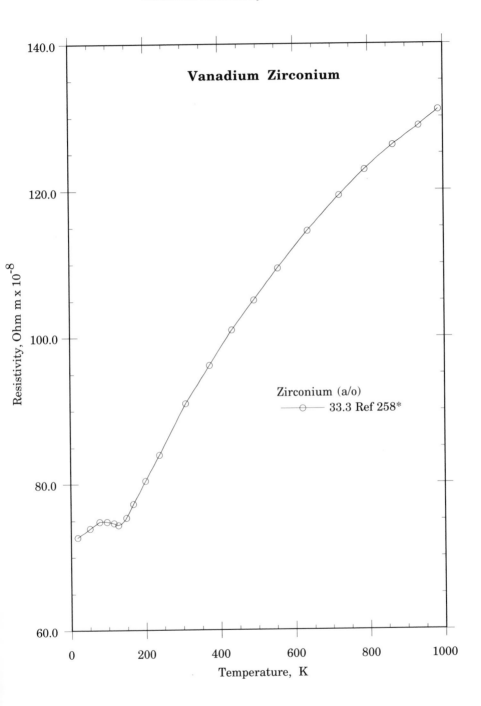

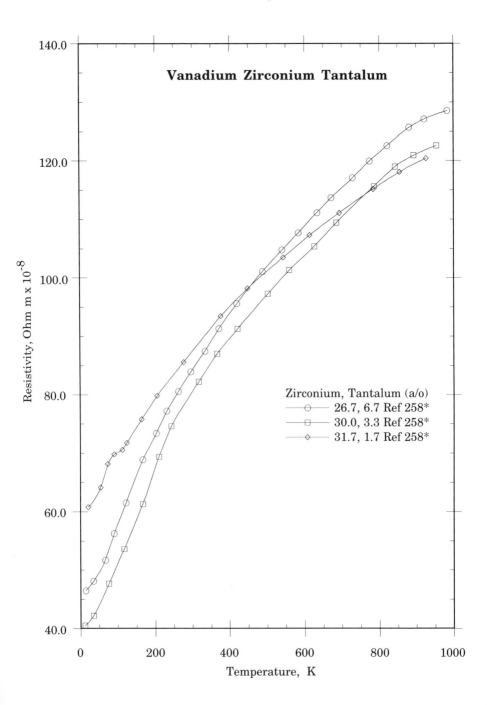

Vanadium Zirconium Tantalum

Resistivity, Ohm m x 10^{-8}

Temperature, K

Zirconium, Tantalum (a/o)
- ○ 26.7, 6.7 Ref 258*
- □ 30.0, 3.3 Ref 258*
- ◇ 31.7, 1.7 Ref 258*

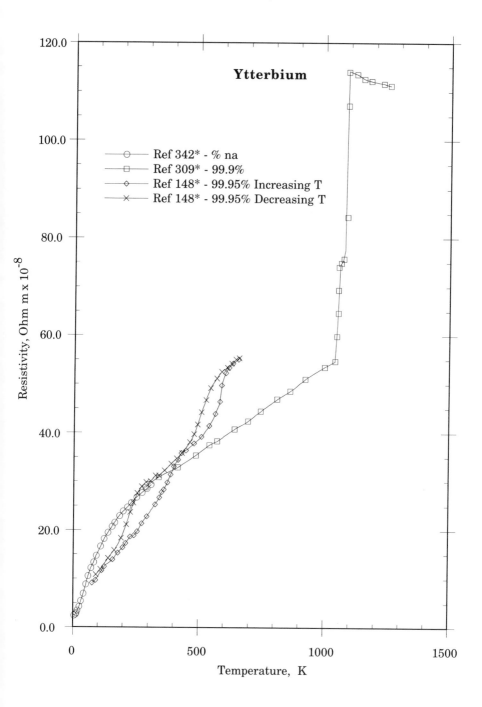

Ytterbium

Ref 342* - % na
Ref 309* - 99.9%
Ref 148* - 99.95% Increasing T
Ref 148* - 99.95% Decreasing T

Resistivity, Ohm m x 10^{-8}

Temperature, K

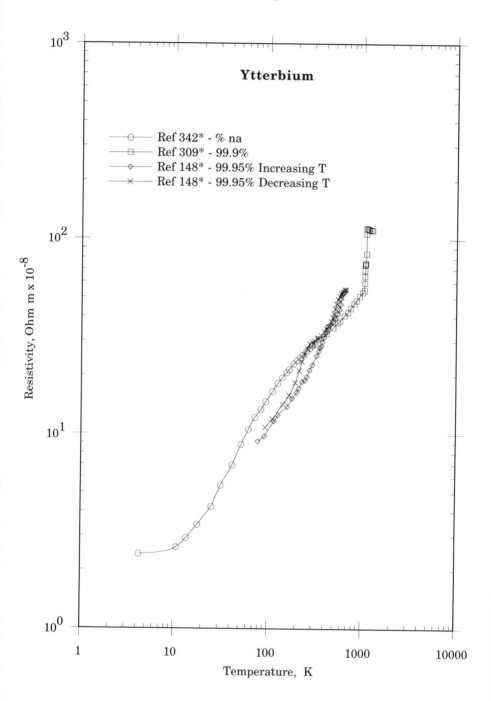

Ytterbium

Ref 342* - % na
Ref 309* - 99.9%
Ref 148* - 99.95% Increasing T
Ref 148* - 99.95% Decreasing T

Resistivity, Ohm m x 10⁻⁸

Temperature, K

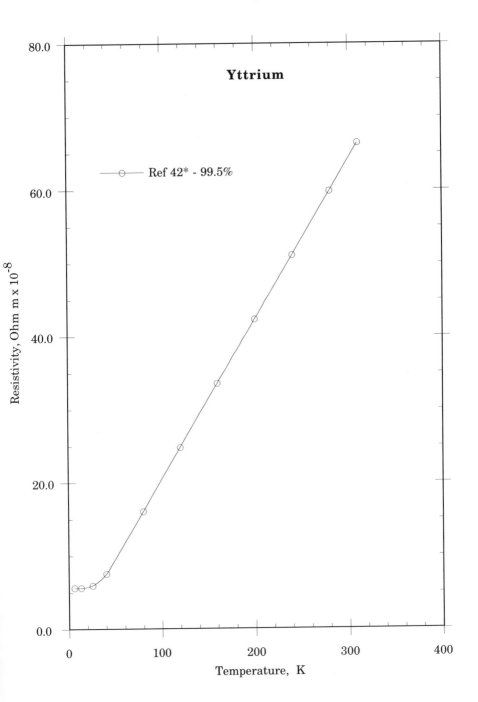

Yttrium

Ref 42* - 99.5%

Resistivity, Ohm m x 10^{-8}

Temperature, K

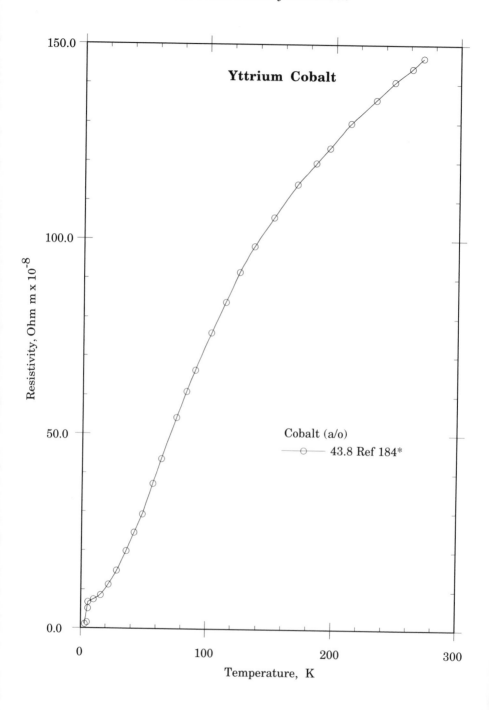

Yttrium Cobalt

Resistivity, Ohm m x 10^{-8}

Temperature, K

Cobalt (a/o)
—○— 43.8 Ref 184*

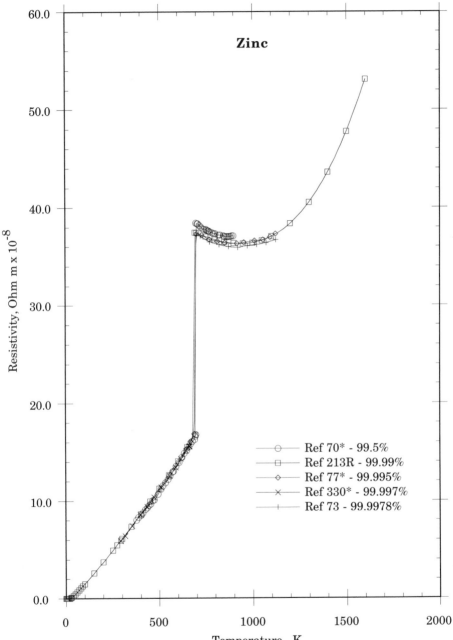

Zinc

Resistivity, Ohm m x 10^{-8}

Temperature, K

Ref 70* - 99.5%
Ref 213R - 99.99%
Ref 77* - 99.995%
Ref 330* - 99.997%
Ref 73 - 99.9978%

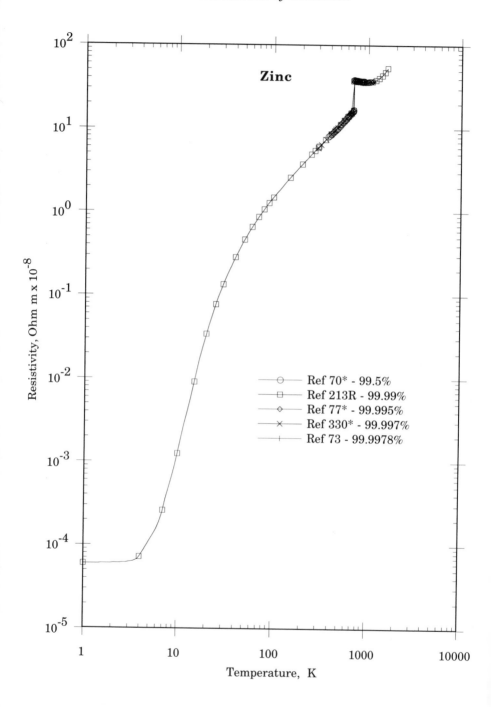

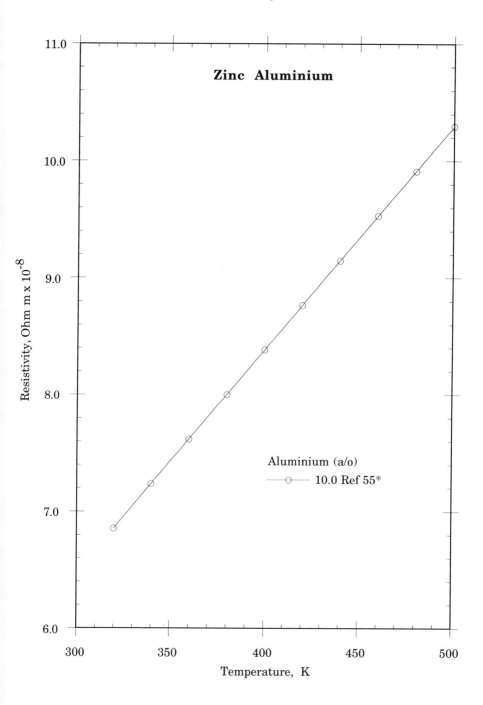

Zinc Aluminium

Aluminium (a/o)
10.0 Ref 55*

Resistivity, Ohm m x 10⁻⁸

Temperature, K

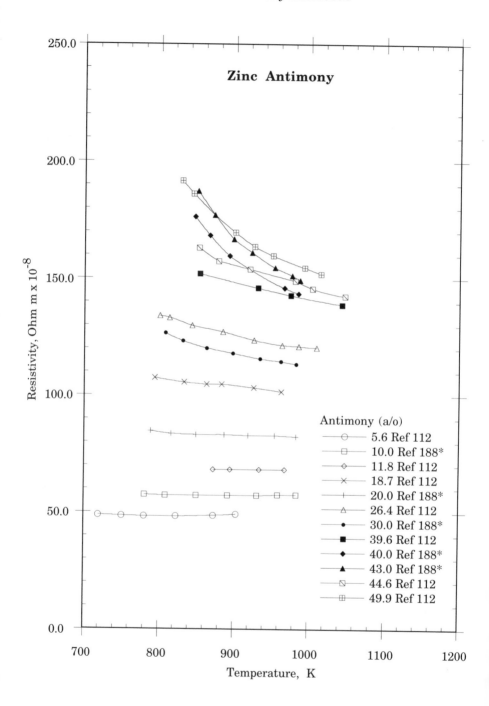

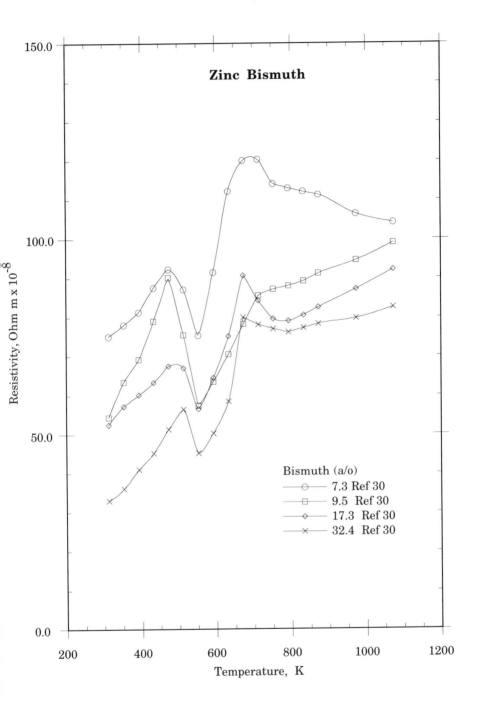

Zinc Bismuth

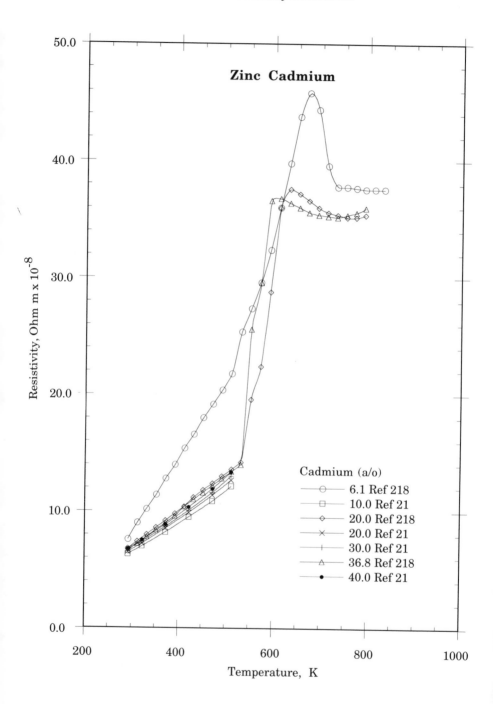

Zinc Cadmium

Resistivity, Ohm m x 10^{-8}

Temperature, K

Cadmium (a/o)
- ⊖ 6.1 Ref 218
- ☐ 10.0 Ref 21
- ◇ 20.0 Ref 218
- ✕ 20.0 Ref 21
- + 30.0 Ref 21
- △ 36.8 Ref 218
- ● 40.0 Ref 21

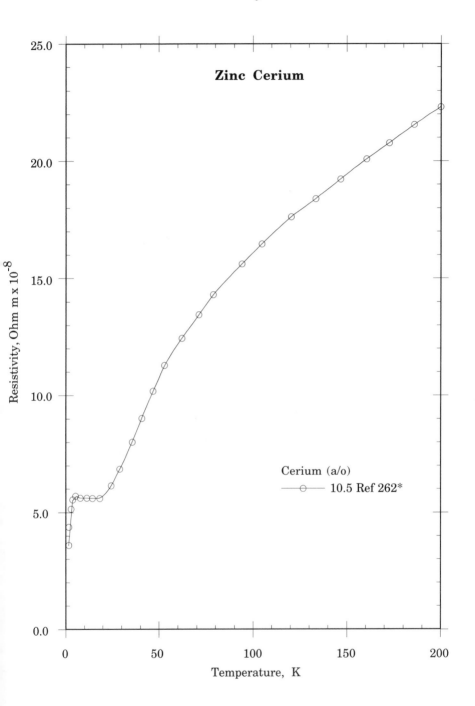

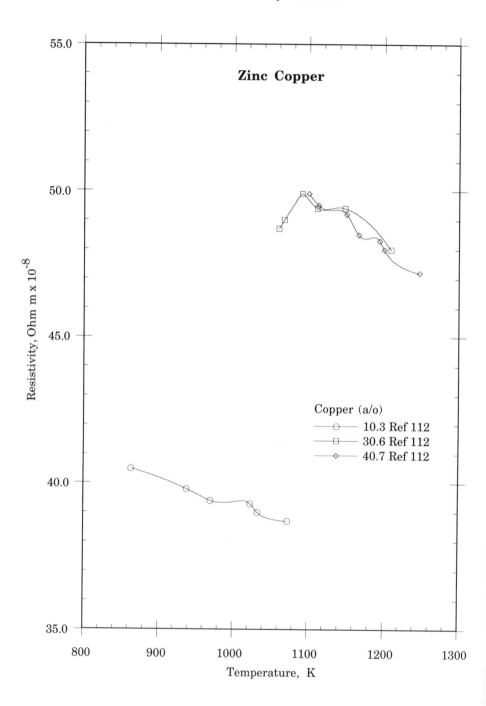

Zinc Copper

Resistivity, Ohm m x 10^{-8}

Temperature, K

Copper (a/o)
— ⊖ — 10.3 Ref 112
— ☐ — 30.6 Ref 112
— ◇ — 40.7 Ref 112

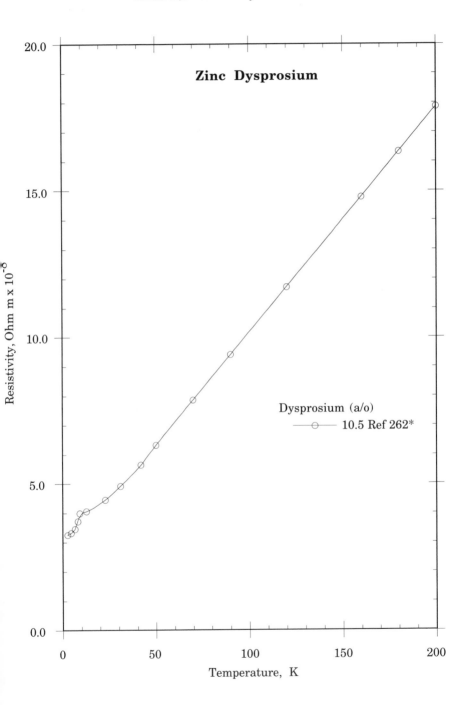

Zinc Dysprosium

Dysprosium (a/o)
10.5 Ref 262*

Resistivity, Ohm m x 10^{-8}

Temperature, K

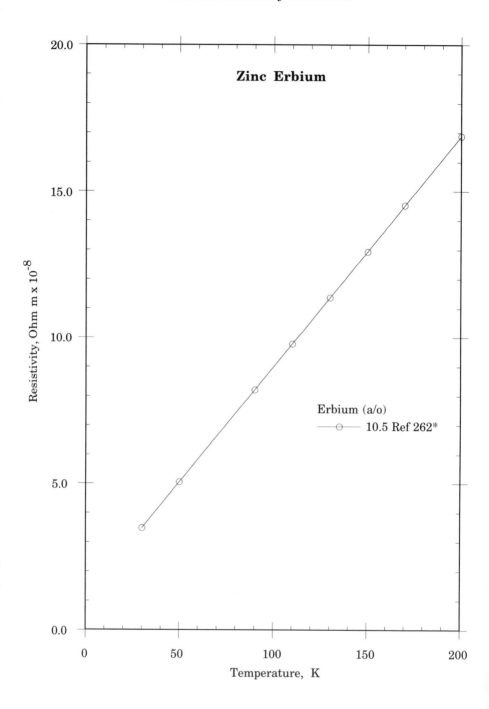

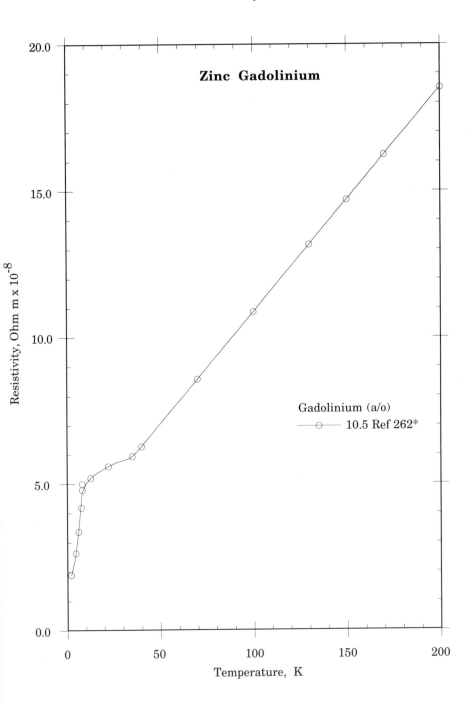

Zinc Gadolinium

Gadolinium (a/o)

——o—— 10.5 Ref 262*

Resistivity, Ohm m x 10^{-8}

Temperature, K

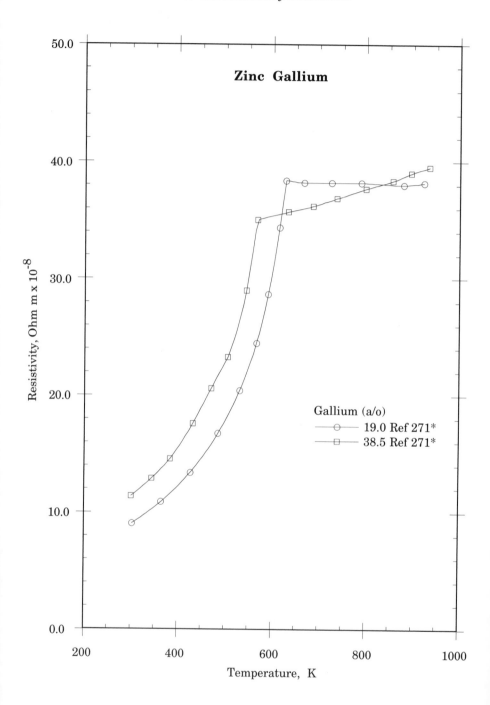

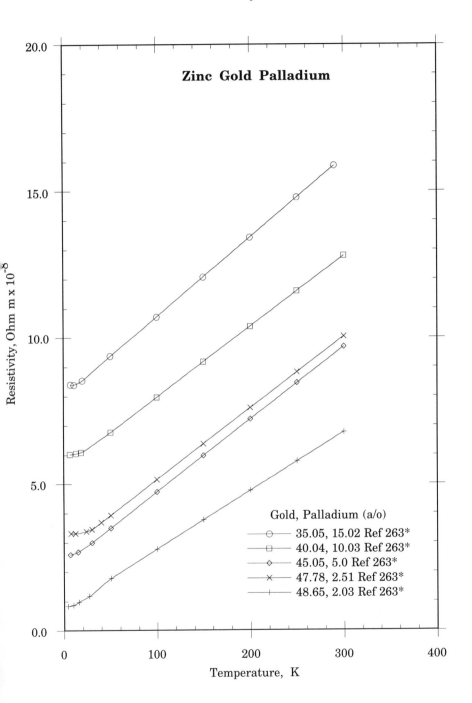

Zinc Gold Palladium

Gold, Palladium (a/o)
— ⊙ — 35.05, 15.02 Ref 263*
— ☐ — 40.04, 10.03 Ref 263*
— ◇ — 45.05, 5.0 Ref 263*
— ✕ — 47.78, 2.51 Ref 263*
— + — 48.65, 2.03 Ref 263*

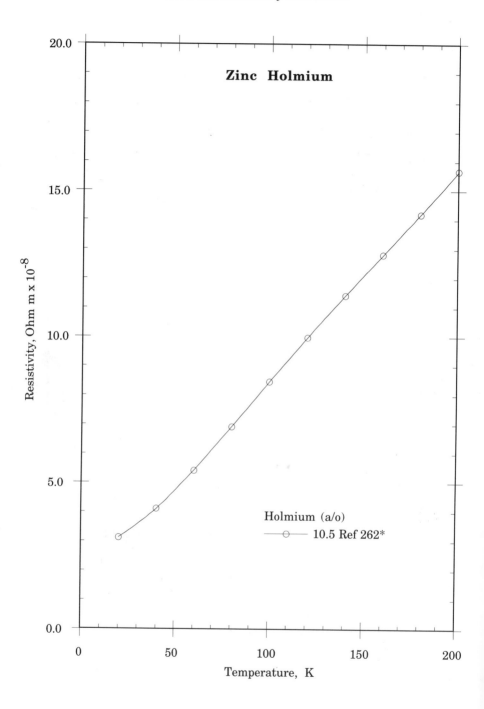

Zinc Holmium

Holmium (a/o)
—⊖— 10.5 Ref 262*

Electrical Resistivity Handbook

If enough material can be gathered, it is hoped that a further volume or supplement may be issued. Emphasis is likely to be placed upon alloys, proprietary named alloys, steels and other commercial materials. If you have data or information on such materials, or suggestions for coverage, please pass this to the editors (via the publishers, using the address overleaf). In addition, if you require further copies of this *Handbook*, or wish to register your potential interest in a future issue or supplement, please complete and return the card.

☐ I am interested in purchasing additional copies of the *Electrical Resistivity Handbook*

☐ I am interested in further issues, supplements or coverage if published

Name _____

Address _____

_____

Electrical Resistivity Handbook

If enough material can be gathered, it is hoped that a further volume or supplement may be issued. Emphasis is likely to be placed upon alloys, proprietary named alloys, steels and other commercial materials. If you have data or information on such materials, or suggestions for coverage, please pass this to the editors (via the publishers, using the address overleaf). In addition, if you require further copies of this *Handbook*, or wish to register your potential interest in a future issue or supplement, please complete and return the card.

☐ I am interested in purchasing additional copies of the *Electrical Resistivity Handbook*

☐ I am interested in further issues, supplements or coverage if published

Name _____

Address _____

_____

Electrical Resistivity Handbook,
The Institution of Electrical Engineers,
Book Publishing Department,
Michael Faraday House, Six Hills Way,
Stevenage, Herts. SG1 2AY, United Kingdom

Electrical Resistivity Handbook,
The Institution of Electrical Engineers,
Book Publishing Department,
Michael Faraday House, Six Hills Way,
Stevenage, Herts. SG1 2AY, United Kingdom

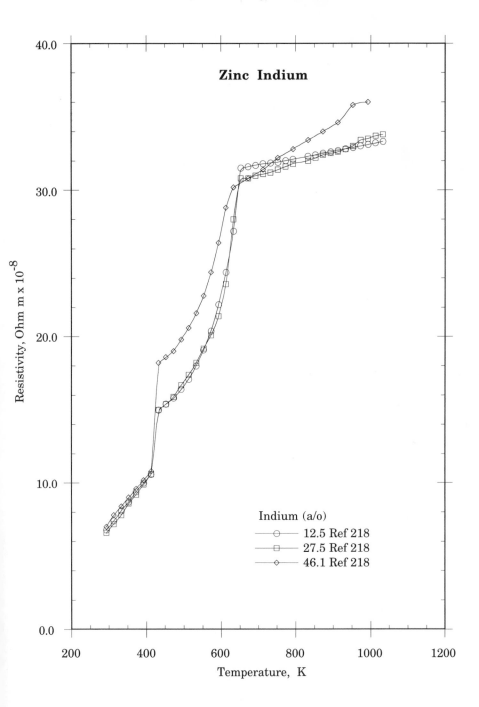

Zinc Indium

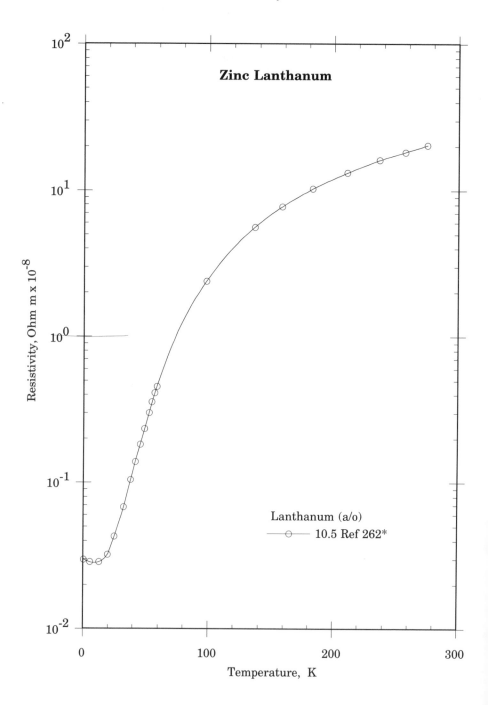

Zinc Lanthanum

Resistivity, Ohm m x 10^{-8}

Temperature, K

Lanthanum (a/o)
⊖ 10.5 Ref 262*

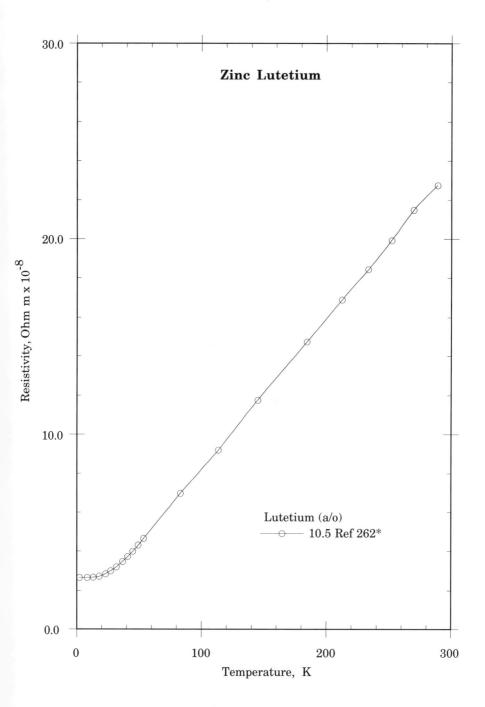

Zinc Lutetium

Resistivity, Ohm m x 10^{-8}

Temperature, K

Lutetium (a/o)
—⊙— 10.5 Ref 262*

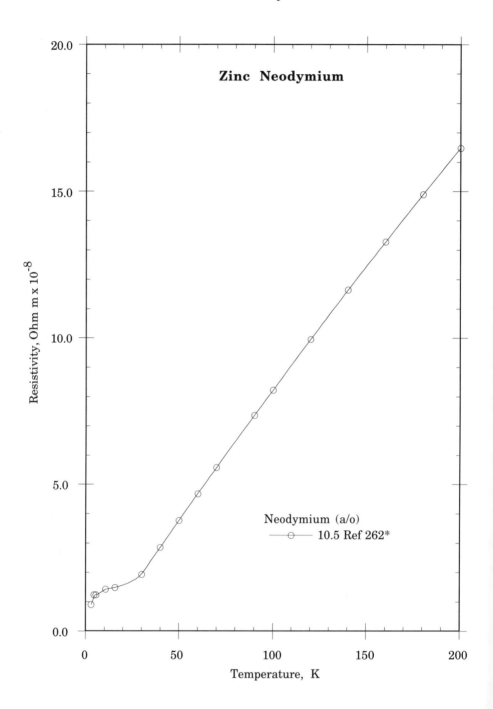

Zinc Neodymium

Resistivity, Ohm m x 10^{-8}

Temperature, K

Neodymium (a/o)
10.5 Ref 262*

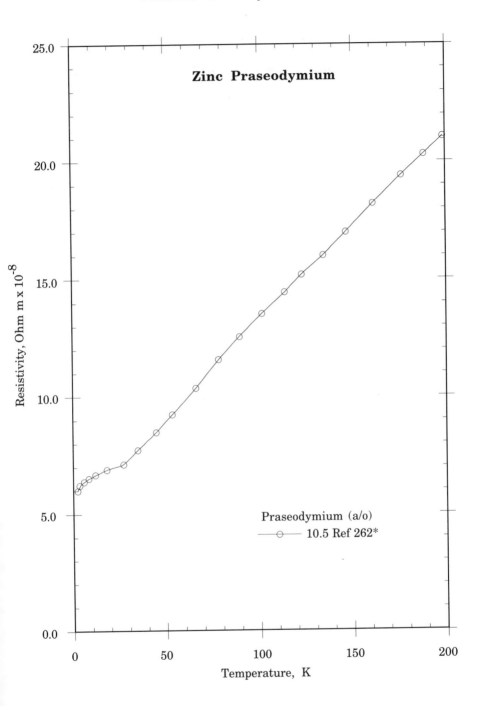

Zinc Praseodymium

Resistivity, Ohm m x 10^{-8}

Temperature, K

Praseodymium (a/o)
—⊙— 10.5 Ref 262*

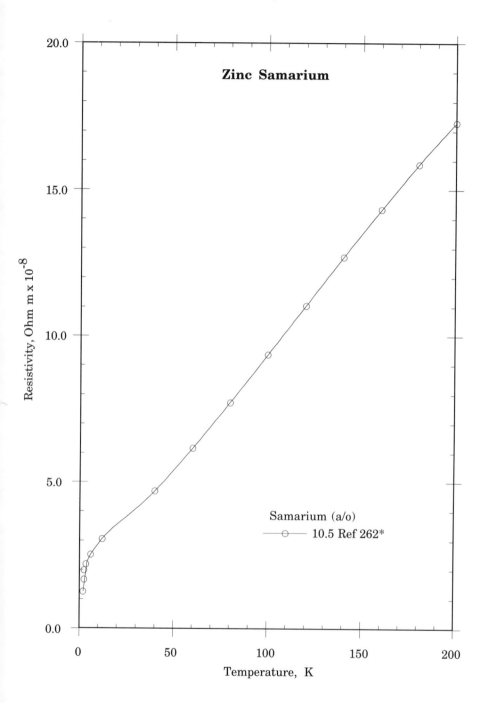

Zinc Samarium

Samarium (a/o)
—o— 10.5 Ref 262*

Resistivity, Ohm m x 10^{-8}

Temperature, K

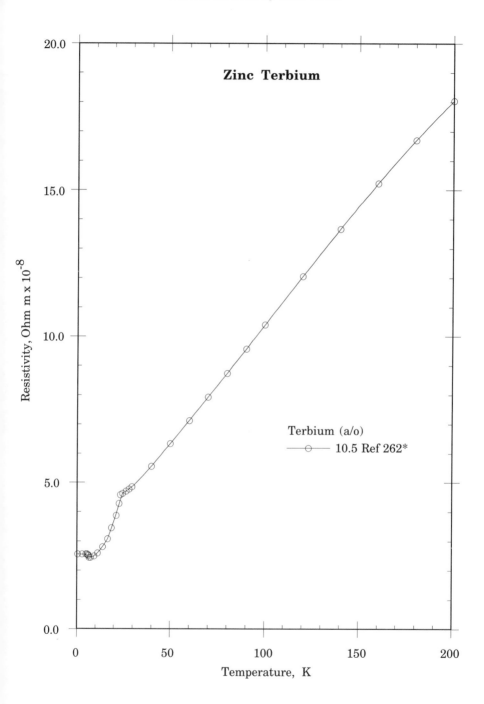

Zinc Terbium

Resistivity, Ohm m x 10^{-8}

Temperature, K

Terbium (a/o)
10.5 Ref 262*

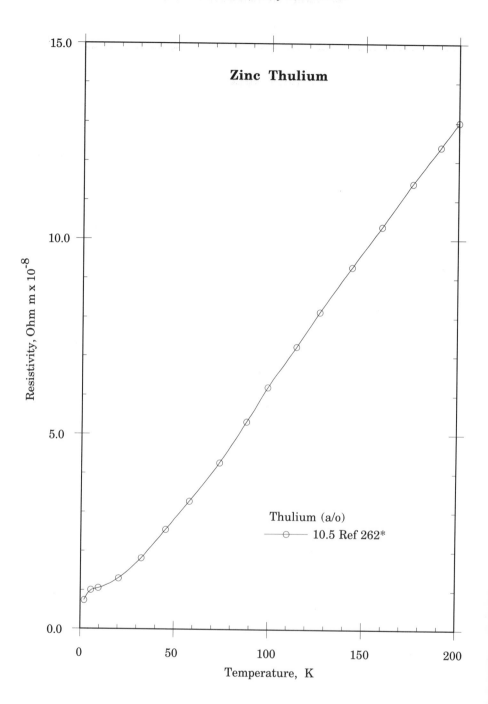

Zinc Thulium

Thulium (a/o)
— ○ — 10.5 Ref 262*

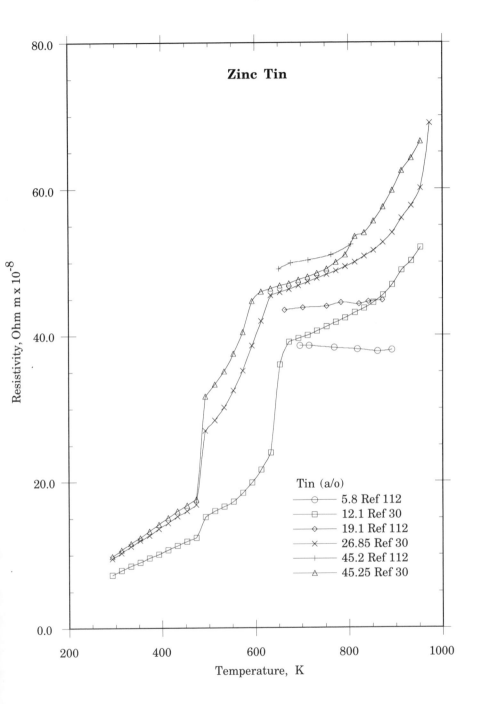

Zinc Tin

Resistivity, Ohm m x 10^{-8}

Temperature, K

Tin (a/o)
- ⊖ 5.8 Ref 112
- ☐ 12.1 Ref 30
- ◇ 19.1 Ref 112
- ✕ 26.85 Ref 30
- + 45.2 Ref 112
- △ 45.25 Ref 30

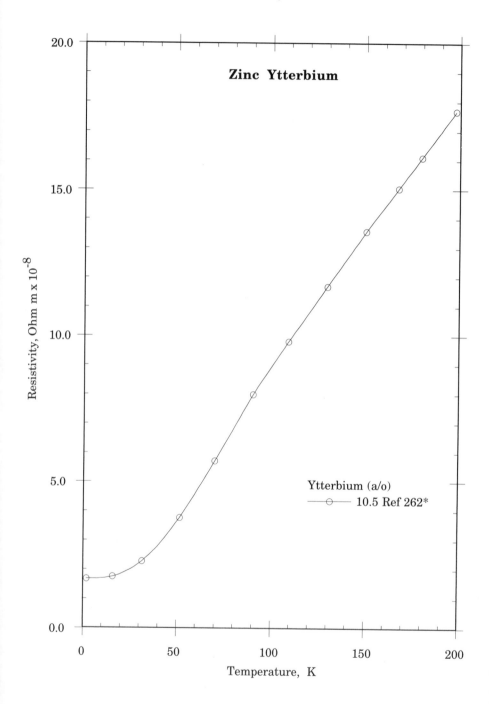

Zinc Ytterbium

Ytterbium (a/o)
10.5 Ref 262*

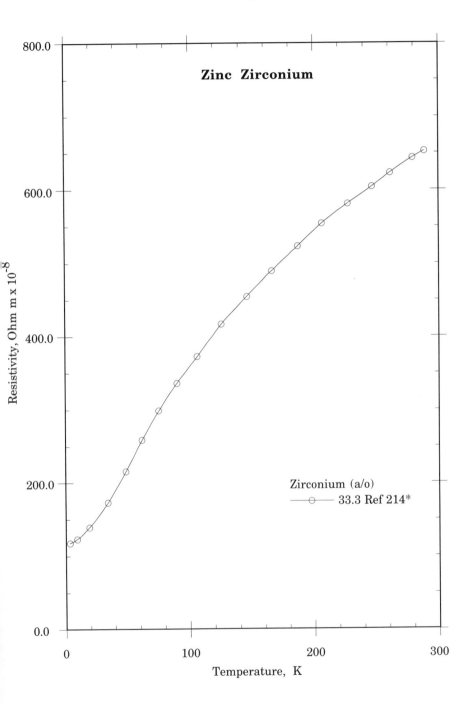

Zinc   Zirconium

Resistivity, Ohm m x 10^{-8}

Temperature, K

Zirconium (a/o)
——⊙—— 33.3 Ref 214*

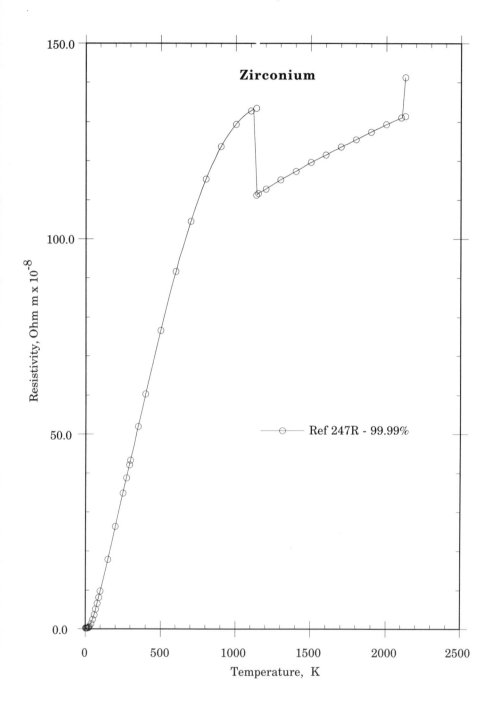

Zirconium

Resistivity, Ohm m x 10^{-8}

Temperature, K

Ref 247R - 99.99%

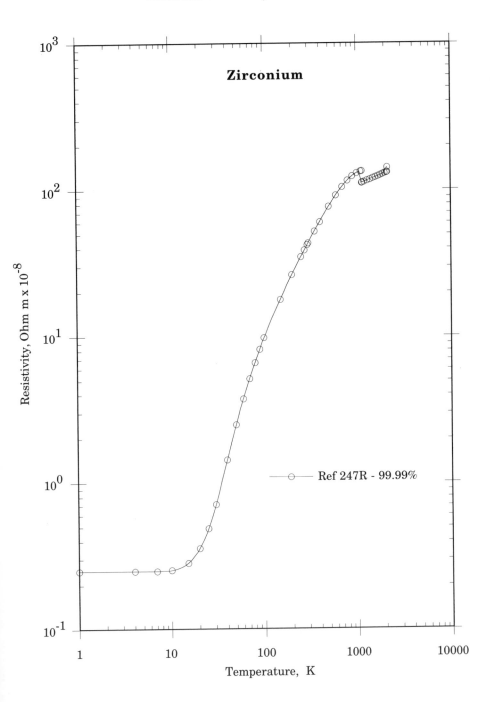

Zirconium

Ref 247R - 99.99%

Resistivity, Ohm m x 10^{-8}

Temperature, K

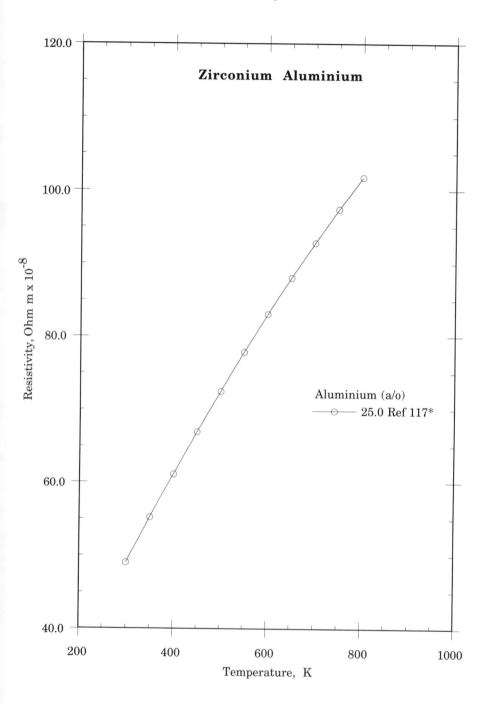

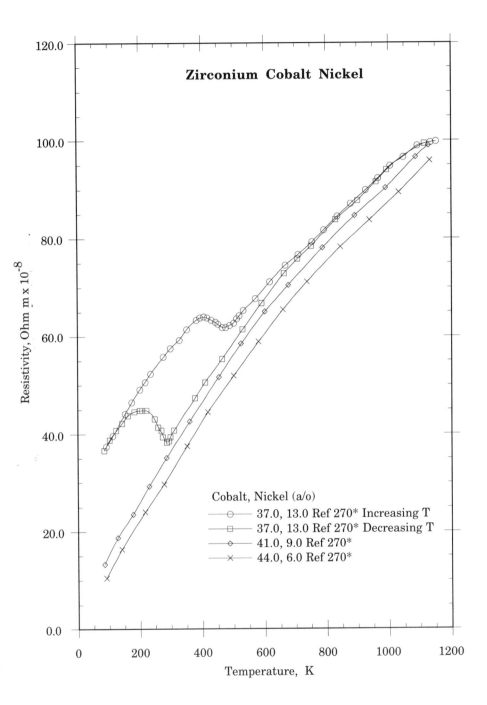

Zirconium Cobalt Nickel

Resistivity, Ohm m x 10^{-8}

Temperature, K

Cobalt, Nickel (a/o)
- ○ 37.0, 13.0 Ref 270* Increasing T
- □ 37.0, 13.0 Ref 270* Decreasing T
- ◇ 41.0, 9.0 Ref 270*
- × 44.0, 6.0 Ref 270*

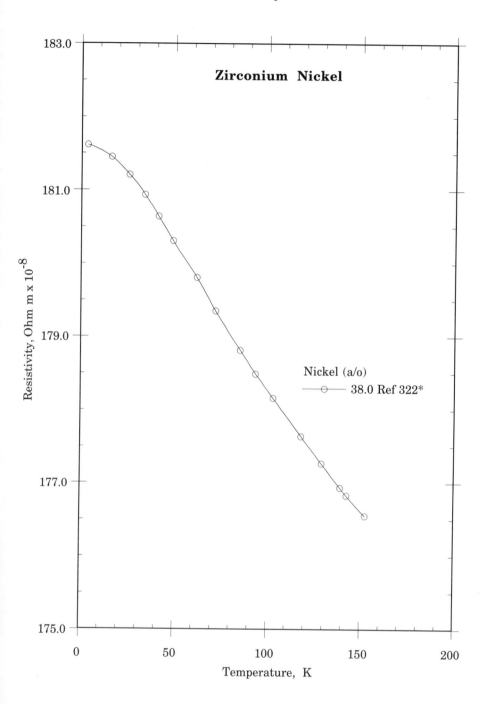

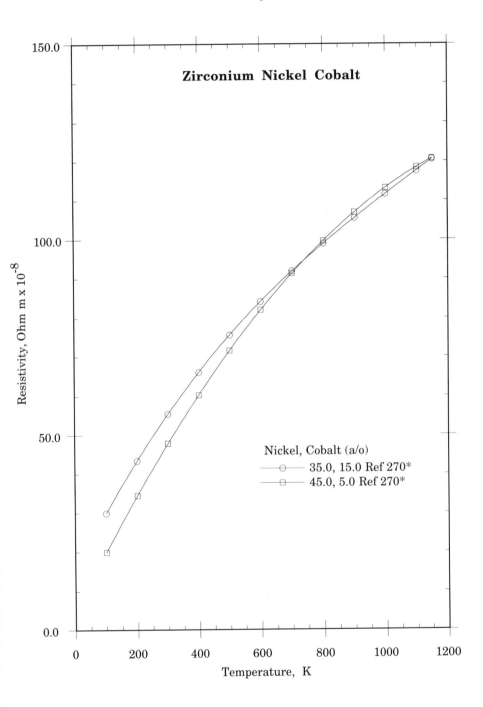

Zirconium Nickel Cobalt

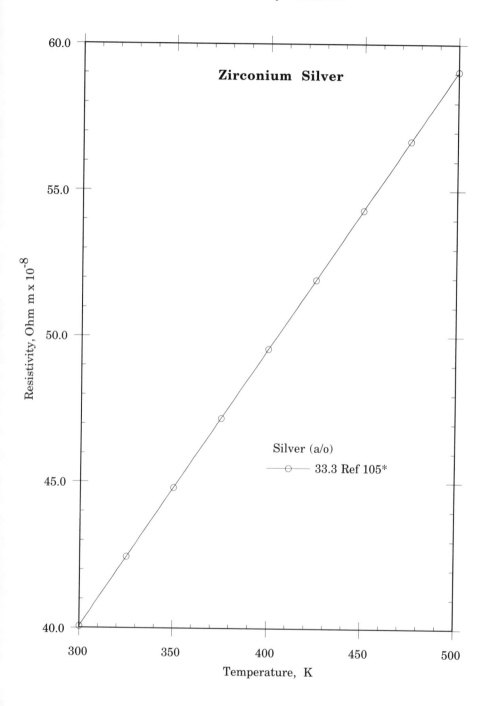

Zirconium Silver

Silver (a/o)

33.3 Ref 105*

Resistivity, Ohm m x 10^{-8}

Temperature, K

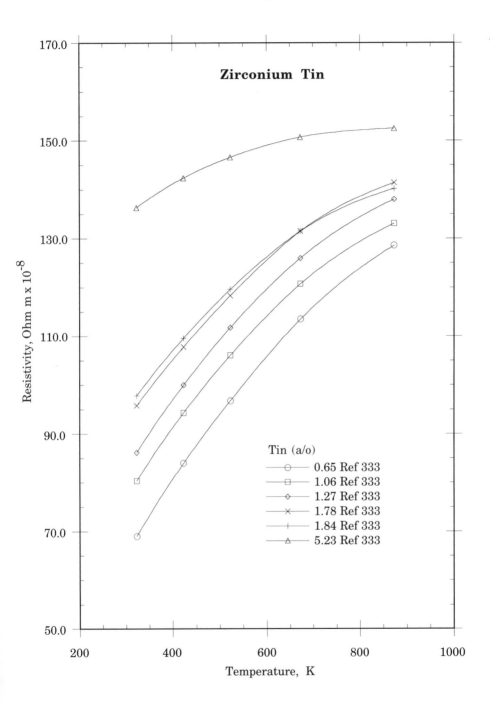

Zirconium Tin

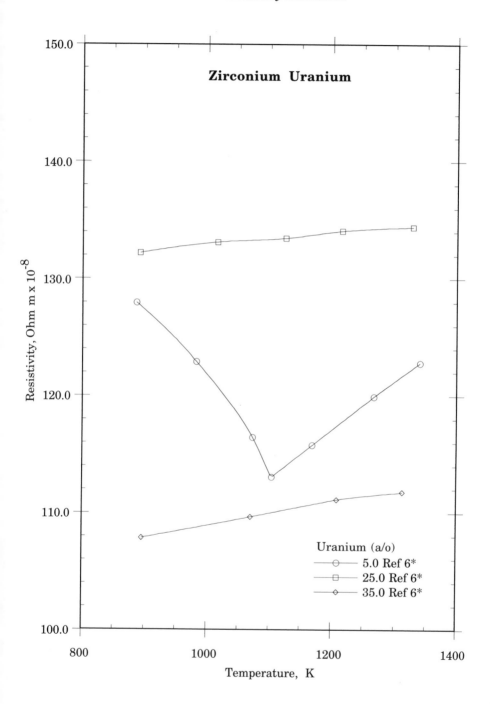

Resistivity references

1 NASU, S., KURASAWA, T., MATSUI, H., TAMAKI, M., and OKUDA, M. (1976): 'Electrical resistivity of uranium monophosphides, uranium mononitrides and uranium carbonitrides at low temperatures', in 'Plutonium and Other Actinides', Blank, H. and Linder, R. (Eds) (North Holland), pp. 515–523

2 GRAY, D.E. (1963): 'Electrical resistivity of some alloys as a function of temperature', in 'American Institute of Physics Handbook', 3rd edition (McGraw-Hill Book Company), pp. 4.13–4.14

3 MAGLICH, K.D., PEROVICH, N., and ZIVOTICH, Z. (1979): 'Thermal diffusivity and electrical resistivity of cobalt', Proc. 16th Int. Conf. Thermal Conductivity, Larsen, D.C. (Ed.) (Plenum Publishing), pp. 325–335

4 VOSKRESENSKII, V.YU., PELETSKII, V.E., and TIMROT, D.L. (1967): 'Apparatus for the absolute measurement of the specific electrical conductivity of a metal above 1000 deg. C', *High Temp.*, **5**, pp. 624–627

5 FLYNN, D.R., and O'HAGAN, M. (1967): 'Measurement of the thermal conductivity and electrical resistivity of platinum from 100 to 900 deg. C', *J. Res. Natl. Bur. Standards*, **71**, pp. 255–284

6 TERECHOV, G.I., SINIAKOVA, S.I., TAGIROVA, R.H., and IVANOV, O.S. (1976): 'Thermoelectric properties of solid solutions based on uranium and thorium', in 'Plutonium and Other Actinides', Blank, H. and Linder, R. (Eds) (North Holland), pp. 535–545

7 BANCHILA, S.N., and FILIPPOV, L.P. (1973): 'Study of the electrical conductivity of liquid metals', *Teplofiz. Vys. Temp.*, **6**, pp. 1301–1305

8 PELETSKII, V.E., DRUZHININ, V.P., and SOBOL, YA G. (1971): 'Thermophysical properties of vanadium at high temperatures', *High Temp.-High Press.*, **3**, pp. 153–159

9 MARTY, W. (1958): 'Determining the specific resistance of solid and molten aluminium and its alloys', *Brown Boveri Rev.*, pp. 549–552

10 SIMMONS, R.O., and BALLUFFI, R.W. (1960): 'Measurement of the high temperature electrical resistivity of aluminium', *Phys. Rev.*, **117**, pp. 62–68

11 PELETSKII, V.E. (1978): 'Electrical resistivity of vanadium in the temperature range 200–2100 K', *High Temp.*, **16**, pp. 57–60

12 CEZAIRLIYAN, A., and McCLURE, J.L. (1974): 'Thermophysical measurements on iron above 1500 K using a transient (subsecond) technique', *J. Res. Natl. Bur. Standards*, **78A**, pp. 1–4

13 BINKELE, L. (1983): 'Zur frage der hochtemperatur-Lorenzzahl bei wolframeine analyse neuer Meßwerte der thermischen und elektrischen leitfähigkeit im temperaturbereich 300 bis 1300 K', *High Temp.-High Press.*, **15**, pp. 525–531

14 BEL'SKAYA, E.A., and PELETSKII, V.E. (1981): 'Electrical resistivity of nickel in the temperature range 100–1700 K', *Teplofiz. Vys. Temp.*, **19**, pp. 525–532

15 KATANO, S., and MORI, N. (1980): 'Resistance minimum in V-doped Cr–4 at.% Co alloys', *J. Phys. Soc. Jpn.*, **48**, pp. 691–692

16 CEZAIRLIYAN, A. (1980): 'Electrical resistivity of molybdenum in the temperature range 1500 to 2650 K', *Int. J. Thermophys.*, **1**, pp. 417–427

17 BERLINCOURT, T.G. (1959): 'Hall effect, resistivity and magnetoresistivity of Th, U, Zr, Ti and Nb', *Phys. Rev.*, **114**, pp. 969–977

18 COLVIN, R.V., LEVGOLD, S., and SPEDDING, F.H. (1960): 'Electrical resistivity of the heavy rare-earth metals', *Phys. Rev.*, **120**, pp. 741–745

19 GRUBE, VON G., MOHR, L., and BORNHAK, R. (1934): 'Elektrische leitfähigkeit und zustandsdiagramm bei binären legierungen—das system magnesium-wismut', *Z. Electrochem.*, **40**, pp. 143–150

20 GRUBE, VON G., HOßKÜHLER, H., and VOGT, H. (1932): 'Elektrische leitfähigkeit und zustandsdiagramm bei binären legierungen—das system lithium-cadmium', *Z. Electrochem.*, **38**, pp. 869–880

21 LE BLANC, VON M., and SCHÖPEL, H. (1933): 'Elektrische leitfähigkeits-messungen an zink-kadmium- und blei-antimon-system unter Berücksichti-gung stabiler gleichgewichtseinstellungen', *Z. Electrochem.*, **39**, pp. 695–701

22 ROSHKO, R.M., and WILLIAMS, G. (1977): 'Comparison of the resistivity of PdCr with canonical spin-glass systems', *Phys. Rev. B*, **16**, pp. 1503–1509

23 GRUBE, VON G., and KLAIBER, H. (1934): 'Elektrische leitfähigkeit und zustandsdiagramm bei binären leigierungen—das system lithium-blei', *Z. Electrochem.*, **40**, pp. 745–754

24 GRUBE, VON G., and MEYER, E. (1934): 'Elektrische leitfähigkeit und zustandsdiagramm bei binären leigierungen—das system lithium-zinc', *Z. Electrochem.*, **40**, pp. 771–774

25 ALSTEAD, J.K., COLVIN, R.V., LEGVOLD, S., and SPEDDING, S.H. (1961): 'Electrical resistivity of lanthanum, praseodymium, neodymium and samarium', *Phys. Rev.*, **25**, pp. 1637–1639

26 GRUBE, VON G., and HILLE, J. (1934): 'Elektrische leitfähigkeit und zustandsdiagramm bei binären legierungen—das system magnesium-thallium', *Z. Electrochem.*, **40**, pp. 101–106

27 KOUVEL, J.S., and HARTELIUS, C.C. (1961): 'Abrupt magnetic transition in MnSn$_2$', *Phys. Rev.*, **123**, pp. 124–125

28 KLAIBER, VON H. (1936): 'Elektrische leitfähigkeit und zustandsdiagramm bei binären legierungen—das system natrium-blei', *Z. Electrochem.*, **42**, pp. 258–264

29 BLUE, M.D. (1961): 'Resistivity and thermoelectric power of transition metals in copper-gold alloys', *Phys. Rev.*, **123**, pp. 1270–1272

30 MAGOMEDOV, A.M. (1978): 'Electric resistance of Zn-Sn and Zn-Bi alloys', *High Temp.*, **16**, pp. 446–450

31 CEZAIRLIYAN, A. (1972): 'High speed (subsecond) simultaneous measure-ment of specific heat, electrical resistivity and hemispherical total emittance of Ta-10(wt.%)W alloy in the range 1500 to 3200 K', *High Temp.-High Press.*, **4**, pp. 541–550

32 MEADEN, G.T. (1963): 'Electronic properties of the actinide metals at low temperatures', *Proc. R. Soc. Lond. A*, **276**, pp. 553–570

33 CHAKRABARTI, D.J., and BECK, P.A. (1971): 'Transport properties of Cr-Al solid solutions', *J. Phys. Chem. Solids*, **32**, pp. 1609–1615

34 ARAJS, S., REEVES, N.L., and ANDERSON, E.E. (1971): 'Antiferromagnetism and electrical transport properties of chromium-aluminium alloys', *J. Appl. Phys.*, **42**, pp. 1691–1692

35 JAN, J.P., and PEARSON, W.B. (1963): 'Electrical properties of $AuAl_2$, $AuGa_2$ and $AuIn_2$'. *Philos. Mag.*, **8**, pp. 279–284

36 FAKIDOV, I.G., and AFANAS'EV, A. (1958): 'The electrical conductivity of a chromium-antimony antiferromagnetic compound', *Phys. Met. & Metallogr.*, **6**, pp. 160–161

37 MAKI, S., and ADACHI, K. (1979): 'Antiferromagnetism and weak ferromagnetism of disordered bcc Cr-Mn alloys', *J. Phys. Soc. Japan*, **46**, pp. 1131–1137

38 MIKHEYEV, V.S., and ALEKSASHIN, V.S. (1962): 'Electrical volume resistivity of alloys of the titanium-chromium system up to temperatures of 1100 deg. C', *Phys. Met. & Metallogr.*, **14**, pp. 62–67

39 GIANNUZZI, A., TOMASCHKE, H., and SCHRÖDER, K. (1970): 'Absolute thermoelectric power and resistivity of Cr-V and Cr-Mn alloys', *Philos. Mag.*, **21**, pp. 479–493

40 ARAJS, S., STELMACH, A.A., and MARTIN, M.C. (1973): 'Magnetic susceptibility and electrical resistivity of TiCo at elevated temperatures', *J. Less-Common Met.*, **32**, pp. 178–180

41 YAO, Y.D. (1978): 'Electrical resistivity of nickel rich nickel-copper alloys between 78 and 700 K', *Annu. Rep. Inst. Phys. Acad.*, **8**, pp. 11–17

42 ALSTAD, J.K., COLVIN, R.V., and LEGVOLD, S. (1961): 'Single crystal and polycrystal resistivity relationships for yttrium', *Phys. Rev.*, **123**, pp. 418–419

43 SHAMASHOV, F.P., NEIMARK, B.E., GREBENNIKOV, R.V., MERKUL'EV, A.N., and POSEL'SKII, V.B. (1971): 'Thermal conductivity and electrical resistivity of a titanium-gadolinium alloy', *High Temp.*, **9**, pp. 1208–1210

44 EROGLU, A., ARAJS, S., MOYER, C.A., and RAO, K.V. (1978): 'Electrical resistivity of chromium-gold alloys between 77 and 700 K', *Phys. Status Solidi*, **87**, pp. 287–291

45 MICHIKAMI, O., and YAMAGUCHI, Y. (1971): 'Electrical resistivity and hardness of Au-Ga alloys', *Jpn. J. Appl. Phys.*, **10**, pp. 660–661

46 OTTER, F.A. (1956): 'Thermoelectric power and electrical resistivity of dilute alloys of Mn, Pd and Pt in Cu, Ag and Au', *J. Appl. Phys.*, **27**, pp. 197–200

47 SINGH, R.L., and MEADEN, G.T. (1972): 'Thermopower and resistivity studies of the AuV and AuCuV systems', *Phys. Rev. B*, **6**, pp. 2660–2668

48 SHIGA, M., and NAKAMURA, Y. (1964): 'Electrical and magnetic study of iron-manganese alloys', *J. Phys. Soc. Jpn.*, **19**, p. 1743

49 FUJIWARA, H., SUEDA, N., and FUJIWARA, Y. (1970): 'Electrical resistivity of alpha-phase iron rich Fe-V alloys', *J. Phys. Soc. Jpn.*, **28**, p. 527

50 MEADEN, G.T., RAO, K.V., and TEE, K.T. (1970): 'Effect of the Néel transition on the thermal and electrical resistivities of Cr and Cr:Mo alloys', *Phys. Rev. Lett.*, **25**, pp. 359–362

51 DOMENICALI, C.A., and OTTER, F.A. (1955): 'Thermoelectric power and electrical resistivity of dilute alloys of silicon in copper, nickel and iron', *J. Appl. Phys.*, **26**, pp. 377–380

52 IKEDA, K., and NAKAMICHI, T. (1975): 'Electrical resistivity of Laves phase compounds containing transition elements: 1. Fe_2A (A = Sc, Y, Ti, Zr, Hf, Nb and Ta)', *J. Phys. Soc. Jpn.*, **39**, pp. 963–968

53 ABALLE, M., REGIDOR, J.J., SISTIAGA, J.M., and TORRALBA, M. (1972):

'Mechanisms of precipitation in lead-antimony alloys', *Z. Metallkd.*, **63**, pp. 564–570

54 IVANOV, G.A., POPOV, A.M., and CHISTYAKOV, B.I. (1963): 'Electrical properties of binary bismuth alloys in a wide temperature range: 1. Sn, Sb and Te solid solutions in bismuth (polycrystals)', *Phys. Met. & Metallogr.*, **16**, pp. 23–29

55 YURKOV, V.A., DUTYSHEVA, N.A., and OKOLYKHINA, L.B. (1966): 'Electrical and thermoelectrical properties of Al-Zn alloys', *Phys. Met. & Metallogr.*, **20**, pp. 33–40

56 KORNILOV, I.I., MIKHEYEV, V.S., and KONSTANTINOV, K.M. (1963): 'Study of the volume resistivity of alloys of the system Ti-Al at temperatures ranging from room to 1200 deg. C', *Phys. Met. & Metallogr.*, **16**, pp. 49–51

57 UL'YANOV, R.A., and TARASOV, N.D. (1964): 'Certain irregularities in the variation in the properties of alloys on niobium base', *Phys. Met. & Metallogr.*, **17**, pp. 60–64

58 ARAJS, S., and COLVIN, R.V. (1964): 'Electrical resistivity of high purity iron from 300 to 1300 K', *Phys. Status Solidi*, **6**, pp. 789–802

59 ARAJS, S., DE YOUNG, T.F., and ANDERSON, E.E. (1970), 'Antiferromagnetism and electrical resistivity of chromium alloys containing ruthenium and osmium', *J. Appl. Phys.*, **41**, pp. 1426–1428

60 VERTOGRADSKII, V.A., and CHEKHOVSKII, V.YA. (1971), 'Electrical resistivity of VR-20 tungsten-rhenium alloy at high temperature', *High Temp.- High Press.*, **3**, pp. 399–401

61 ARAJS, S. (1961): 'Electrical resistivities of nickel-niobium solid solutions', *J. Appl. Phys.*, **32**, pp. 97–99

62 YAO, Y.D., ARAJS, S., and RAO, K.V. (1977): 'Magnetic phase transition in nickel rich nickel-titanium alloys', *Chin. J. Phys.*, **1**, pp. 1–5

63 WOODARD, D.W., and CODY, G.D. (1964): 'Anomalous resistivity of Nb_3Sn', *Phys. Rev. A*, **136**, pp. 166–168

64 KNAPPWOST, VON A. (1953): 'Magnetische und resistometrische untersuchungen an stoffen mit negativem volumensprung in der umgebung des schmelzpunktes', *Z. Electrochem.*, **57**, pp. 618–624

65 VERTOGRADSKII, V., and CHEKHOVSKII, V.YA (1973): 'Resistivity of VR-5 and VR-10 tungsten-rhenium alloys at 1200–3000 K', *High Temp.*, **11**, pp. 386–387

66 ZAICHENKO, V.M., MINTS, R.G., and CHEKHOVSKOI, V.YA (1976): 'Electrical resistivity of tungsten-rhenium alloys in the region of a concentration anomaly', *High Temp.*, **14**, pp. 263–267

67 WHITE, G.K., and WOODS, S.B. (1958): 'The thermal and electrical resistivity of bismuth and antimony at low temperatures', *Philos. Mag.*, **3**, pp. 342–359

68 MOOIJ, J.H. (1973): 'Electrical conduction in concentrated disordered transition metal alloys', *Phys. Status Solidi*, **17**, pp. 521–530

69 CHEN, T., ROGOWSKI, G., and WHITE, R.M. (1978): 'On the properties and electronic structure of MnSb, CoSb and NiSb', *J. Appl. Phys.*, **49**, pp. 1425–1427

70 BUSCH, G., and TIÈCHE, Y. (1963): 'Résistivité électrique et effet Hall de métaux et semiconducteurs fondus', *Phys. Condens. Matter*, **1**, pp. 78–104

71 WILLIAMS, W., and STANFORD, J.L. (1973): 'Resistivity studies of antiferromagnetic Mn rich alloys containing up to 2.5 at. % V, Cr, Fe and Ru', *Phys. Rev. B*, **7**, pp. 3244–3250

72 POGA, M., BRADEA, I., IVANCIU, O., NICULESCU, D., and GRUCEANU, E. (1975): 'Synthesis and electrical behaviour of Ga_2Mg_5 type structure compounds', *Mater. Res. Bull.*, **10**, pp. 1349–1353

73 SCALA, E., and ROBERTSON, W.D. (1953): 'Electrical resistivity of liquid metals and of dilute liquid metal solutions', *J. Met.*, **5**, pp. 1141–1147

74 ARAJS, S. (1968): 'Electrical resistivity and antiferromagnetism in binary chromium alloys with molybdenum and tungsten', *J. Appl. Phys.*, **39**, pp. 673–674

75 SOUSA, J.B., PINTO, R.S., AMADO, M.M., MOREIRA, J.M., and BRAGA, M.E. (1979): 'Universal behaviour of thermal and electrical transport coefficients near the Néel temperature of $Cr_{99.94}Al_{0.06}$', *Solid State Commun.*, **31**, pp. 209–212

76 ARAJS, S. (1970): 'Electrical resistivity of dilute binary chromium alloys with nickel and manganese', *Phys. Status Solidi B*, **1**, pp. 499–503

77 ROLL, VON A., and MOTZ, H. (1957): 'Der elektrische widerstand von metallischen schmelzen', *Z. Metallkd.*, **48**, pp. 272–280

78 KUME, K. (1967): 'Anomalies due to s-d interaction in dilute alloys', *J. Phys. Soc. Jpn.*, **23**, pp. 1226–1234

79 TOTH, R.S., ARROTT, A., SHINOZAKI, S.S., WERNER, S.A., and SATO, H. (1969): 'Studies of Au_4X ordered alloys: electron and neutron diffraction, resistivity and specific heat', *J. Appl. Phys.*, **40**, pp. 1373–1375

80 SMITH, J.H., and WELLS, P. (1969): 'Antiferromagnetism in Au_5Mn_2', *J. Phys. C*, **2**, pp. 356–360

81 RICKER, VON T., and PFLÜGER, E. (1966): 'Die temperaturbhängigkeit des widerstandes und andere elektrische eigenschaften von palladium-silber- und palladium-rhodium-legierungen', *Z. Metallkd.*, **57**, pp. 39–45

82 FORT, D., and HARRIS, J.R. (1975): 'The physical properties of some palladium alloy hydrogen diffusion membrane materials', *J. Less-Common Met.*, **41**, pp. 313–327

83 BORELIUS, G., and LARSSON, L.E. (1956): 'Resistometric and calorimetric studies on precipitation in aluminium-silver alloys', *Ark. Fys.*, **11**, pp. 137–163

84 VERTOGRADSKII, V.A., and CHEKHOVSKOI, V.YA (1972): 'Electrical resistivity and thermal conductivity of alloys of the tungsten-molybdenum system', *High Temp.-High Press.*, **4**, pp. 621–626

85 TSCHIRNER, VON H.U. (1969): 'Messung des elektrischen widerstandes von flüssigem Pb, Sn, In, Bi und Sb nach der drehfeldmethode', *Z. Metallkd.*, **60**, pp. 46–49

86 ROLL, VON A., FELGER, H., and MOTZ, H. (1956): 'Elektrodenlose messung der elektrischen leitfähigkeit mit der drehfeldmethode', *Z. Metallkd.*, **47**, pp. 707–713

87 TAKEUCHI, S., and ENDO, H. (1962): 'The electric resistivity of the metals in the molten state', *Trans. Jpn. Inst. Met.*, **3**, p. 30

88 ADAMS, P.D., and LEACH, J.S. (1967): 'Resistivity of liquid lead-tin alloys', *Phys. Rev.*, **156**, pp. 178–183

89 AGTE, C., ALTERHUM, B., BECKER, K., HEYNE, G., and MOERS, K. (1931): 'Physical and chemical properties of rhenium', *Z. Chemie*, **196**, pp. 129–159

90 GAMES, G.B., and SIMS, C.T. (1957): 'Electrical resistivity and thermoelectric potential of rhenium metal', *Trans. ASTM.*, **57**, pp. 759–769

91 SPEDDING, F.H., DAANE, A.H., and HERRMANN, K.W. (1957): 'Electrical resistivities and phase transformations of lanthanum, cerium, praseodymium and neodymium', *J. Met.*, **9**, pp. 895–897

92 LITSCHEL, H., and POP, I. (1985): 'Temperature dependence of the electrical resistivity of the nickel-platinum alloy system', *J. Phys. & Chem. Solids*, **46**, pp. 1421–1425

93 IYER, V.K., and ASIMOW, R.M. (1967): 'The electrical resistivity of gold-silver alloys', *J. Less-Common Met.*, **13**, pp. 18–23

Resistivity references

94 HABERMANN, C.E., and DAANE, A.H. (1964): 'The high temperature resistivities of dysprosium, holmium and erbium', *J. Less-Common Met.*, **7**, pp. 31–36

95 TAKEUCHI, S., and ENDO, H. (1962): 'The electric resistivity of the metals in the molten state', *Trans. Jpn. Inst. Met.*, **3**, pp. 30–35

96 OSONO, H., KAWATA, S., ENDO, T., and KINO, T. (1978): 'Resistometric study of the configuration of Ag atoms in aluminium–1.03 at.% silver alloy', *Trans. Jpn. Inst. Met.*, **19**, pp. 69–76

97 DOMENICALI, C.A., and CHRISTENSON, E.L. (1961): 'Effect of transition metal solutes on the electrical resistivity of copper and gold between 4 and 1200 K', *J. Appl. Phys.*, **32**, pp. 2450–2456

98 LOBANOV, V.V., KLYUYEVA, I.B., KAGAN, G.YE., RYABOV, R.A., GUK, YU.N., and GEL'D, P.V. (1980): 'Influence of atomic ordering on the behaviour of hydrogen in the alloy Pd-20 W', *Phys. Met. & Metallogr.*, **47**, pp. 176–179

99 LEGVOLD, S., SPEDDING, F.H., BARSON, F., and ELLIOTT, J.F. (1953): 'Some magnetic and electrical properties of gadolinium, dysprosium and erbium metals', *Rev. Mod. Phys.*, **25**, pp. 129–130

100 SMITH, J.H. (1968): 'Magnetic properties of manganese-gold alloys and the effect of thermal and mechanical treatment', *J. Appl. Phys.*, **39**, pp. 675–677

101 ARUTYUNOV, A.V., and FILIPPOV, L.P. (1970): 'Thermal properties of rhenium at high temperatures', *Teplofiz. Vys. Temp.*, **8**, pp. 1095–1097

102 DESAI, P.D., JAMES, H.M., and HO, C.Y. (1984); 'Electrical resistivity of aluminium and manganese', *J. Phys. & Chem. Ref. Data*, **13**, pp. 1131–1172

103 GASSER, J.G., MAYOUFI, M., and BELLISSENT-FUNEL, M.C. (1989): 'Electrical resistivity and structure of liquid Al-Ge alloys', *J. Phys. Cond. Matter*, **1**, pp. 2409–2425

104 TAKEUCHI, A.Y., and DA CUNHA S.F. (1985): 'Electrical resistivity of the pseudo-binary system $Ce(Fe_{1-x}Al_x)_2$', *J. Magn.. & Magn. Mater.*, **49**, pp. 257–264

105 RAUB, C.J., and WOPERSNOW, W. (1983): 'Die elektrische leitfähigkeit einiger silber-aluminium und kobaltlegierungen', *J. Less-Common Met.*, **90**, pp. 153–159

106 WILLIAMS, R.K., GRAVES, R.S., and WEAVER, F.J. (1987); 'Effects of temperature and composition on the thermal and electrical conductivities of Ni_3Al', *J. Appl. Phys.*, **61**, pp. 1486–1492

107 GRUBE, VON G., and KNABE, R. (1936): 'Elektrische leitfähigkeit und zustandsdiagramm bei binären legierungen—das system Palladium-Chrom', *Z. Electrochem.*, **42**, pp. 793–804

108 COSMA, I., LUPSA, I., and VANCEA, M. (1975): 'Studiul rezistivitatii aliajelor Ni-Cu-Al', *Studdi Cercetari de Fiz.*, **27**, pp. 113–116

109 YAKHMI, J.V., GOPALAKRISHNAN, I.K., and GROVER, A.K. (1984): 'Electrical resistivity studies on the Heusler alloys $Co_2T_{1-x}Al_{1+x}$, (T = Ti or Zr)', *Phys. Status Solidi.*, **85**, pp. 89–92

110 UEMURA, O., and IKEDA, S. (1973): 'Electrical resistivity, magnetic susceptibility and heat of mixing of liquid silver-germanium alloys', *Trans. Jpn. Inst. Met.*, **14**, pp. 351–354

111 VERTOGRADSKII, V.A. (1977): 'Thermal and electric conductivity of platinum at high temperatures', *High Temp.*, **15**, pp. 178–180

112 MATUYAMA, Y. (1927): 'On the electrical resistance of molten metals and alloys', *Sci. Rep. Tokohu Univ.*, **16**, pp. 447–474

113 KÉITA, M., STEINEMANN, S., KÜNZI, H.U., and GÜNTHERODT, H.J. (1977):

'Determination of the diffusion coefficient in liquid metal alloys from measurements of the electrical resistivity', in *Inst. Phys. Conf. Liq. Met.*, pp. 655–662

114 JACKSON, C.B. (Ed.) (1955): 'Liquid Metals Handbook Sodium (NaK) Supplement', (Atomic Energy Commission USAEC Rep. No. TID 5277), pp. 28–30

115 SHPIL'RAIN, E.E., KAGAN, D.N., BARKHATOV, L.S., and ZHMAKIN, L.I. (1976): 'Experimental study of the specific electrical conductivity of molten aluminium oxide at temperatures up to 3000 K', *High Temp.*, **14,** pp. 843–846

116 ROMANOVA, A.V., and PERSION, Z.V. (1972): 'Electrical conductivity of aluminium-copper alloys in the liquid state', *Ukv. Fiz. Zh.*, **17,** pp. 1747–1749

117 SCHULSON, E.M., and TURNER, R.B. (1978): 'The electrical resistivity of ordered Zr_3Al', *Phys. Status Solidi*, **50,** pp. 83–86

118 GRATZ, E., GROSSINGER, R., OESTERREICHER, H., and PARKER, F.T. (1981): 'Resistivity and magnetization in disordered crystalline compound series $R(Al_xM_{1-x})_2$ (R = rare earth; M = Cu, Co, Fe)', *Phys. Rev. B*, **23,** pp. 2542–2547

119 MUIR, W.B., BUDNICK, J.I., and RAJ, K. (1982): 'Electrical resistivity and the band structure of $Fe_3Si_{1-x}Al_x$ alloys', *Phys. Rev. B*, **25,** pp. 726–729

120 CHELKOWSKI, A., TALIK, E., and WNETRZAK, G. (1983): 'Magnetic susceptibility and electric resistivity investigation in $Gd_{1-x}Dy_xAl_2$', *Solid State Commun.*, **46,** pp. 759–761

121 SMIT, P., and ALBERTS, H.L. (1987): 'Electrical resistivity of dilute $(Cr_{1-x}Al_x)_{95}Mo_5$ alloys', *J. Phys. Chem. Solids*, **48,** pp. 887–893

122 KOCH, C.C., and LOVE, G.R. (1967): 'The electrical resistivity of technetium from 8 to 1700 K', *J. Less-Common Met.*, **12,** pp. 29–35

123 TAYLOR, R.E., KIMBROUGH, W.D., and POWELL, R.W. (1971): 'Thermophysical properties of tantalum, tungsten and tantalum-10 wt.% tungsten at high temperatures', *J. Less-Common Met.*, **24,** pp. 369–382

124 YAMAMOTO, K. (1970): 'The phase relation of the electrical resistivity of Au-Mn alloys in Au rich side', *J. Phys. Soc. Jpn.*, **28,** pp. 1374–1375

125 HEIMANN, J., KACZMARSKA, K., KWAPULINSKA, E., SLEBARSKI, A., and CHELKOWSKI, A. (1982): 'Electrical resistivity and magnetic properties of the $Re_{1-x}Gd_xAl_2$ Laves phase (for Re: La, Lu, Y)', *J. Magn. & Magn. Mater.*, **27,** pp. 187–194

126 TYLER, W.W., WILSON, A.C., and WOLGA, G.J. (1953): 'Thermal conductivity, electrical resistivity and thermoelectric power of uranium', *J. Met.*, **5,** pp. 1238–1239

127 IKEDA, K., SUZUKI, K., and TANOSAKI, K. (1983): 'Electrical resistivity in ferromagnetic Ni-based alloys containing Al, Si and Ge', *J. Magn. & Magn. Mater.*, **40,** pp. 232–240

128 YAKHMI, J.V., GOPALAKRISHNAN, I.K., and IYER, R.M. (1983): 'On the electrical resistivity and Néel temperature of dilute Cr-Ir alloys', *J. Less-Common Met.*, **91,** pp. 327–331

129 PECHERSKAYA, V.I., and BOLSHUTKIN, D.N. (1982): 'Electrical resistance and thermal expanson of Fe-18Cr-(10 to 25)Ni alloys', *Phys. Status Solidi*, **74,** pp. 285–290

130 JEN, S.U., and YAO, Y.D. (1987): 'Temperature dependence of the electrical resistivity and deviations from Matthiessen's rule of Pd-Er alloys', *J. Phys. Chem. Solids*, **48,** pp. 571–574

131 SEYDEL, U., and FUCKE, W. (1977): 'Sub-microsecond pulse heating measurements of high temperature electrical resistivity of the 3d- transition metals Fe, Co and Ni', *Z. Naturforsch. A*, **32A,** pp. 994–1002

132 ZINOV'EV, V.E., ABEL'SKII, SH., SANDAKOVA, M.I., PETROVA, L.N., and GEL'D, P.V. (1973): 'Matthiessen rule and high temperature electrical resistivity of solid solutions of silicon in iron', *Sov. Phys. JETP*, **36**, pp. 1174–1176

133 GRUBE, VON G., and KÄSTNER, H. (1936): 'Elektrische leitfähigkeit und zustandsdiagramm bei binären legierungen—das system palladium-kobalt', *Z. Elektrochem.*, **42**, pp. 156–159

134 GRUBE, VON G., and SCHMIDT, A. (1936): 'Elektrische leitfähigkeit und zustandsdiagramm bei binären legierungen—das system natrium-thallium', *Z. Elektrochem.*, **42**, pp. 201–209

135 GRUBE, VON G., BAYER, G., and BUMM, H. (1936): 'Elektrische leitfähigkeit und zustandsdiagramm bei binären legierungen—das system palladium-mangan', *Z. Elektrochem.*, **42**, pp. 805–815

136 ARAJS, S., and DUNMYRE, G.R. (1967): 'Electrical resistivity of polycrystalline praseodymium between 2 and 325 K', *J. Less-Common Met.*, **12**, pp. 162–166

137 FEITSMA, P.D., HALLERS, J.J., WERFF, F.V.D., and VAN DER LUGT, W. (1975): 'Electrical resistivities and phase separation of liquid lithium-sodium alloys', *Physica*, **79B**, pp. 35–52

138 ASHELBY, D.W., BROOKS, S., LEACH, J.S.L., and MOON, J.R. (1975): 'Some measurements of the electrical resistivities of dilute liquid amalgams containing alkali and alkaline earth elements', *J. Phys. F*, **5**, pp. 2105–2108

139 STEINLEITNER, G., FREYLAND, W., and HENSEL, F. (1975): 'Electrical conductivity and excess volume of the liquid alloy system Li-Bi', *Ber. Bunsen. Phys. Chem.*, **79**, pp. 1186–1189

140 OOMI, G., and WOODS, S.B. (1985): 'The electrical resistance of some lithium-magnesium alloys near the martensitic transformation', *Solid State Commun.*, **53**, pp. 223–225

141 SAVCHENKO, V.A., and SHPIL'RAIN, E.E. (1973): 'Effect of small additions of rubidium on electrical resistance of liquid sodium', *High Temp.*, **11**, pp. 606–609

142 SAVCHENKO, V.A., and SHPIL'RAIN, E.E. (1974): 'Effect of small additions of cesium on electrical resistance of liquid sodium', *High Temp.*, **12**, pp. 782–784

143 CHI, T.C. (1979): 'Electrical resistivity of alkali elements', *J. Phys. & Chem. Ref. Data*, **8**, pp. 339–438

144 CHI, T.C. (1979): 'Electrical resistivity of alkaline earth elements', *J. Phys. & Chem. Ref. Data*, **8**, pp. 439–497

145 SAVCHENKO, V.A., and SHPIL'RAIN, E.E. (1977): 'Experimental investigation of the electrical resistivity of a liquid alloy of lithium and sodium', *High Temp.*, **15**, pp. 1123–1125

146 KURIYAMA, K., KAMIJOH, T., and NOZAKI, T. (1980): 'Anomalous electrical resistivity in LiAl near critical composition', *Phys. Rev. B*, **22**, pp. 470–471

147 YAHAGI, M., and IWAMURA, K. (1977): 'Temperature dependences on electrical resistivity and thermal expansion in the intermetallic compound LiIn', *Phys. Status Solidi*, **39**, pp. 189–192

148 VEDERNIKOV, M.V., BURKOV, A.T., DVUNITKIN, V.G., and MOREVA, N.I. (1977): 'The thermoelectric power, electrical resistivity and Hall constant of rare earth metals in the temperature range 80–1000 K', *J. Less-Common Met.*, **52**, pp. 221–245

149 CALAWAY, W.F. (1982): 'The electrical resistivity of liquid lithium', *J. Less-Common Met.*, **86**, pp. 305–319

150 YATSENKO, S.P., ZHAKUPOV, SH.R., and CHUNTONOV, K.A. (1979): 'Magnetic susceptibility and electrical conductivity of sodium alloys with gallium', *Fiz. Met. & Metalloved.*, **48**, pp. 927–930

151 BAUHOFER, W., and SIMON, A. (1977): 'Der elektrische widerstand von kalium/cäsium-legierungen', *Z. Naturforsch. A*, **32A**, pp. 1275–1280

152 MITCHELL, M.A., and SUTULA, R.A. (1987): 'The density, electrical resistivity and Hall coefficient of Li-B alloys', *J. Less-Common Met.*, **57**, pp. 161–175

153 BURKOV, A.T., and VEDERNIKOV, M.V. (1984): 'Temperature dependences of the thermoelectric power and electrical resistivity of praseodymium and neodymium in solid and liquid states', *Sov. Phys.-Solid State*, **26**, pp. 2211–2212

154 LAUBITZ, M.J. (1969): 'Transport properties of pure metals at high temperatures. II. Silver and gold', *Can. J. Phys.*, **47**, pp. 2633–2644

155 BATH, A., GASSER, J.G., BRETONNET, J.L., BIANCHIN, R., and KLEIM, R. (1980): 'Electrical resistivity and the thermoelectric power of liquid Ge-Sb and Pb-Sb alloys', *J. de Phys. Colloq.*, **41**, pp. 519–523

156 SAVCHENKO, V.A., and SHPIL'RAIN, E.E. (1969): 'Influence of a small admixture of sodium on the resistivity of lithium', *High Temp.*, **7**, pp. 769–780

157 FREEDMAN, J.F., and ROBERTSON, W.D. (1961): 'Electrical resistivity of liquid sodium, liquid lithium and dilute liquid sodium solutions', *J. Chem. Phys.*, **34**, pp. 411–415

158 FEI-XIANG, L., FENG, L., GUO-HUA, T., and WAN-RONG, C. (1988): 'The behaviour of electrical resistivity in amorphous $Co_{100-x}B_x$ alloys', *Mater. Sci. & Eng.*, **99**, pp. 227–229

159 KEDVES, F.J., HORDOS, M., and GERRELY, L. (1972): 'Temperature dependence of impurity resistivity in dilute Al-based 3d transition metal alloys between 78 and 930 K', *Solid State Commun.*, **11**, pp. 1067–1071

160 KAPPLER, J., and WUCHER, J. (1973): 'Résistivités électriques à hautes températures des systèmes $(Cr_{1-x}Mn_x)Be_2$ et $(V_{1-x}Cr_x)Be_2$', *C. R. Acad. Sc. Paris*, **277**, pp. 337–340

161 BABIC, E., SAUB, K., and MAROHNIC, Z. (1980): 'Electrical resistivities of amorphous $(Fe_xNi_{100-x})_{0.75}B_{0.25}$ alloys', *Fizika*, **12**, pp. 215–219

162 POWELL, R.W., TYE, R.P., and WOODMAN, M.J. (1967): 'The thermal conductivity and electrical resistivity of polycrystalline metals of the platinum group and of single crystals of ruthenium', *J. Less-Common Met.*, **12**, pp. 1–10

163 KETTLER, W., WERNHARDT, R., and ROSENBERG, M. (1982): 'Electrical resistivity and thermopower of amorphous $Fe_xCo_{80-x}B_{20}$ alloys', *J. Appl. Phys.*, **53**, pp. 8248–8250

164 DOLOGAN, V., and DOLOGAN, E. (1985): 'Temperature dependence of the electrical resistivity of some amorphous ferromagnetic alloys', *Rev. Roum. Phys.*, **30**, pp. 531–534

165 REDDY, P.V., and AKHTAR, D. (1987): 'Electrical resistivity and thermoelectric power of $Fe_{74}Co_{10-x}Cr_xB_{16}$ metallic glasses', *J. Mater. Sci. Lett.*, **6**, pp. 1344–1346

166 DOLOGAN, V., and DOLOGAN, E. (1987): 'Electrical resistivity and crystallization of glassy $Fe_{100-x}B_x$, $Fe_{80-x}Co_xB_{20}$, $Fe_{78}B_{22-x}Si_x$ and $Fe_{100-x}B_{x-y}Si_y$ alloys', *Rev. Roum. Phys.*, **32**, pp. 1077–1090

167 YAO, Y.D., ARAJS, S., and LIN, S.T. (1980): 'Electrical resistivity of amorphous $Fe_{80}B_{20-x-y}Si_xC_y$ alloys', *Annu. Rep. Inst. Acad. Sin.*, **10**, pp. 55–60

168 BARANDIARÁN, J., GÓMEZ-SAL, J., FERNÁNDEZ, J., LÓPEZ-SÁNCHEZ, R., and NIELSEN, O.V. (1987): 'Temperature dependence of the electrical

resistivity in $[Co_{1-x}(Fe_{0.5}Ni_{0.5})_x]_{75}Si_{15}B_{10}$ metallic glasses', *Phys. Status Solidi A*, **99**, pp. 243–249

169 JESSER, R., and LABORDE, O. (1984): 'Magnétisme de $CrBe_{12}$ en champs intenses et résistivité électrique de $CrBe_{12}$ et de VBe_{12}', *J. Phys.*, **45**, pp. 1819–1826

170 SHTERNER, S.R., and DOVGOPOL, S.P. (1983): 'Density, electric resistance and short range order of Co-B and Ni-B melts', *Ukr. Fiz. Zh.*, **28**, pp. 858–861

171 HUST, J.G., and GIARRATANO, P. J. (1975): 'Thermal conductivity and electrical resistivity standard reference materials: electrolytic iron, SRM's 734 and 797 from 4 to 1000 K', (National Bureau of Standards NBS-SP-260-50)

172 MATULA, R.A. (1979): 'Electrical resistivity of copper, gold, palladium and silver', *J. Phys. & Chem. Ref. Data*, **8**, pp. 1147–1298

173 CEZAIRLIYAN, A., and MILLER, A.P. (1977): 'Heat capacity and electric resistivity of titanium in the range 1500 to 1900 K by a pulse heating method', *High Temp.-High Press.*, **9**, pp. 319–324

174 CEZAIRLIYAN, A., RIGHINI, F., and McCLURE, J. L. (1974): 'Simultaneous measurements of heat capacity, electrical resistivity and hemispherical total emittance by a pulse heating technique: Vanadium, 1500 to 2100 K', *J. Res. Natl. Bur. Standards A*, **78A**, pp. 143–147

175 KOPCANSKY, P., and ZENTKO, A. (1983): 'Temperature dependence of the electrical resistivity of amorphous Pd-Si alloy', *Czech. J. Phys. Sect. B*, **B33**, pp. 243–246

176 MEDVEDEV, A.I., STOLZ, A.K., AND GELD, P.V. (1985): 'Electrical resistivity of solid solutions of iron, chromium and cobalt in V_3Si and V_3Ge', *Metallofizika Kiev*, **7**, pp. 613–618

177 IKEDA, K. (1986): 'Electrical resistivity in strongly disturbed Ni-based alloys', *J. Magn. & Magn. Mater.*, **49**, pp. 161–171

178 COOK, J.G., and LAUBITZ, M.J. (1978): 'The transport properties of the cubic alkaline earths Ca, Sr and Ba', *Can. J. Phys.*, **56**, pp. 161–174

179 JENSEN, C.L., and ZALESKY, M.P. (1980): 'Electrical resistivity of scandium-hydrogen alloys from 295 to 4.2 K', *J. Less-Common Met.*, **74**, pp. 197–205

180 ZINON'EV, V.E., CHUPINA, L.I., and GEL'D, P.V. (1973): 'Transport properties of scandium at high temperatures', *Sov. Phys.-Solid State*, **14**, pp. 2416–2418

181 TSCHIRNER, H.U., PAZDRII, I.P., and WEIKART, R. (1983): 'Messung des spezifischen widerstandes flüssiger legierungen der system Ge-Sb, Ge-Te, Sb-Te und Ge-Sb-Te', *Wiss. Z. Tech. Hochsch.*, **25**, pp. 219–225

182 ROTTMAN, C., and VAN ZYTVELD, J.B. (1979): 'Electronic properties of high purity liquid calcium and strontium', *J. Phys. Fluids*, **9**, pp. 2049–2056

183 LOGUNOV, A.V., and PETRUSHIN, N.V. (1975): 'Electrical resistivity of niobium-titanium alloy at high temperatures', *High Temp.*, **13**, pp. 84–86

184 SARKISSIAN, B.V.B. (1986): 'Resistivity, thermopower and susceptibility of Y_9Co_7 and local-mode effects', *J. Phys. F*, **16**, pp. 755–768

185 GRATZ, E., BAUER, E., SECHOVSKY, V., and CHMIST, J. (1986): 'Magnetic properties, electrical resistivity and thermopower of $(Ho_xY_{1-x})Co_2$ $(0 \le x \le 1)$', *J. Magn. & Magn. Mater.*, **54-7**, pp. 517–518

186 GRATZ, E., PILLMAYR, N., BAUER, E., and HILSCHER, G. (1987): 'Temperature and concentration dependence of the electrical resistivity in $(RE, Y)Co_2$ (RE = Rare earth element)', *J. Magn. & Magn. Mater.*, **70**, pp. 159–161

187 BURKOV, A.T., VEDERNIKOV, M.V., IOFFE, A.F., and GRATZ, E. (1988): 'Thermopower and electrical resistivity of YCo_2 at high temperatures', *Solid State Commun.*, **67**, pp. 1109–1111

188 BATH, A., and KLEIM, R. (1982): 'Electrical resistivity and thermoelectric power of liquid Zn-Sb alloys', *J. Phys. F*, **12**, pp. 2975–2984

189 McELROY, D.L., WILLIAMS, R.K., and WEAVER, F.J. (1979): 'The physical properties of V(Fe,Co,Ni)$_3$ alloys from 300 to 1600 K', in *Proc. 16th Int. Conf. Thermal conductivity*, Larsen, D.C. (Ed.) (Plenum Publishing), pp. 337–341

190 RAO, K., RAPP, O., JOHANNESSON, C., ASTRÖM, H., BUDNICK, J., BURCH, T., and NICULESCU, V. (1977): 'Electrical resistivity of Fe dissolved in PdAg matrix alloys', *Physica B & C*, **86-8**, pp. 831–832

191 YAO, Y.D., and ARAJS, S. (1981): 'Electrical resistivity of Fe-rich Fe-Pd alloys between 77 and 1250 K', *Phys. Status Solidi*, **64**, pp. 95–104

192 WU, Y.C., CHEN, J., and FANG, S.H. (1973): 'Determination of Curie temperature, short range order parameter and number of nearest neighbours of Pd atoms to Fe atoms in Pd-Fe alloys by resistivity measurement', *Chin. J. Phys.*, **11**, pp. 130–135

193 SAKURAI, J., TAGAWA, Y., and KOMURA, Y. (1985): 'Electrical resistivity and lattice constant of $(Ce_{1-x}La_x)Ni_2$', *J. Magn. & Magn. Mater.*, **52**, pp. 205–207

194 KOUROV, N.I., TSIOVKIN, YU.N., and VOLKENSHTEYN, N.V. (1983): 'Electrical resistivity of $Pt_3Mn_xFe_{1-x}$ alloys with interacting magnetic order parameters', *Phys. Met. & Metallogr.*, **55**, pp. 955–959

195 LEVY, R.A., and PAYNE, J.A. (1975): 'Resistive anomalies in $(Pd_{1-x}Ag_x)_{0.99}Fe_{0.01}$ ternary alloys', *Phys. Lett. A*, **53**, pp. 329–330

196 GASSER, J.G., and KLEIM, R. (1976): 'Electrical resistivity of liquid alloys of manganese with antimony, tin and indium as a function of temperature and composition', in *3rd. Int. Conf. on Liq. Metals*, pp. 352–358

197 ZINOV'EV, V.E., ZAGREBIN, L.D., PETROVA, L.N., and SIPAILOV, V.A. (1984): 'Electrical resistance and thermophysical properties of solid solutions of germanium in iron', *Sov. Phys. J.*, **27**, pp. 474–478

198 KOUROV, N.I., TSIOVKIN, YU.N., and KARPOV, YU.G. (1987): 'Concentration dependence of resistivity in alloys $(Pd_xPt_{1-x})_3Fe$', *Phys. Met. & Metallogr.*, **64**, pp. 199–202

199 HAUG, R., KAPPEL, G., and JAEGLE, A. (1980): 'Electrical resistivity and magnetic susceptibility studies of the system Mn_5Si_3-Fe_5Si_3', *J. Phys. & Chem. Solids*, **41**, pp. 539–544

200 OKADA, T., and OHNO, S. (1986): 'Electrical conductivities of liquid Mn and Mn-Te alloys', *J. Phys. Soc. Jpn.*, **55**, pp. 599–605

201 YAHIA, J., and THOBE, J.P. (1972): 'The temperature dependence of the resistivity of liquid gallium to 1000 deg. C', *Can. J. Phys.*, **50**, pp. 2554–2556

202 GRATZ, E., SASSIK, H., NOWOTNY, H., STEINER, W., and MAIR, G. (1979): 'Electrical resistivity, thermopower and X-ray structure investigations on $RECo_2$ (RE = Nd, Tb, Dy, Ho, Er) as a function of temperature', *J. Phys. Colloq.*, **C5**, pp. 186–187

203 TOMILO, ZH.M. (1980): 'Thermophysical properties of scandium-titanium alloys', *J. Eng. Phys.*, **38**, pp. 337–341

204 ADAMYAN, V.YE., KARAGEBAKYAN, G.G., MELIKYAN, M.A., and ABOVYAN, E.S. (1983): 'Temperature dependences of magnetic susceptibility and electrical resistivity in the system of solid solutions $GdZn_xMg_{1-x}$', *Phys. Met. & Metallogr.*, **55**, pp. 36–40

205 LUTSKAYA, L.F. (1986): 'Thermo-emf and electrical resistivity in Fe_3Si, Mn_3Si and $(Fe_{1-x}Mn_x)_3Si$ solid solutions', *Inorg. Mater.*, **22**, pp. 1812–1814

206 HAUSER, E., RAY, S., and TAUC, J. (1979): 'Electrical resistivity of liquid gold-silicon alloys', *Solid State Commun.*, **29**, pp. 821–823

Resistivity references

207 IKEDA, K. (1987): 'Electrical resistivity and ferromagnetism in Ni-Pd alloys', *J. Appl. Phys.*, **62**, pp. 4499–4503

208 IKEDA, K., ARAKI, Y., and TANOSAKI, K. (1984): 'Electrical resistivity of Ni-rich Ni-Mn alloys', *J. Phys. Soc. Jpn.*, **53**, pp. 3545–3552

209 NABEREZHNYKH, V.P., MOROZ, T.T., and BELOV, B.F. (1985): 'Electrical resistivity of liquid Ni-P alloys', *Phys. Met.*, **6**, pp. 1055–1060

210 SEYDEL, U., FUCKE, W., and MÖLLER, B. (1977): 'The electrical resistivity of exploding Ni wires in fast RCL circuits', *Z. Nat. Forsch. A*, **32**, pp. 147–151

211 BÖTTGER, CH., and HESSE, J. (1989): 'The electrical resistivity of re-entrant spin glass alloys ($Fe_{0.65}Ni_{0.35})_{1-x}Mn_x$ in the temperature range 4 to 280 K', *Z. Phys. Condens. Matter*, **75**, pp. 485–493

212 IKEDA, K., and LI, X. (1989): 'Electrical resistivity and ferromagnetism in Ni-Ga alloys', *J. Less-Common Met.*, **152**, pp. 261–273

213 DESAI, P.D., CHU, T.K., JAMES, H.M., and HO, C.Y. (1984): 'Electrical resistivity of selected elements', *J. Phys. & Chem. Ref. Data*, **13**, pp. 1069–1096

214 OGAWA, S. (1976): 'Electrical resistivity of weak itinerant ferromagnetic $ZrZn_2$', *J. Phys. Soc. Jpn.*, **40**, pp. 1007–1009

215 OHNO, S., OKAZAKI, H., and TAMAKI, S. (1974): 'Electrical resistivity of liquid Sb alloy', *J. Phys. Soc. Jpn.*, **36**, pp. 1133–1136

216 SUGAWARA, T., TAKANO, M., and TAKAYANAGI, S. (1974): 'Anomalous resistivity of dilute MoMn, MoFe and MoCo alloys', *J. Phys. Soc. Jpn.*, **36**, pp. 451–455

217 YAKHMI, J.V. (1983): 'Electrical resistivity of β Co_xGa_{1-x}', *J. Phys. F*, **13**, pp. 659–664

218 MAGOMEDOV, A.M. (1979): 'Electrical resistivity of cadmium-zinc and indium-zinc melts', *High Temp.*, **17**, pp. 271–276

219 VORONTSOV, B.S., DOVGOPOL, S.P., and GEL'D, P.V. (1979): 'Electrical resistivity of Co-Ga alloys at high temperatures', *Sov. Phys. J.*, **21**, pp. 1572–1576

220 YAKHMI, J.V., GOPALAKRISHNAN, I.K., and IYER, R.M. (1984): 'Electrical resistivity and the magnetic phase transitions of Cr-Mn alloys', *J. Phys. F*, **14**, pp. 923–929

221 BURGER, J.P., DAOU, J.N., and VAJDA, P. (1984): 'Electrical resistivity of rare earth dihydrides and dideutrides', *J. Less-Common Met.*, **103**, pp. 381–388

222 CEZAIRLIYAN, A., and MILLER, A.P. (1983): 'Heat capacity and electrical resistivity of nickel in the range 1300–1700 K measured with a pulse heating technique', *Int. J. Thermophys.*, **4**, pp. 389–396

223 IKEDA, K., and TANOSAKI, K. (1984): 'Origin of anomalous resistivity in Ni-Cu alloys', *Trans. Jpn. Inst. Met.*, **25**, pp. 447–457

224 IKEDA, K. (1988): 'Electrical resistivity of nickel, cobalt and their alloys', *Trans. Jpn. Inst. Met.*, **29**, pp. 183–190

225 MATVEYEV, V.A., FEDOROV, G.V., and VOLKENSHTEYN, N.V. (1976): 'Kinetic properties of palladium-nickel alloys. 1. Resistivity', *Phys. Met. & Metallogr.*, **42**, pp. 42–56

226 REDDY, B.K., and GOEL, T.C. (1975): 'Total and spectral emittances and electrical resistivity of nickel at high temperatures', *Ind. J. Pure & Appl. Phys.*, **13**, p. 138

227 MON'KIN, V.D., UKHOV, V.F., VATOLIN, N.A., and GEL'CHINSKIY, B.R. (1978): 'Electrical resistivity of Au-Sn, Au-Bi, Au-Ga and Au-Ni alloys in the liquid states', (Ivz. Akad. Nauk. SSR Met), pp. 41–44

228 PAL'GUYEV, YE.E., KURANOV, A.A., SYUTKIN, P.N., and SIDORENKO, F.A. (1976): 'Anomalous temperature dependence of the resistivity of ordering palladium-iron alloys', *Phys. Met. & Metallogr.*, **42**, pp. 46–50

229 TAKEZAWA, T., and YOKOYAMA, T. (1975): 'The temperature dependence of electrical resistivity at 70 at.% Pd-Fe alloy', *J. Jpn. Inst. Met.*, **39**, pp. 550–551

230 IKEDA, K., LI, X., TANOSAKI, K., and NAKAZAWA, K. (1989): 'Electrical resistivity and ferromagnetic properties in Ni-Zn alloys', *Mat. Trans. JIM*, **30**, pp. 1–9

231 CYWINSKI, R. (1979): 'Electrical resistivity anomalies of highly resistive crystalline alloys', *J. Phys. F*, **9**, pp. L29–L33

232 SEYDEL, U., and FUCKE, W. (1980): 'Electrical resistivity of liquid Ti, V, Mo and W', *J. Phys. F*, **10**, pp. L203–L206

233 CLAUS, H., and MYDOSH, J.A. (1974): 'Electrical resistivity and initial susceptibility of weakly ferromagnetic V-Fe alloys', *Solid State Commun.*, **14**, pp. 209–212

234 OKADA, T., and OHNO, S. (1987): 'Electrical conductivities of liquid transition metal-Te alloys', *J. Phys. Soc. Jpn.*, **56**, pp. 1092–1100

235 ISINO, M., and MUTO, Y. (1985): 'Electrical resistivity of binary vanadium alloys', *J. Phys. Soc. Jpn.*, **54**, pp. 3839–3847

236 PAN, V.M., VULAKH, I.E., KASATKIN, A.L., and SHEVCHENKO, A.D. (1978): 'Origin of the anomalous temperature dependences of the electrical resistivity and magnetic susceptibility of ZrV_2 and HfV_2', *Sov. Phys. Solid State*, **20**, pp. 1437–1438

237 MOORE, J.P., WILLIAMS, R.K., and GRAVES, R.S. (1977): 'Thermal conductivity, electrical resistivity and Seebeck coefficient of high purity chromium from 280 to 1000 K', *J. Appl. Phys.*, **48**, pp. 610–617

238 MOYER, C.A., ARAJS, S., and EROGLU, A. (1980): 'Electrical resistivity and antiferromagnetism of chromium-palladium alloys between 77 and 700 K', *Phys. Rev. B*, **22**, pp. 3277–3281

239 KATANO, S., MÔRI, N., and NAKAYAMA, K. (1980): 'Resistivity minimum in antiferromagnetic Cr-Co alloys', *J. Phys. Soc. Jpn.*, **48**, pp. 192–199

240 ARAJS, S., AIDUN, R., and MOYER, C.A. (1980): 'Antiferromagnetism and electrical resistivity of dilute chromium-germanium alloys', *Phys. Rev. B*, **22**, pp. 5366–5368

241 CEZAIRLIYAN, A., and MILLER, A.P. (1977): 'Melting point, normal spectral emittance (at the melting point), and electrical resistivity (above 1900 K) of titanium by a pulse heating method', *J. Res. Natl. Bur. Standards*, **82**, pp. 119–122

242 RAMAKRISHNAN, S., NIGAM, A.K., and CHANDRA, G. (1986): 'Resistivity and magnetoresistance studies on superconducting Al_5V_3Ga, V_3Au and V_3Pt compounds', *Phys. Rev. B*, **34**, pp. 6166–6171

243 GASSER, J.G., MAYOUFI, M., GINTER, G., and KLEIM, R. (1984): 'Electrical resistivity of bismuth and bismuth-germanium alloys in the liquid state', *J. Non-Cryst. Solids*, **61-2**, pp. 1237–1242

244 PASHAYEV, B.P., PALCHAYEV, D.K., CHALABOV, R.I., and REVELIS, V.G. (1974): 'Electrical resistance of gallium-indium alloys in the solid and liquid phases', *Fiz. Met. & Metalloved.*, **37**, pp. 67–70

245 GINTER, G., GASSER, J.G., and KLEIM, R. (1986): 'The electrical resistivity of liquid bismuth, gallium and bismuth-gallium alloys', *Philos. Mag. B*, **54**, pp. 543–552

246 MORI, H., YASHIMA, H., and SATO, N. (1985): 'A new dense Kondo system:

$Ce_yLa_{1-y}Ge_2$. Resistivity, specific heat and susceptibility studies', *J. Low Temp. Phys.*, **58**, pp. 513-531

247 DESAI, P.D., JAMES, H.M., and HO, C.Y. (1984): 'Electrical resistivity of vanadium and zirconium', *J. Phys. & Chem. Ref. Data*, **13**, pp. 1097-1130

248 PALSTRA, T.T.M., MYDOSH, J.A., and NIEUWENHUYS, G.J. (1983): 'Study of the critical behaviour of the magnetization and electrical resistivity in cubic $La(Fe, Si)_{13}$ compounds', *J. Magn. & Magn. Mater.*, **36**, pp. 290-296

249 HAUG, R., KAPPEL, G., and JAÉGLÉ, A. (1979): 'Electrical resistivity studies of the system Mn_5Ge_3-Mn_5Si_3', *Phys. Status Solidi A*, **55**, pp. 285-290

250 MARCHENKO, V.A. (1973): 'Temperature dependence of the electrical resistivity of V_3Si', *Sov. Phys. Solid State*, **15**, pp. 1261-1262

251 ZINOV'YEV, V.Y., PETROVA, M.I., SANDAKOVA, M.I., and GEL'D, P.V. (1974): 'Influence of silicon on the high temperature resistivity of Fe-Si solid solutions', *Fiz. Met. & Metalloved.*, **37**, pp. 65-69

252 ARAJS, S., RAO, K.V., and ANDERSON, E.E. (1975): 'Electrical resistivity and antiferromagnetism of chromium-platinum alloys', *Solid State Commun.*, **16**, pp. 331-333

253 YAO, Y.D., ARAJS, S., and ANDERSON, E.E. (1975): 'Electrical resistivity of nickel rich nickel-chromium alloys between 4 and 300 K', *J. Low Temp. Phys.*, **21**, pp. 369-376

254 VAN ZYTVELD, J.B. (1984): 'Electrical resistivity of liquid chromium', *J. Non-Cryst. Solids*, **61-2**, pp. 1085-1090

255 YAKHMI, J.V., GOPALAKRISHNAN, I.K., and IYER, R.M. (1985): 'Some anomalous aspects of resistivity behaviour in dilute chromium alloy systems', *J. Appl. Phys.*, **57**, pp. 3223-3225

256 MORTON, N., JAMES, B.W., WOSTENHOLM, G.H., and NUTTALL, S. (1975): 'The thermal and electrical conductivities of niobium-65% titanium alloys', *J. Phys. F*, **5**, pp. 2098-2104

257 WESTLAKE, D.G., and MILLER, J.F. (1980): 'Resistivity due to hydrogen in transition metal alloys', *J. Phys. F*, **10**, pp. 859-863

258 OOTA, A., KAMATANI, Y., and NOGUCHI, S. (1981): 'Electrical resistivity of C-15 compounds $Zr_{1-x}Ta_xV_2$', *Jpn. J. Appl. Phys.*, **20**, pp. 2411-2417

259 TAKEDA, S., OHNO, S., and TAMAKI, S. (1976): 'Electrical resistivities of dilute 3d-transition solutes in liquid tellurium', *J. Phys. Soc. Jpn.*, **40**, pp. 113-117

260 KEFIF, B., and GASSER, J.G. (1987): 'Electrical resistivity of liquid rare earth-transition alloys of nickel with lanthanum and cerium', in Proc. Nato Adv. Stdy Inst., Passo della Mendola, Trentino, Italy (Martinus Nijhoff), pp. 427-431

261 KITA, Y., AND MORITA, Z. (1984): 'The electrical resistivity of liquid Fe-Ni, Fe-Co and Ni-Co alloys', *J. Non-Cryst. Solids*, **61-2**, pp. 1079-1084

262 OLIVIER, M., SIEGRIST, T., and McALISTER, S.P. (1987): 'Magnetic susceptibility and electrical resistivity of some Th_2Zn_{17}-rare earth zinc phases', *J. Magn. & Magn. Mater.*, **66**, pp. 281-290

263 KROMPHOLZ, K., and WEISS, A. (1980): 'Hall effect and electrical conductivity of gold-zinc and gold-platinum-zinc alloys', *Z. Metallkd.*, **71**, pp. 497-506

264 MITCHELL, M.A., and GOFF, J.F. (1975): 'Effect of molybdenum and vanadium on the lattice thermal conductivity and Lorenz number of chromium', *Phys. Rev. B*, **12**, pp. 1858-1867

265 CHIU, J.C.H. (1976): 'Deviations from linear temperature dependence of the electrical resistivity of V-Cr and Ta-W alloys', *Phys. Rev. B*, **13**, pp. 1507-1514

266 TEOH, W., ARAJS, S., ABUKAY, D., and ANDERSON, E.E. (1976): 'Electrical

resistivity of iron-vanadium alloys between 78 and 1200 K', *J. Magn. & Magn. Mater.*, **3**, pp. 260–263

267 VALIANT, J. C., and FABER, T.E. (1974): 'The resistivity and thermoelectric power of liquid alloys containing tellurium', *Philos. Mag.*, **29**, pp. 571–583

268 POKORNY, M., and ÅSTRÖM, H.U. (1976): 'Temperature dependence of the electrical resistivity of liquid gallium between its freezing point (29.75 deg.) and 752 deg. C', *J. Phys. F*, **6**, pp. 559–565

269 THÖRMER, K., COUFAL, H., FRITSCH, H., and DILETTI, H. (1976): 'Measurement of the specific resistance of Cu-Ga alloys at high temperatures', *Phys. Lett.*, **56A**, pp. 489–490

270 CARVALHO, E.M., and HARRIS, I.R. (1985): 'Electrical resistivity studies of phase transitions in the system $Zr_{50}Co_{50-x}Ni_x$ $(0 \le x \le 50)$', *J. Less-Common Met.*, **106**, pp. 129–141

271 ISMAILOV, M.A., MAGOMEDOV, A.M.A., and PASHAYEV, B.P. (1976): 'Electrical resistivity of Ga-Sn and Ga-Zn melts', (Izv. Akad. Nauk, SSSR Met.), pp. 45–47

272 RASSOKHIN, V.A., VOLKENSHTEIN, N.V., ROMANOV, A.P., and PREKUL, A.F. (1974): 'Localized spin fluctuations and singularities of the electric resistance of Ti_x-V_{1-x}', *Zh. Eksp. & Teor. Fiz.*, **66**, pp. 348–353

273 GALKIN, V. YU., and TUGUSHEVA, T.E. (1988): 'Relationship between an anomalous temperature dependence of the electrical resistivity and resonance impurity scattering in dilute Cr-Fe alloys', *Sov. Phys. Solid State*, **30**, pp. 487–491

274 HÜBER, G., and PLOUMBIDIS, D. (1984): 'Electrical resistivity and NMR investigations of the V-Cr system at high temperatures', *Phys. Status Solidi B*, **126**, pp. K101–K104

275 HODGKINSON, R.J., EASTMAN, N.J., FAVARON, J., and OWEN, H. (1982): 'Contact and contactless electrical conductivity measurements on the liquid semiconductor systems In-Te and Sb-Te', *J. Phys. C*, **15**, pp. 4147–4153

276 NEWPORT, R.J., HOWE, R.A., and ENDERBY, J.E. (1982): 'The structure and electrical properties of liquid semiconductors: II. Electron transport in liquid Ni-Te alloys', *J. Phys. C*, **15**, pp. 4635–4640

277 HIEMSTRA, C., KEEGSTRA, P., MASSELINK, W.T., and ZYTVELD, J.B. (1984): 'Electrical resistivities of solid and liquid Pr, Nd and Sm', *J. Phys. F*, **14**, pp. 1867–1875

278 SHIRAKAWA, K., KANEKO, T., YOSHIDA, H., and MASUMOTO, T. (1986): 'Electrical resistivity of amorphous and crystalline $(Co_{1-x}Mn_x)_2B$ alloys', *J. Magn. & Magn. Mater.*, **54-57**, pp. 267–268

279 COOK, J.G. (1982): 'Transport properties of solid and liquid Cs', *Can. J. Phys.*, **60**, pp. 1759–1769

280 COOK, J.G., VAN DER MEER, M.P., and BROWN, D.J. (1982): 'Measurement of the transport properties of liquid Rb using automated equipment', *Can. J. Phys.*, **60**, pp. 1311–1316

281 COOK, J.G. (1981): 'Electron-electron scattering and the transport properties of liquid potassium', *Can. J. Phys.*, **59**, pp. 25–34

282 KURISU, M., TANAKA, H., KADOMATSU, H., and FUJIWARA, H. (1987): 'Pressure study on electrical and magnetic properties of CeMg', *J. Phys. Soc. Jpn.*, **56**, pp. 1127–1131

283 ITOH, Y., KADOMATSU, H., KURISU, M., and FUJIWARA, H. (1987): 'Electrical resistivity of dense Kondo system CePt under pressure', *J. Phys. Soc. Jpn.*, **56**, pp. 1159–1164

284 AKASOFU, T., TAKEDA, S., and TAMAKI, S. (1983): 'Compound forming

effect in the electrical resistivity of liquid Pb-Te alloys', *J. Phys. Soc. Jpn.*, **52**, pp. 2485–2491

285 KAKINUMA, F., and OHNO, S. (1981): 'Electrical resistivities and magnetic susceptibilities of dilute rare earth solutes in liquid tellurium', *J. Phys. Soc. Jpn.*, **50**, pp. 1951–1957

286 KAKINUMA, F., OKADA, T., and OHNO, S. (1986): 'Electronic properties of liquid S-Te mixtures', *J. Phys. Soc. Jpn.*, **55**, pp. 284–294

287 RUPP, B., ROGEL, P., PILLMAYR, N., HILSCHER, G., SCHAUDY, G., and FELNER, I. (1990): 'Magnetic, specific heat and resistivity measurements of alloys $CePd_{2-x+y}Mn_xSi_{2-y}$ ($0 \leq x \leq 2$, $-0.1 \leq y \leq 0.1$)', *Phys. Rev. B*, **41**, pp. 9315–9322

288 SATOW, T., UEMURA, O., MATSUMOTO, K., and OKAMURA, S. (1987): 'Electronic properties of liquid Sb-Se and Bi-Se alloys', *Phys. Status Solidi B*, **140**, pp. 233–242

289 KAKINUMA, F., and OHNO, S. (1981): 'Electrical resistivities and magnetic susceptibilities of dilute rare earths in liquid indium', *J. Phys. Soc. Jpn.*, **50**, pp. 3644–3649

290 ISHIKAWA, M., and TAKABATAKE, T (1986): 'New dense Kondo compound: $CeCu_{1.6}Si_{1.4}$', *J. Phys. Soc. Jpn.*, **55**, pp. 40–42

291 MATSUNAGA, S., and TAMAKI, S. (1983): 'Compound forming effect in the resistivity of liquid Na-Pb alloys', *J. Phys. Soc. Jpn.*, **52**, pp. 1725–1729

292 OKADA, T., KAKINUMA, F., and OHNO, S. (1983): 'Electronic properties of liquid Cu-Te, Ag-Te, Au-Te and Pd-Te alloys', *J. Phys. Soc. Jpn.*, **52**, pp. 3526–3535

293 KAKINUMA, F., TANABE, H., OHNO, S., ITOH, F., and SUZUKI, K. (1980): 'Electronic behaviour of dilute liquid Te alloys', *J. Phys. Soc. Jpn.*, **49**, pp. 1034–1038

294 ABDELMOHSEN, N., LABIB, H.H.A., ABOU EL ELA, A.H., and ELSAYED, S.N. (1989): 'Electrical properties and thermal conductivity of $AgTlTe_2$ in the solid and liquid phases', *Appl. Phys. A*, **48**, pp. 251–253

295 ABOU EL ELA, A.H., ABDELGHANI, A., and LABIB, H.H.A. (1982): 'Electrical conductivity and thermoelectric power of liquid selenium doped with thallium and indium', *Appl. Phys. A*, **27**, pp. 161–165

296 ABOU EL ELA, A.H., ABDELMOHSEN, N., and LABIB, H.H.A. (1981): 'Electrical properties of $CuTlSe_2$ in the liquid state', *Appl. Phys. A*, **26**, pp. 171–173

297 CHO, S.A., GÓMEZ, J.A., ICHASO, A., and VÉLEZ, M.H. (1985): 'Electrical resistivity of In-Sb liquid system and its relation to thermodynamic behaviour', *J. Mater. Sci.*, **20**, pp. 620–623

298 DAVEY, T.C., and BAKER, E.H. (1983): 'Electrical conductivity studies of some chalcogenide glasses over a large temperature range', *J. Mater. Sci.*, **18**, pp. 717–720

299 ZAPADAEVA, T.E., and PETROV, V.A. (1982): 'Emissivity and electrical conductivity of stoichiometric titanium nitride at high temperatures', *High Temp.*, **20**, pp. 57–60

300 SATO, N., WOODS, S.B., KOMATSUBARA, T., OGURO, I., KUNII, S., and KASUYA, T. (1983): 'Transport properties of CeB_6', *J. Magn. & Magn. Mater.*, **31-34**, pp. 417–418

301 EL-ZAIDIA, M.M., NASSER, A.M., and EL-MOUSALLAMY, M. (1981): 'Electrical conductivity and bond concentration of liquid $Se_{80}Te_{20-x}S_x$', *J. Phys. D*, **14**, pp. 2275–2278

302 TAKEUCHI, A.Y., and DA CUNHA, S.F. (1981): 'Electrical resistivity of CeFe$_2$', *J. Phys. F*, **11**, pp. L241–L244

303 IKEDA, K., SATO, T., and CHIBA, A. (1990): 'Electrical resistivity of the Ni base alloys containing Fe', *Mater. Trans. Jpn. Inst. Met.*, **31**, pp. 93–97.

304 NEWPORT, R.J., DUPREET, B.C., ENDERBY, J.E., and HOWE, R.A. (1981): 'The resistivity and thermoelectric power of liquid Ag-Pd alloys', *J. Phys. F*, **11**, pp. 2539–2548

305 TAKEDA, S., AKASOFU, T., TSUCHIYA, Y., and TAMAKI, S. (1983): 'Compound forming effect for the electronic properties of liquid Sn-Te alloys', *J. Phys. F*, **13**, pp. 109–117

306 BARNES, A.C., and ENDERBY, J.E. (1988): 'The structure and electrical properties of liquid copper selenide', *Philos. Mag. B*, **58**, pp. 497–512

307 BARNES, A.C., LAUNDY, D., and ENDERBY, J.E. (1987): 'The electrical properties of liquid Mn-Te alloys', *Philos. Mag. B*, **55**, pp. 497–506

308 KOUBAA, W., ANNO, L., and GASSER, J.G. (1990): 'Electronic properties of liquid gallium-germanium alloys interpreted with Bachelet, Hamann and Schlutter pseudopotential', *J. Non-Cryst. Solids*, **117-8**, pp. 312–315

309 CATE, J.T., ZWART, J., and VAN ZYTVELD, J.B. (1980): 'Electrical resistivity and thermopower of europium and ytterbium in the solid and liquid phases', *J. Phys. F*, **10**, pp. 669–676

310 BENAZZI, N., GASSER, J.G., and TERZIEFF, P. (1990): 'Anomalous concentration dependence of the resistivity of liquid Ni-Sb alloys', *J. Non-Cryst. Solids*, **117-8**, pp. 391–394

311 POWELL, R.W., and TYE, R.P. (1963): 'The promise of platinum as a high temperature thermal conductivity reference material', *Brit. J. Appl. Phys.*, **14**, pp. 662–666

312 LAUBITZ, M.J., and MATSUMURA, T. (1973): 'Transport properties of the ferromagnetic metals. I. Cobalt', *Can. J. Phys.*, **51**, pp. 1247–1256

313 SATO, T., and SAKATA, M. (1983): 'Magnetic and electrical properties of CrGe and Cr$_{11}$Ge$_8$', *J. Phys. Soc. Jpn.*, **52**, pp. 1807–1813

314 VASIL'YEVA, R.P., PANAKHOV, T.M., and GUSEYNOV, F.S. (1983): 'Anomalies of the electrical resistivity in Ni$_3$Fe alloyed with Nb, Ta and V', *Phys. Met. & Metallogr.*, **55**, pp. 155–157

315 OHNO, S., OKADA, T., and TOGASHI, M. (1987): 'Electrical conductivities of liquid In$_2$Te$_3$ and InTe$_4$ alloys containing transition metal solutes', *J. Phys. Soc. Jpn.*, **56**, pp. 3616–3621

316 OHNO, S. (1986): 'Electrical properties of transition metal solutes in liquid Se-Te mixtures', *J. Phys. Soc. Jpn.*, **55**, pp. 295–306

317 HALL, R.O., and KAKINUMA, F. (1978): 'The thermal conductivity of α-uranium between 5 and 100 K', *J. Low Temp. Phys.*, **4**, pp. 283–290

318 OHNO, S., and KAKINUMA, F. (1987): 'Electronic properties of liquid Se-Te mixtures containing noble metal solutes', *J. Phys. Soc. Jpn.*, **56**, pp. 283–290

319 KAZANDZHAN, B.I., and MATVEEV, V.M. (1980): 'Electronic properties of alloys of the binary systems Cu-Te, Cu-Se and Cu-S in the liquid state', *High Temp.*, **18**, pp. 61–67

320 MERLO, F., and CANEPA, F. (1987): 'Electrical resistivity of some RAg equiatomic compounds (R = La, Ce, Y)', *J. Phys, F*, **17**, pp. 2373–2376

321 ABOU EL ELA, A.H., and ABDELMOHSEN, N. (1982): 'Electrical properties of AgTlSe$_2$ semiconductor in the liquid state', *Fizika*, **14**, pp. 15–20

322 RISTIC, R., BABIC, E., and SAUB, K. (1989): 'Temperature and concentration dependence of the electrical resistivity of ZrCu and ZrNi glassy alloys', *Fizika*, **21**, pp. 216–219

323 KAZANDZHAM, B.I., MATVEEV, V.M., and UMAROV, A.M. (1988): 'Electrical conductivity and thermo-emf of melts of the system Tl-Te-Se', *High Temp.*, **26**, pp. 59-64

324 ZHOROV, G.A. (1972): 'Electrical resistivity and emissivity of some transition metals and alloys in the high temperature range', *High Temp.*, **10**, pp. 1202-1204

325 GALLOB, R., JÄGER, H., and POTTLACHER, G. (1985): 'Recent results on thermophysical data of liquid niobium and tantalum', *High Temp.-High Press.*, **17**, pp. 207-213

326 KOTKATA, M.F., EL-ELA, A.A., MAHMOUD, E.A., and EL-MOUSLY, M.K. (1982): 'Electrical transport and structural properties of Se-Te semiconductors', *Acta Phys. Acad. Sci. Hung.*, **52**, pp. 3-13

327 COLVIN, R.V., and ARAJS, S. (1963): 'Electrical resistivity of scandium', *J. Appl. Phys.*, **34**, pp. 286-290

328 GOFF, J.F. (1970): 'Lorenz number of chromium', *Phys. Rev. B*, **1**, pp. 1351-1362

329 IVANOV, G.A., and LEVITSKIY, YU.T. (1967): 'Electrical properties of bismuth at temperatures ranging from 300 to 540 K', *Fiz. Met. & Metalloved.*, **24**, pp. 253-259

330 KUVANDIKOV, O.K., CHEREMUSHKINA, A.V., and VASIL'YEVA, R.P. (1972): 'Hall effect, resistance, Nernst-Ettingshausen effect and thermo emf in zinc and cadmium', *Fiz. Met. & Metalloved.*, **34**, pp. 867-869

331 TAYLOR, M.A. (1962): 'The electrical resistivity of dilute solutions of transition metals in chromium', *J. Less-Common Met.*, **4**, pp. 476-478

332 ARAJS, S., and COLVIN, R.V. (1964): 'Electrical resistivity of α-uranium from 2 to 300 K', *J. Less-Common Met.*, **7**, pp. 54-66

333 POWELL, R.W., and TYE, R.P. (1961): 'The thermal and electrical conductivities of zirconium and of some zirconiuum alloys', *J. Less-Common Met.*, **3**, pp. 202-215

334 TAYLOR, R.E., and FINCH, R.A. (1964): 'The specific heats and resistivities of molybdenum, tantalum and rhenium', *J. Less-Common Met.*, **6**, pp. 283-294

335 PROVOW, D.M., and FISHER, R.W. (1964): 'The electrical resistivity of solid and molten uranium-chromium eutectic', *J. Less-Common Met.*, **6**, pp. 313-321

336 BETT, R., SPIRLET, J.C., and MÜLLER, W. (1984): 'The electrical resistivity of proactinium,' *J. Less-Common Met.*, **102**, pp. 41-52

337 YAO, Y.D., HO, L.T., and YOUNG, C.T. (1980): 'The electrical resistivity of polycrystalline samarium and of Sm-3 at.% Dy between 15 and 300 K', *J. Less-Common Met.*, **69**, pp. 355-359

338 ES-SAID, O.S., and MERCHANT, H.D. (1984): 'Electrical resistivity of bismuth and its dilute alloys', *J. Less-Common Met.*, **102**, pp. 155-166

339 YARBROUGH, D.W., WILLIAMS, R.K., and GRAVES, R.S. (1979): 'Transport properties of concentrated Ag-Pd and Cu-Ni alloys from 300-1000 K', in Proc. 16th Int. Conf. Thermal Conductivity, Larsen, D.C. (Ed.) (Plenum Publishing), pp. 319-324

340 WATANABE, K., and FUKAI, Y. (1980): 'Electrical resistivity due to interstitial hydrogen and deuterium in V, Nb, Ta and Pd', *J. Phys. F*, **10**, pp. 1795-1801

341 WHITE, G.K., and WOODS, S.B. (1959): 'Electrical and thermal resistivity of the transition elements at low temperatures', *Trans. Roy. Soc. London*, **251A**, pp. 273-302

342 CURRY, M.A., LEGVOLD, S., and SPEDDING, F.H. (1960): 'Electrical resistivity of europium and ytterbium', *Phys. Rev.*, **117**, pp. 953-954

343 PANOVA, G.Kh., and SAMOILOV, B.N. (1968): 'Effect of isotopic composition on the electrical conductivity of cadmium', *Sov. Phys. JETP*, **26,** pp. 888-890

344 GÜNTHERODT, H.J., HAUSER, E., and KÜNZI, H.U. (1974): 'Electrical resistivity of solid and liquid Gd and Tb', *Phys. Lett.*, **47A,** pp. 189-190